PERGAMON POLICY STUDIES ON INTERNATIONAL DEVELOPMENT

Renewable Resource Utilization for Development

Robert P. Morgan
Larry J. Icerman

with
Elizabeth S. Andrews
James A. Chapman
J.P.R. Falconer
Horrace Herring
Charles G. Kick, III
Eugene B. Shultz, Jr.
Holly G. Winger

Published in cooperation with the
Center for Development Technology
Washington University, St. Louis

WASHINGTON UNIVERSITY IN ST. LOUIS
PER VERITATEM VIS

Pergamon Press
NEW YORK • OXFORD • TORONTO • SYDNEY • PARIS • FRANKFURT

Pergamon Press Offices:

U.S.A.	Pergamon Press Inc., Maxwell House, Fairview Park, Elmsford, New York 10523, U.S.A.
U.K.	Pergamon Press Ltd., Headington Hill Hall, Oxford OX3 0BW, England
CANADA	Pergamon Press Canada Ltd., Suite 104, 150 Consumers Road, Willowdale, Ontario M2J 1P9, Canada
AUSTRALIA	Pergamon Press (Aust.) Pty. Ltd., P.O. Box 544, Potts Point, NSW 2011, Australia
FRANCE	Pergamon Press SARL, 24 rue des Ecoles, 75240 Paris, Cedex 05, France
FEDERAL REPUBLIC OF GERMANY	Pergamon Press GmbH, Hammerweg 6, Postfach 1305, 6242 Kronberg/Taunus, Federal Republic of Germany

This material is based upon work supported by the Agency for International Development under Contract No. AID/DSAN-C-0061. Any opinions, findings, and conclusions or recommendations expressed in this book are those of the authors and do not necessarily reflect the views of the Agency for International Development.

Library of Congress Cataloging in Publication Data

Morgan, Robert P 1934-
Renewable resource utilization for development.

(Pergamon policy studies on international development)
Includes bibliographical references and indexes.
1. Underdeveloped areas--Renewable energy sources.
2. Underdeveloped areas. I. Icerman, L., joint author. II. Andrews, Elizabeth, S., joint author.
III. Title. IV. Series.
TJ163.2.M66 1981 333.7 80-28321
ISBN 0-08-026338-0

Printed in the United States of America

To Our Parents

Contents

List of Figures

List of Tables

Acknowledgments

The work reported in this book was carried out during the fall of 1978 and the winter and spring of 1979 under U.S. Agency for International Development (AID) sponsorship as part of U.S. preparation for the 1979 United Nations Conference on Science and Technology for Development (UNCSTD). The book closely follows the final report prepared for AID in June 1979, and is reasonably current through that date. In preparing the book for wider publication, we have made some minor modifications to reflect recent developments and to emphasize our belief that the contents will continue to be of value even though UNCSTD is now history. Much of it should be relevant to the 1981 UN Conference on New and Renewable Sources of Energy (UNCNRSE) and to the growing interests of governments, enterprises, and individuals in renewable resources for energy and other purposes.

We are grateful to many individuals who aided us in this work. The study was carried out in collaboration with Volunteers in Technical Assistance (VITA). VITA's role in the project as subcontractor included consultation in the selection of study areas, provision of information, assistance in convening of study panels, and reviewing the final report. VITA Project Manager, Eduardo Perez, was cooperative, enthusiastic, and responsive. The participants in the VITA-CDT panels, listed in Appendix A, contributed their time, ideas, and expertise. Approximately 150 individuals took time out to answer requests for information. Many of these were in developing countries and were extremely busy doing important work. John Ashworth of the U.S. Solar Energy Research Institute (SERI) responded in timely fashion on two occasions with helpful information. Paul Smulders of the Technische Hugeschool, Eindhoven provided a useful critique of the Wind Energy chapter.

We were fortunate in having two AID program monitors, William Eilers and John Fry, who went out of their way to provide us with information, encouragement, and assistance. William Anderson, Alan Poole, and W. H. Scofield of AID provided useful comments on the draft final report. Ming Ivory of AID supplied data on AID programs in appropriate technology.

The staff of our Center for Development Technology (CDT) was responsive to our usual, unreasonable demands. We wish particularly to thank Donna Williams, Mary Baetz, Paula Markham, and Emily Pearce for typing and retyping numerous drafts and getting the draft AID final report and AID final report together on time; Su Fagerberg and Barbara Needham of Principia College typed Chapter 4. Ava Fried of our Center helped with the data gathering in the initial stages of the project. Michael Klock and Professor Virginia Walbot of Washington University provided useful information for Chapter 5. Our colleague, Professor John Kardos, was in integral part of the study team and provided helpful comments and advice. Georgia Valaoras skillfully drew many of the figures. We are also grateful to Pergamon Press for their strong editorial and production support.

We wish to thank Thomas Fox, former Executive Director of VITA, for his early, enthusiastic support of the project and of the concept of CDT-VITA collaboration. Thanks go also to Laurel Druben, formerly of VITA, for her efforts in connection with the wind energy survey reported in Chapter 1. Other VITA personnel who supplied us with information and/or who read and commented on the draft report include W.R. Breslin, William Magrath, Brij Mathur, Mark Mullenax, and Marvin Munk.

While gratefully acknowledging the assistance of these and other individuals who have gone unnamed, the responsibility for the contents, accuracy, and wisdom of the work rest solely with the authors.

Introduction

BACKGROUND AND STUDY OBJECTIVES

The 1970s have seen a steep rise in prices of nonrenewable sources of energy, coupled with growing concern about the availability and desirability of these sources in both the short and long terms. As a result, there is growing interest in many nations in the use of renewable resources not only for energy but also for food, feed, soil conditioning, and many other uses of importance to people. Among these renewable resources are solar energy, in both its direct and indirect manifestations, and agricultural and timber products, by-products, and wastes.

The primary objective of this investigation, undertaken by the Center for Development Technology (CDT) at Washington University in St. Louis, in collaboration with Volunteers in Technical Assistance (VITA), a leading U.S. appropriate technology organization, was to define U.S. initiatives in the area of appropriate, light-capital technology for renewable resource utilization which might be supported by the U.S. in connection with preparations for the 1979 United Nations Conference on Science and Technology for Development (UNCSTD). By initiatives we mean steps, policies, and programs that the U.S. government might take, adopt, or support to aid developing countries in utilizing appropriate technology for renewable resources for the benefit of people - in particular the poor majority - in those countries. Study objectives also include identification of potential contributions of U.S. universities, research institutes, private industry, volunteer organizations, and appropriate technology groups in contributing to these initiatives, as well as identification of desirable mechanisms for linking U.S. resources to developing country organizations in order to implement promising initiatives.

In somewhat of a departure from previous UNCSTD studies oriented toward appropriate technology,(1) we chose to organize our study project around five "technology" or "product" areas within the broad spectrum of appropriate technology for renewable resource utilization which appear to us to have potential for bringing about improvements in the life of the rural poor in developing countries. These five areas are summarized in table I.1. We believe that the material contained in this study, which is current through the spring of 1979, fills a need for systematic analysis and evaluation of technologies for utilizing renewable resources for international development. Work on the first three study areas of table I.1. is relevant to the 1981 U.N. Conference on New and Renewable Sources of Energy (UNCNRSE) and to the growing worldwide interest in renewable energy.

Table I.1. Principal Project Study Areas (By Technology)

1. Wind Energy for Agricultural and Other Uses (water lifting, rotary power, electricity)

2. Processes for Direct Biomass Utilization for Cooking (more efficient cookstoves, ovens)

3. Direct Solar Energy Utilization for Drying (grain crop drying, timber drying)

4. Useful Food and Feed Products from Agricultural By-Products and Wastes, with Emphasis on Rice Bran Utilization

5. Useful Non-Food and Non-Feed Products Based on Natural Fibers and Agricultural and Timber Wastes (building components, other products)

CHOICE OF STUDY AREAS

Initially, a list of eleven technologies was compiled, after discussions with VITA personnel, in two main subject areas: 1) utilization of agricultural residues, non-wood fibers, and timber wastes; and 2) solar-based energy technolgies. This list was further narrowed down to the topics in table I.1. The five study areas in table I.1 were chosen because they were areas in which:

- Volunteers in Technical Assistance (VITA) has received substantial numbers of requests from individuals in developing countries for information that would appear to be useful at the grass roots level in developing countries.
- CDT and VITA staff agreed that these areas had potential for helping to meet individual and/or community needs at reasonable cost. Both staffs include individuals with Peace Corps and other overseas experience.
- There was a good match with the interests, skills, and knowledge of the people who were available to the project. In Topics 4 and 5 of table I.1, we were able to obtain significant inputs from consultants who had been involved with CDT in work on those areas.
- Substantial progress could be made on several topics within the limited time and resources available to us. We eliminated the topic of biogas generation from consideration because, although important, it appears to constitute a topic which we could not do justice to in the limited time available; it also has been the subject of a National Research Council study.(2)
- There was believed to be a range of research and development activity required. We selected topics which we hoped would illustrate a variety of types of U.S. initiatives, from support for primarily extension efforts on one hand, to more intensive R&D on the other.

We recognize that it is not possible to focus exclusively on hardware and expect such a study to lead to beneficial change, since many of the areas we are talking about involve significant social, economic, political, and environmental factors. On the other hand, the technologies and products which we consider appear not to have been examined in any kind of systematic way and therefore the data base - the knowledge of what they are needed for, what is available, how well they work - is lacking. Thus, we chose primarily to examine specific technology or product areas but we did so with some consideration of the non technological factors which strongly govern the success or failure of innovative efforts in appropriate technology. Interdisciplinary field investigations and more detailed economic comparisons of alternative technologies are required to shed further light on the status of the technologies under study and their potential for widespread use.

DEFINITIONS OF APPROPRIATE TECHNOLOGY

Definitions of appropriate technology (A.T.) vary. One such definition was provided by a group of A.T. practitioners who met to define appropriate technology inputs to U.S. UNCSTD preparations. According to the Winrock participants: (3)

> Appropriate Technology does not offer an absolute prescription but rather a process of choosing from among a broad range of options. The AT approach seeks to optimize solutions wherever possible through reliance on problem-solving capabilities of local people as well as a sensitivity to environmental and cultural impacts.
>
> In the process of choosing an appropriate technology, economic priorities include: maintenance of balance of material, labor, and capital supply; availability to the widest range of people; utilization of local materials; minimization of unemployment; gains for both producer and consumer. Environmental priorities include: renewable energy resources, natural resource conservation, recycling, replenishable raw materials, minimal pollutants, waste minimization, harmony with nature. Social priorities include: enhanced quality of life, creative work, elimination of alienation, maximum participation in decision-making, and respect for cultural values and practices.
>
> Characteristics of appropriate technology include: can be simple, intermediate, or high technology. Choice based on dialogue and mutual respect between those who plan, produce, consume, and are affected by technologies chosen. Relates to existing skills or those acquirable by training. Tends to distribute political and economic power more equitably. Efficient use of all resources and maintenance of low total-life costs. Understandable by and accessible to the end users.

Another definition of appropriate technology is given in the draft report of the Technical/Official Level Meeting to the Ministerial Level Meeting of the International Forum on Appropriate Industrial Technology sponsored in November 1978 by UNIDO and the government of India. According to the UNIDO report: (4)

> there (is) a need for reorientation in industrial strategy in most developing countries in order to extend the benefits of industrialization to all sections of the population, particularly poorer communities resident in rural areas, while sustaining overall growth. The degree of reorientation, the strategy to achieve it, and the choice of industries would vary with specific country situations, factor endowments and development objectives. The use of appropriate industrial technology (is) considered an essential element in any reorientation of industrial strategies and programs.
>
> The concept of appropriate technology was viewed as being the technology or technologies contributing most to economic, social and environmental objectives, bearing in mind the resource endowments and the conditions of application in each country. Appropriate technology was stressed as being a dynamic concept which must be responsive to changing conditions in each economy.
>
> While the concept is generally applicable and in some cases implies modern and large-scale technologies, it was felt that additional emphasis should be laid on small-scale, low-cost and relatively simple technologies that could be used by and for the benefit of the rural poor.

The definitions of appropriate technology presented above are but two of many. One of us (Morgan) has addressed the definitional problems elsewhere,(5) as has Evans, who along with Adler has compiled a series of case histories of appropriate technologies for development.(6) We chose in this project to employ an approach that attempts to avoid the pitfalls inherent in arguing about definition. We have focused on a specific set of technologies which we believe can have some positive impact on low-income populations, mostly the rural poor, who appear to be a major focus of most appropriate technology definitions. Some of these technologies will be more affordable to individuals than others. Some may require community or village-level investment. In all such cases, we are dealing with renewable resources and with prospects for meeting needs or bringing about improved income and employment through upgrading of traditional technologies or use of new technologies. While we do not claim that these technologies will immediately improve the lot of the rural poor or in all cases necessarily be immediately affordable by them, they do appear to us to be technologies that are worthy of more attention and effort than has been the case to date.

Technologies for renewable resource utilization are of growing interest to the United States, which once made extensive use of windmills and biomass in its own development, and other industrialized countries which currently are oil importers. Cooperative international efforts could result in developments which are of mutual benefit to all countries.

DATA GATHERING AND ANALYSIS

Data-gathering activities for this study included:

- Visits, telephone conversations, and correspondence with staff members of Volunteers in Technical Assistance (VITA) to obtain information from their document center and project files in the subject areas of interest. In several of the topical areas, VITA served as a major source of information for our study.
- Examination of journals and other literature containing articles with key words in the topical areas of interest.
- Mail requests for information on all topics of interest to some 221 institutions and individuals in the U.S. and other countries. As of April 1979, 69 responses had been received for an overall response rate of 31 percent. Of these responses, 24 were from the U.S. and 45 from other countries corresponding to response rates of 30 and 32 percent, respectively. Follow-up letters in the specific areas of wood stoves, solar technology, and wind devices resulted in additional responses.
- Participation by one of us (Morgan) in the UNIDO Forum on Appropriate Industrial Technology in New Delhi, India, November 1978. Forum documentation was extensive and provided useful unputs.
- The convening of VITA-CDT Expert Panels in each of three areas: 1) wind energy; 2) cookstoves; 3) solar drying. These one-day panels were convened during the period from January through March 1979. Panel reports are included in their entirety as Appendix A, along with the names of panel participants and preparatory materials for the panel. Panel members were chosen based on their interest, expertise, and experience in subjects related to the specific topics being considered. The panel members provided useful inputs to the project from their practical insights and experiences. They also reviewed a preliminary set of initiatives prepared by the CDT staff which served as the basis for the initiatives listed in tables 6.1 through 6.3 of chapter 6.
- Requests for information on current AID activities relevant to the project areas. Visits by both CDT and VITA

staff to the AID UNCSTD Coordinator's office provided useful information.

Data analysis and chapter preparation were carried out by the following individuals:

Chapter 1.	Wind Energy. E. Andrews, L. Icerman, R. Morgan
Chapter 2.	Cookstoves. H. Winger, L. Icerman
Chapter 3.	Solar Drying. J. Chapman, H. Herring, L. Icerman, R. Morgan
Chapter 4.	Rice Bran Utilization. E. Shultz, R. Morgan
Chapter 5.	Waste Utilization. C. Kick, J. Falconer, R. Morgan, L. Icerman
Chapter 6.	Initiatives. R. Morgan, L. Icerman
Appendix A.	Panel Reports. Drafted by L. Icerman with feedback from panel members

ORGANIZATION OF THE BOOK

Chapters 1 through 5 are devoted to the five principal study areas listed in table I.1. Although the content varies somewhat from chapter to chapter, each considers the need for the technology or project in question, a state-of-the-art review, and a discussion of social, economic, and environmental factors. Initiatives and implementation considerations are included in some of these chapters. Chapter 6 focuses on recommended initiatives, organizational capability, and mechanisms for implementation, as well as summaries of the review of the technologies examined in the other chapters. Thus, this work can serve both as a guide to possible future actions (chapter 6) and as a source of information on specific technologies (chapters 1-5).

The Appendix contains summaries of the panel meetings in three topical areas, a list of organizations, also by topic, involved in renewable resource utilization, and conversion factors for some of the currencies cited in the study. We recommend that the summaries of the panel deliberations be read as an integral part of the book, in conjunction with the corresponding topical chapters (chapters 1-3).

List of Acronyms

AAAS	American Association for the Advancement of Science (United States)
AATP	Arusha Appropriate Technology Project (Tanzania)
AID	Agency for International Development (United States)
AIT	Asian Institute of Technology (Thailand)
ASRCT	Applied Scientific Research Corporation of Thailand
ASTRA	Cell for the Application of Science and Technology to Rural Areas (India)
AT	Appropriate Technology
ATDA	Appropriate Technology Development Association (India)
ATI	Appropriate Technology International (United States)
CARE	Cooperative for American Relief Everywhere (United States)
CASD	Center for Applied Studies in Development (Fiji)
CBRI	Central Building Research Institute (India)
CDT	Center for Development Technology (United States)
CEMAT	Centro Mesoamericano de Estudios Sobre Tecnologia Apropiada (Guatemala)
CENICAFE	Centro Nacional de Investigaciones de Cafe (Colombia)
CEPEC	Centro de Pesquises do Cacov (Brazil)
CFTRI	Central Food Technological Research Institute (India)
CIAT	Centro Internacional de Agricultura Tropical (Colombia)

CONACYT	Mexican Science and Technology Council
CSA	Community Services Administration (United States)
CSIR	Council of Scientific and Industrial Research (India)
CSIRO	Commonwealth Scientific and Industrial Research Organization (Australia)
DOE	United States Department of Energy
DRI	Denver Research Institute (United States)
ESCAP	Economic and Social Commission for Asia and the Pacific
FAO	Food and Agricultural Organization (United Nations)
FORPRIDECOM	Forest Products Research Industries Development Commission (Philippines)
FUNDAEC	Fundación para la Aplicación y Enseñanza de las Ciencias (Colombia)
GRET	Groupe de Recherches sur les Techniques Rurales (France)
HERL	Hyderabad Engineering Research Laboratories (India)
IATA	Instituto de Agroquimica y Tecnologia de Alimentos (Spain)
ICADA	Investigaciones Cientificas Asociadas del Altiplano (Guatemala)
ICAITI	Instituto Centroamericano de Investigación y Technologia Industrial (Guatemala)
IDCA	International Development Cooperation Agency (United States)
IDRC	International Development Research Center (Canada)
IITA	International Institute of Tropical Agriculture (Nigeria)
IMAT	International Mechanism for Appropriate Technology
IPAT	Interdisziplinäre Projektgruppe für Angepasste Technologie (Federal Republic of Germany)
IPIRI	Indian Plywood Industries Research Institute
IRRI	International Rice Research Institute (Philippines)
ISTC	Institute for Scientific and Technological Cooperation (United States)
ISTI	International Science and Technology Institute (United States)
ITDG	Intermediate Technology Development Group (United Kingdom)
ITINTEC	National Institute for Development Technology (Peru)
MFM	Meals for Millions (United States)
NAL	National Aeronautical Laboratory (India)

NAS	National Academy of Sciences (United States)
NCAT	National Center for Appropriate Technology (United States)
NCAT	National Centre for Alternative Technology (United Kingdom)
NHC	National Housing Corporation (Philippines)
NRC	National Research Council (United States)
NSF	National Science Foundation (United States)
ODC	Overseas Development Council (United States)
OMB	United States Office of Management and Budget
ORP	Organization of the Rural Poor (India)
RERIC	Renewable Energy Resources Information Center (Thailand)
SATIS	Socially Appropriate Technology Information System
SCF	Save the Children Federation (United States)
SEATEC	Southeast Asia Technology, Ltd. (Thailand)
SERI	Solar Energy Research Institute (United States)
SIDA	Swedish International Development Authority (Sweden)
SWD	Steering Committee for Wind Energy in Developing Countries (Netherlands)
TCC	Technology Consultancy Center (Ghana)
TOOL	Technische Ontwikkeling Ontwikkelingslander (Netherlands)
TRANET	Transnational Network for Appropriate/Alternative Technologies (United States)
UN	United Nations
UNCNRSE	United Nations Conference on New and Renewable Sources of Energy
UNCSTD	United Nations Conference on Science and Technology for Development
UNDP	United Nations Development Program
UNEP	United Nations Environment Program
UNICAMP	Universidade Estadual de Campinas (Brazil)
UNICEF	United Nations Children's Fund
UNIDO	United Nations Industrial Development Organization
USDA	United States Department of Agriculture
VCOAD	Voluntary Committee on Overseas Aid and Development (Great Britain)
VFA	Village Forestry Association (Republic of Korea)
VIA	Volunteers in Asia (United States)
VITA	Volunteers in Technical Assistance (United States)
WHO	World Health Organization

1 Wind Energy Utilization

WIND ENERGY APPLICATIONS AND NEEDS

Wind energy systems have considerable potential to supply some of the village-level energy needs in developing countries. Although apparently no well-coordinated effort has been made to improve windmill designs, to reduce windmill manufacturing costs, and to identify desirable matches between wind energy technologies and developing countries' applications, a number of performance characteristics of wind energy conversion systems are well suited to many of the basic energy needs in developing countries. Some of these characteristics are:(1)

1. renewability - wind energy is a renewable resource which is not subject to the supply and demand fluctuations of international energy markets;
2. technological complexity - wind machines are available in a wide range of technical designs and, hence, should be easy to match to the available technical skills and experience in developing countries;
3. scale - wind energy systems are well matched to many of the power and energy levels needed in developing countries;
4. distribution - wind is a distributed resource available at the point of application so that remote villages may be provided power without the use of a distribution network;
5. applications - wind energy technologies are well suited for water pumping, mechanical power, and electrical generation, all of which are applications of interest to developing countries;
6. environmental control - small-scale applications of wind power have very few negative environmental impacts on the surrounding area; and

7. costs - wind machines typically have relatively high capital and energy production costs but small-scale light-capital designs may provide a cost-effective power supply over the system lifetime.

Wind Energy Resources

Detailed and careful quantitative evaluations of site-specific wind energy resources are usually considered as very important inputs to assessments of favorable windmill sites. The importance of these evaluations increases as the technical sophistication of candidate windmill designs increases and, hence, may be overemphasized for many applications in developing countries. In cases where small-scale, simply constructed and operated, low power output wind machines are appropriate for applications such as low-lift water pumping, crude qualitative descriptions of the local wind regime may be sufficient. An adequate wind survey for many potential applications in developing countries may include, for example, qualitative assessments of significant, and ideally predictable, diurnal and seasonal variations in wind patterns.(2) Another possibly useful technique is the Beaufort scale(3) which assigns a qualitative description of the observable effects of wind to a numerical scale correlated to known ranges of wind speeds.

If the apparent need for quantitative wind resource data is likely to delay, defer, or eliminate possible installation of windmills in developing countries, then serious consideration should be given to the adequacy of quickly collected qualitative wind data on which the system design can be based. Most of the rural applications of wind energy technologies in the United States were almost certainly sited on qualitative wind data alone and proved to be quite successful. A similar approach in developing countries should lead to a no less successful siting program because the power and scheduling requirements should be at least as flexible for most wind system applications in developing countries, even though the current economic conditions in developing countries may be less favorable than the conditions that existed in the United States during the period of rapid growth of windmill installations.

Wind Energy Applications

Water pumping, mechanical power, and electricity generation are the three principal applications of wind energy technologies in developing countries. One of the main requirements for small power sources in developing countries is for pumping agricultural and domestic water supplies. For example, a major motivation for rural electrification projects in India is to provide power to drive irrigation pumps. In fact, as much as 80 percent of the electricity distributed to rural villages in India is used for noncommercial, nondomestic purposes, primarily irrigation water pumping.(4)

Water lifting and pumping needs in developing countries are commonly supplied by animal power, fossil-fuels, electricity, and human labor. Use of windmills could free the energy sources currently dedicated to water pumping for other uses and could free people, usually women and children, from the often time consuming and difficult task of providing water supplies. Both of these changes should offer possibilities for improvements in the quality of life in rural villages in developing countries.

Specific power requirements for water pumping depend upon the specific application (e.g., domestic use, stock watering, irrigation), the depth of the water source or the pumping distance, and the operating characteristics of the pump. Power requirements of no more than 100 watts should be required for most domestic water supplies, whereas as much as one kilowatt may be required for typical irrigation applications of 1 to 2.5 hectares.(5) For example, because considerable quantities of water are pumped in India by bullock pairs, a minimum power requirement for the replacement of this traditional power source is about 250 watts.(6) Based on these power requirements, a variety of windmill designs is capable of providing the basic water pumping needs in developing country rural areas.

Electricity generation applications require, in general, more technically sophisticated windmill designs, higher capital investments, and more careful analysis of and hardware matching with the local wind regime than water pumping installations. In addition, storage facilities of some type may be required for many of the preferred uses of electricity. Wind-powered electric generators could be used to provide domestic lighting and refrigeration, to power communications equipment, to supply small-scale industrial power needs, and to drive electric pumps for water pumping (see table 1.1).

Typical power requirements, in kilowatts, for small rural villages in Tanzania are: lighting and domestic use, 4 to 6; refrigeration and cooling, 3 to 4; farm and workshop machines, 5 to 10; irrigation 10 to 20; communication equipment, $\leq$ 1.(7) These estimates are comparable to reported values of 4 to 6 kilowatts for a communal outlet ($\simeq$ 2 kilowatts) and domestic lighting in 40 to 50 homes in Fiji.(8) However, to supply even very modest estimates of power requirements, a fairly sophisticated windmill or set of windmills would be required to provide a significant fraction of the electricity needs. Therefore, windmills for electricity generation will probably have

Table 1.1. Rural Energy Needs and Possible Windmill Applications

Sector	Primary energy needs	Windmill applications	Additional possible energy needs
Domestic	lighting	electrical	small appliances
	heating	electrical or mechanical	
	water	mechanical	
Agricultural	cattle watering	mechanical or electrical	aquaculture
	irrigation	mechanical or electrical	water aeration
	drying	mechanical	
	refrigeration	electrical	
Industrial	grinding	mechanical	sawing
	refrigeration	electrical	salt production
	rotational power	mechanical	
Community	communication	electrical	lighting

Source: Based in part on: Working Group XI, "Appropriate Technology for Energy for Rural Requirements: Draft Report," International Forum on Appropriate Industrial Technology, New Delhi, India, November 20-25, 1978, Table 1.

only limited applications for supplying special needs for electricity such as communications in developing countries which could still improve the quality of life and support the achievement of development goals in rural villages.

Windmills may also be used to provide mechanical power for applications such as grain grinding or milling, cane crushing, oilseed pressing, air circulation for crop drying, refrigeration, sawing of wood, and water aeration.(9) Many other needs for mechanical power in small industries could also be provided by windmills. However, even if windmills can meet the power requirements, difficulties may arise as the result of shaft speed requirements or the need for continuous power supplies by many processes that utilize mechanical power sources.

Materials Availability

The local availability of materials and/or components needed for windmill construction in developing countries may heavily influence the preferred windmill design choice for a specified application. A wide variety of materials including bamboo, woven grasses, cloth, Dacron, wood, steel, aluminum, fiberglass, plastics, concrete, and bricks may be used in windmill construction,(10) many of which are widely available in developing countries. However, the availability of precision components (e.g., gears, chains, crankshafts) is likely to be more of a problem than the access to raw materials. Likewise, difficulty in obtaining repair parts for commercial windmill models installed in developing countries is often a significant impediment to the successful operation of a wind energy project.

Skill Requirements

Lack of the required manual skills and process capabilities for windmill fabrication may limit the choice of windmill designs that could be used to satisfy the power needs in some areas in developing countries. In general, windmills with higher operating rotational speeds and power outputs, such as wind turbines for electricity generation, require more sophisticated skills and process capabilities than, for example, windmills designed primarily for low-lift water pumping. According to Sherman, skills and process capabilities, listed in order of decreasing importance for typical windmill water pumping applications in developing countries, include: (1) rough carpentry, (2) good carpentry, (3) metal drilling and cutting, (4) metal welding, (5) machine lathe operation, (6) fine carpentry, (7) fiberglass reinforced plastic molding, (8) steel casting, (9) gear making, and (10) hydraulics fitting.(11)

Windmill project failures in developing countries are frequently the result of inadequate technical capability. In many cases, a lack of trained personnel for proper design choice and installation of windmills and/or the lack of skills needed to maintain and repair windmills have contributed to project failures. The quality and availability of technical support may influence heavily the daily performance of a windmill, as well as the total percentage of time the windmill is operational. The value of technical expertise and experience to the successful construction, operation, and maintenance of windmill installations in developing countries should not be underestimated.

Windmill Costs

Capital costs for windmills vary widely with design and differ significantly on a per unit rated power output basis over the range of available power outputs (see table 1.2). For a given

Table 1.2. Estimates of Windmill Capital Costs

	Rated power capacity and capital costs	
Windmill design	less than 4 kw*	4 to 11 kw
Commercially available mechanical power	\$5,000 to \$10,000/kw	\$1,300 to \$2,600/kw
Electrical power	\$3,000 to \$6,000/kw_e	\$1,000 to \$2,000/kw_e
Village-constructed mechanical power	\$1,300 to \$2,000/kw	

*A kilowatt (kw) is equal to 1.34 horsepower. These values are a range of costs representative of a number of windmill designs and operating characteristics. More detailed cost estimates expressed in terms of unit capacities should include the associated rated windspeed and the average wind speed at a typical siting location. Another useful approach to the reporting of windmill costs, especially when the objective is to highlight materials costs, is to estimate costs in terms of unit swept area of the rotor.

Source: Board on Science and Technology for International Development, Energy for Rural Development: Renewable Resources and Alternative Technologies for Developing Countries, Washington, D.C., National Academy of Sciences, 1976, p. 34.

capital cost, the unit cost of energy (mechanical or electrical) is influenced significantly by the local wind regime, the duty cycle of the application, and the anticipated lifetime of the windmill; therefore, reliable cost comparisons of various designs and applications are not easily obtainable and simple cost comparisons are often misleading.(12) The cost of delivered energy for a specified design and wind regime is a function of the operating mode of the windmill (i.e., the average efficiency and the match between the wind regime and the dry cycle) and the operating and maintenance costs, as well as the capital costs and machine lifetime.

Windmill installation costs also vary widely with the level of owner and/or operator participation in the construction process. The costs of commercially available windmills will, in terms of initial capital expenditures, be considerably higher than comparable designs which are built locally. A number of sources of high costs for imported wind-powered water pumps in developing countries has been identified:(13)

- weight and bulk of windmills which result in high shipping costs;
- small sales volumes which encourage high profit margins and exclude any opportunities for cost reductions through mass production;
- high materials requirements which may reflect early design considerations based on relatively inexpensive materials costs; and
- lack of windmill designs which are well matched to water pumping from deep water sources or water pumping in regions without high winds.

However, considerable potential for cost reductions appears to exist for the local construction of most windmill designs, especially designs with a relatively low level of technological sophistication (e.g., sail wing designs for low-lift water pumping), through trade-offs between efficiency, durability, and initial capital costs. Cost reductions for delivered energy may also be achieved by careful design of the windmill rotor and associated equipment in order to achieve high efficiency operation while keeping the materials costs low.(14)

In general, the costs of all windmill designs are expensive when compared to annual family incomes of a few hundred dollars. Because of these high initial costs, windmill technologies are not currently available to many poor people in developing countries. Individuals may, in some special cases, be able to afford windmills if inexpensive or free materials and/or labor are available along with suitable windmill designs or if external financial assistance is available. Capital investment requirements for most windmill designs may even pose more difficult financial burdens on purchasers than does acquiring the funds to purchase diesel engines and the required nonrenewable fuel supplies. This situation is a difficult problem to overcome even when the total costs (capital and operating) over the entire operational lifetime are less for the wind energy system than for the diesel system.(15) Accordingly, for most applications in developing countries, financial resources for the purchase or construction of a windmill may need to be pooled by a number of individual users, acquired by some type of user cooperative or village organization, or made available through low-interest loan programs supported by the government or other institutions.

WIND ENERGY TECHNOLOGIES

Energy in the Wind

The kinetic energy $\dot{E}$ in the wind passing through the swept area of a windmill rotor per unit time is proportional to the cube of the wind speed, viz.

$$\dot{E} = (1/2)\ A\,\rho\,v^3 \tag{1}$$

where A is the rotor area, ρ is the air density, and v is the wind speed. The instantaneous power output P of a windmill may then be expressed as

$$P = C_p\,\dot{E} = C_p\,(1/2)\ A\rho v^3 \tag{2}$$

where C_p is the power coefficient which is a function of operating conditions and can be shown from aerodynamic propeller theory to have a maximum value of 0.593. The instantaneous power available from a real windmill installation in the form of mechanical or electrical power will be less than that given by equation (2) as the result of inefficiencies in power transmission and/or electricity generation.

Windmill Operation

The ability of a windmill to start under load is a function of the design tip-to-wind speed ratio. This ratio is

$$\mu = \omega r/v \tag{3}$$

where ω is the angular frequency of the rotor and r is the rotor diameter. The ability of a windmill to start under load decreases as the design tip-to-wind speed increases (i.e., very roughly proportional to the inverse of the square of the tip-to-wind speed ratio).(16) Performance for off-design conditions of the tip-to-wind speed ratio results in substantial reductions in the instantaneous power coefficient for a specified windmill design (see fig. 1.1).

A convenient way to characterize the wind power potential of a candidate site is the power duration curve (see fig. 1.2)

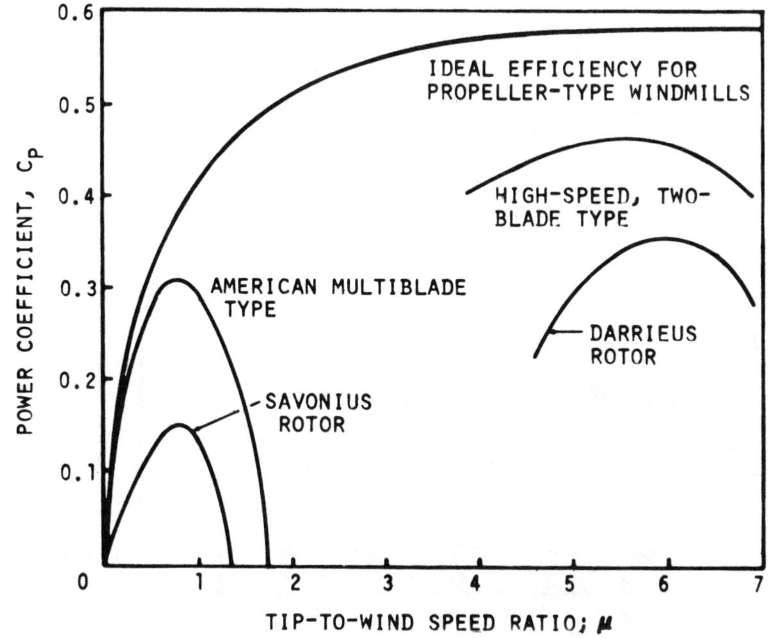

Fig. 1.1. Performance Characteristics of Representative Windmill Designs

Source: Adapted from R. E. Wilson and P. B. S. Lissaman, "Applied Aerodynamics of Wind Power Machines," Oregon State University, Corvallis, Oregon, 1974.

which may be determined easily from equation (2) and from a velocity duration curve generated from wind speed frequency data. The cross-hatched area under the power duration curve represents the annual energy output for a specified wind energy conversion system. Three important values of the wind velocity are defined on this curve: (1) v_c is the cut-in wind speed below which the windmill will not be self-starting; v_r is the rated wind speed that corresponds to the wind speed for which full design power is reached; v_f is the furling wind speed above which the windmill must cease operation to avoid structural damage. Full-load operation may not be obtainable across the full range of wind velocities between v_r and v_f unless the windmill has some mechanism for blade and/or yaw control. However, for most designs the power level will remain approximately constant over this wind speed regime.

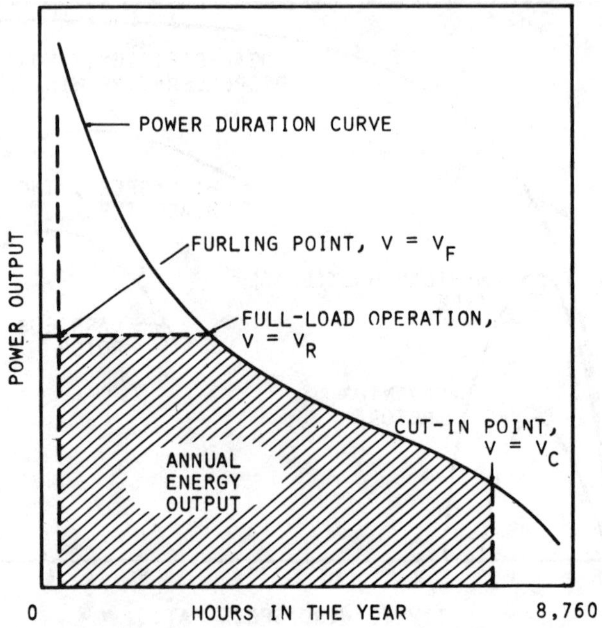

Fig. 1.2. Typical Power Duration Curve for a Wind Energy Conversion System

The use of power duration curves is a much better approach to estimating annual power production potential of a given site than using, for example, mean annual wind speeds. The distribution of wind speeds around the annual mean speed has a very pronounced effect on the wind power potential.(17)

Power Matching Considerations

Power duration curves are useful for estimating annual or, perhaps, monthly wind power potential independent of application considerations. Attractive windmill installations must have a good match between the power duration curve and the power demand characteristics of the particular application. This match becomes much less critical if the application task can be performed adequately with intermittent power delivery or if energy storage systems are incorporated in the system or the application. In addition, the instantaneous power output characteristics should be well matched to the components in the wind system (e.g., transmission, pumps, generators) as well as to the instantaneous power demand levels of the application. Careful power matching will improve wind energy system performance considerably, especially for seasonal applications such as irrigation.

Basic Windmill Designs

Windmill designs are commonly divided into two categories: (1) vertical axis (i.e., rotor mounted on a vertical axis); and (2) horizontal axis (i.e., rotor mounted on a horizontal axis). Examples of vertical-axis windmills include the Savonius rotor, Darrieus rotor, and sail designs. Horizontal windmill designs include the multivane, sail, and medium- and high-speed (propeller type) models. Vertical-axis designs have the advantage of not requiring yaw control mechanisms to follow changes in the wind direction. Horizontal-axis windmills may be designed to operate in an upwind (i.e., rotor receiving wind in advance of the tower or supporting structure) or a downwind (i.e., rotor receiving wind behind the supporting structure) mode. Downwind, horizontal-axis windmills do not require yaw control mechanisms to follow the wind. Upwind, horizontal-axis windmills typically employ tails for yaw control.

Two performance parameters are often used to describe windmill operation, namely, the power coefficient and the tip-to-wind speed ratio, defined previously. In general, windmills of advanced design have high power coefficients, high tip-to-wind speed ratios, operate in high wind speed ranges, and have low ratios of blade area to rotor swept area (i.e., solidity ratio). Low solidity ratios are characteristic of windmills with low starting torques and a small number of blades. Windmills with high solidity ratios and low design tip-to-wind speed ratios are most often employed in water pumping applications, whereas electricity generation applications typically utilize windmills with low solidity ratios and high design tip-to-wind speed ratios. In addition, the required level of sophistication in blade design and fabrication is, in general, lower for windmills with high solidity ratios and low design tip-to-speed wind speed ratios than for the reverse case. Thus, from technical considerations alone, water pumping appears to be the most attractive application for wind energy use in developing countries. A summary of typical design and performance parameters for the major windmill designs is given in table 1.3.

Savonius rotor

The Savonius rotor has received attention as a simply constructed windmill with a large potential for applications in developing countries (see fig. 1.3). Several organizations have published construction plans (e.g., VITA and Brace Research Institute) and numerous examples of this design are operating in developing countries. Although several windmill

Table 1.3. Typical Design and Performance Parameters for Major Windmill Designs Based on Past Windmill Installations

Windmill design	Rated power, kw	Rotor diameter, m	Operating velocities, m/sec			Tip-to-wind speed ratio	Power coefficient	Applications		Estimated lifetime, years
			cut-in	rated	furling			primary	secondary	
Vertical axis										
Savonius rotor	0.03 to 0.37	0.9 to 12	0.9 to 4	4.2 to 8.9	-*	≤ 1.75	0.14 to 0.23	water pumping	mechanical power, electricity generation	-
Sail	-	≤ 10	-	-	-	-	-	water pumping	mechanical power	5
Darrieus rotor	1 to 230	3.7 to 24.4	3.1 to 5.8	6.9 to 14.8	17.9 to 35.8	2 to 9	0.2 to 0.4	electricity generation		-
Horizontal axis										
Sail	0.06 to 5	3.4 to 10	1.6 to 2.9	2.2 to 5.4	4. to 9	0.84 to 4.5	≤ 0.45	water pumping	electricity generation, mechanical power	10 to 20
Multivane	0.04 to 7	1.4 to 9.1	1.8 to 4.5	4.4 to 12.5	8.9 to 33.5	≤ 3.5	0.15 to 0.28	water pumping	mechanical power, electricity generation	5 to 40
Medium-speed	0.1 to 45	1.0 to 12.2	2.2 to 5	4.3 to 11.2	9 to 18	1.8 to 4.5	0.15 to 0.35	water pumping	electricity generation	5 to 40
High-speed	0.4 to 1,250	0.61 to 53.4	2.7 to 6.3	5.8 to 16.5	11.1 to 67	2 to 16	0.18 to 0.47	electricity generation	water pumping	5 to 20

*No entry indicates that no information on this parameter has been located in the literature.

Principal Sources:

1. "A Guide to Commercially-Available Wind Machines," Wind systems Program, Rockwell International, Rocky Flats, Colorado, April 1978.
2. M. M. Sherman, "Practical Application of Wind Powered Water Pumps," Economic and Social Commission for Asia and the Pacific, Bangkok, Thailand, February 1977.

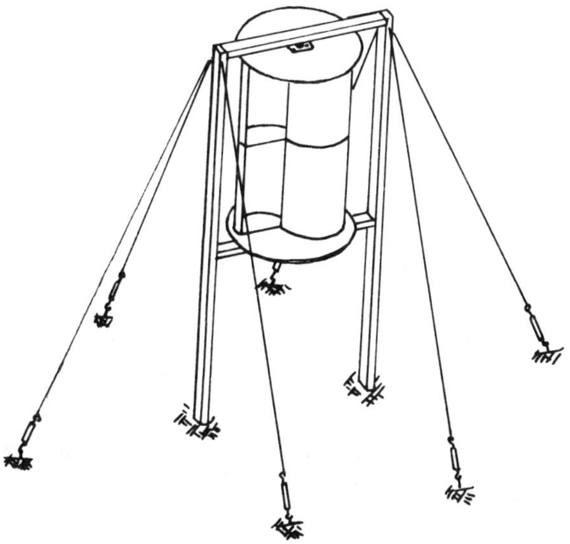

Fig. 1.3 Schematic Diagram of a Savonius Rotor Windmill

Source: A. Bodek, "How to Construct a Cheap Wind Machine for Pumping Water," Do-It-Yourself Leaflet No. 5, Brace Research Institute, Brace Experiment Station, St. James, Barbados, West Indies, February, 1965.

projects in developing countries appear to be experiencing some level of success with this design,(18) limitations of the Savonius rotor have been reported for many types of applications.(19)

Savonius windmills are vertical-axis designs with usually two, though sometimes as many as four, curved wings or blades. In past applications in developing countries, the blades have been fabricated out of steel sheets, oil drums, aluminum, or cloth. This design produces low shaft speeds, medium torque, and overall low power output. Overspeed controls are often avoided completely because the inherent high drag forces created by the rotor limit the achievable rotational speeds. In some designs, additional drag brakes, triggered for operation at high wind speeds, have been employed. Bell cranks, V-belts, and direct drive systems have been used with Savonius windmills for power transmission mechanisms. The low rotational speed and medium torque characteristics of the Savonius design make this machine well suited to low power water pumping and mechanical drive operations. The efficiency of this design is low but the ability to accept wind from any direction, the simplicity of construction, and the widespread availability of the required materials are technical advantages that should make Savonius rotors suitable for some wind energy applications in developing countries.

Vertical-axis sail designs

Information on the performance of vertical-axis sail windmills (see fig. 1.4) is not readily available, in spite of the reported extensive use of this design in China.(20) However, a report by the Economic and Social Commission for Asia and the Pacific (ESCAP) noted work done in China during the 1950s to improve the windmill design and work being done in Thailand in 1976 on adapting the design.(21) The technical sophistication required to build this type of windmill is relatively low.(22) Wood and cloth are commonly used as the primary materials in this design, which should facilitate the fabrication of these windmills for appropriate applications in developing countries.

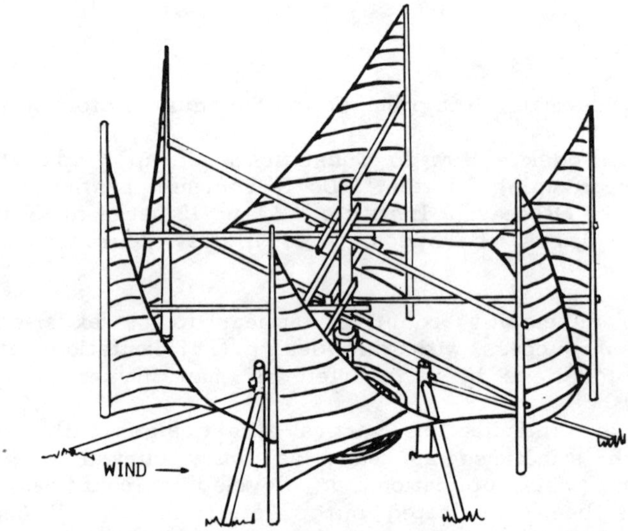

Fig. 1.4. Schematic Diagram of a Vertical-Axis Sail Windmill

Source: Economic and Social Commission for Asia and the Pacific, Proceedings of the Meeting of the Expert Working Group on the Use of Solar and Wind Energy, Energy Resources Development Series No. 16, New York, United Nations, 1976.

The vertical-axis sail windmills typically have six to eight sails, constructed from cloth or bamboo mat, that are arranged about a central shaft by means of spokes radiating perpendicularly from the center shaft to the tops and bottoms of the sails. Each sail is allowed to pivot about a vertical axis as the windmill turns. The individual sails turn automatically to catch the full force of the wind and then turn to offer the least drag resistance on the remainder of the turn as the result of being restrained by ropes and being acted upon by wind forces. Power transmission has commonly been achieved through the use of wooden gears. The primary use of the vertical axis sail windmill has been for low-lift water pumping.

Horizontal-axis sail designs

There are three principal types of horizontal-axis sail windmills: (1) Cretan or Greek design fitted with two-dimensional triangular sails; (2) sail wing design with double thickness sails that enclose the spar in the leading edge to form a three-dimensional rotor blade; (3) sail trouser design which is similar to the sail wing design but has the spar located along the centerline of the blade. The Cretan type (see fig. 1.5) has been used most extensively, particularly on the island of Crete.(23) The other two designs are less well established and represent efforts to obtain the advantages of aerodynamically shaped blades without the use of expensive materials and complex fabrication techniques. All three designs have relatively low materials and construction requirements compared to other windmill types and the potential for village-level construction by users, artisans, and small-shop operators in developing countries has been demonstrated.(24)

Horizontal-axis sail windmills typically have 3 to 12 sails, with the Cretan-type usually using 6 to 12 sails. The sails are commonly made of cloth, although Dacron and bamboo mat have also been used, and are sometimes coated with preservatives which also decrease permeability of the sails. Manual furling of the sails is by far the most frequent method of overspeed control, although manual brakes and feathering have also been employed. Tail vanes are usually employed for yaw control although, in some cases, these windmills have been constructed to permanently face in one direction. Power transmission has been achieved by crankshafts, gears (sometimes constructed from wood), and open-link chains.

Sail windmills exhibit a range of operating characteristics. Cretan-type models are primarily high torque, low rotational speed wind machines designed to operate in low to moderate wind speeds and to produce small amounts of power. Sail wing and sail trouser designs are moderate torque, moderate rotational speed wind machines designed to operate in moderate to high speed winds and to produce moderate amounts of power.

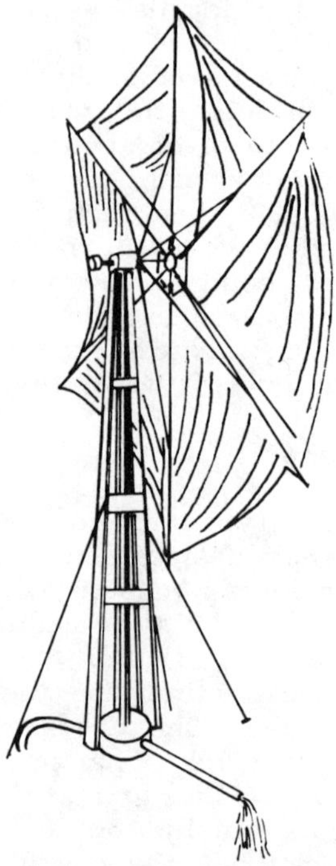

Fig. 1.5. Schematic Diagram of a Cretan-Type Horizontal-Axis Windmill

Source: Economic and Social Commission for Asia and the Pacific, Proceedings of the Meeting of the Expert Working Group on the Use of Solar and Wind Energy, Energy Resources Development Series No. 16, New York, United Nations, 1976.

The most common application of these designs to date has been for water pumping and other mechanical drive needs, even though sail wing and sail trouser models may have limited applications for electricity generation.

Multivane windmills

The multivane windmill is probably the most common windmill design in use today (see fig. 1.6). Extensive use of this design was made in the United States for pumping domestic water supplies and for stock watering by the late 1800s, and many of these windmills are still being used in rural areas. The design has also been adopted widely in the United Kingdom, Australia, and South America.(25) Because many of the

Fig. 1.6 A Multivane Windmill Built in the United States in the 1880s.

Source: New Mexico Energy Institute, "Selecting Water-Pumping Windmills," NMEI 11-0-6M, New Mexico State University, Las Cruces, New Mexico, January 1978.

windmills in the United States were built by farmers with little or no training,(26) substantial opportunities for village-level construction should exist. In addition, several development organizations, technical assistance agencies, universities, and church and other charitable organizations are currently involved in disseminating existing plans, building prototypes, and implementing demonstration projects of the multivane windmill design which require a level of technical sophistication expected to be widely available at the village level.

Multivane windmills usually have 16 to 32 blades, although designs with from 6 to 120 have been built. The blades are normally made out of steel but other materials including cloth, wood, and aluminum have been used. Overspeed control has been achieved by the use of a tail located slightly off the rotor axis to push the windmill out of high winds, by blade pitch changes, by the tendency of the rotor to create high drag forces in high winds, and by the use of a rope to manually pull the rotor out of the windstream. A tail is also commonly used to direct the rotor in the direction of the oncoming wind. Several types of power transmission mechanisms, including gears, crankshafts, V-belts with rotary cams, and rim belts, have been used with this type of windmill.

Multivane windmills have a high starting torque, as a result of the high solidity, and operate at low rotational speeds. Thus, the design is particularly well suited to drive water pumps, which commonly have high starting torque requirements. These windmills are suitable for driving many different types of pumps; however, piston-type pumps are the design in most common use. Although water lifting is the principal application for multivane windmills, electricity generation has been achieved, in some cases, by using a step-up transmission to produce higher shaft speeds.

Medium-speed designs

The category of medium-speed windmills is used to refer to those windmills that have a horizontal axis but are not typical sail, multivane, or high-speed windmills. Medium-speed windmills offer performance compromises between the other three categories. These windmill designs have low to medium solidities, usually with two to six blades, and low to medium starting torques. The torque and rotational speed characteristics of these windmills are suitable for water pumping, especially low- to medium-head applications, but may be used with step-up gearing for the generation of electricity. Blades for these windmills are usually constructed from curved steel plate although aluminum, wood, cloth, Dacron, and bamboo mat have also been used. Both manual and automatic systems have been utilized for overspeed control and tail vanes and motor-driven controls have been used for yaw control. Power transmission is typically accomplished by gears (sometimes constructed from wood), crankshafts, pulley and belt systems, or cams. In general, materials costs for medium-speed windmill designs should be lower than for multivane windmills, and less sophisticated construction skills and tools can be utilized for these designs than are required for high-speed, propeller-type windmills. For those applications in developing countries where a compromise between sail, multivane, and high-speed windmill performance characteristics is desirable, medium-speed windmills may be an attractive design choice.

High-speed designs

High-speed windmills, also known as propeller-type windmills and wind turbines, are available commercially in a wide variety of models and sizes (see fig. 1.7). These windmills require sophisticated designs, durable construction materials, and precision engineering to sustain high rotational speeds and to

Fig. 1.7 A Washington University Experimental 10-kw_e High-Speed Propeller-Type Windmill Constructed from Components Supplied by Astral/Wilcon. Photograph courtesy of Nancy H. Morgan.

withstand the strains of operating in locations with medium or high winds. The need for precision aerodynamic blade shaping is probably beyond the technical sophistication of traditional village-level fabrication activities. However, a two-bladed wind machine representing compromise between efficiency, durability, and simple construction has been developed into a workable water pumping windmill design in

Thailand. Although the efficiency advantages of high-speed designs are easily lost by imprecise machine construction, the use of new materials, particularly for blade fabrication, and new construction techniques may offer opportunities for village-level manufacture and/or fabrication of high-speed, propeller-type windmill designs. For the foreseeable future, however, high-speed designs will be much more accessible in developing countries in the form of commercial models being produced with sophisticated manufacturing techniques.

High-speed windmills typically have one to three, but sometimes as many as six, rotor blades made of wood, aluminum, or fiberglass. Overspeed control is commonly achieved by automatic feathering of the blades, although other methods that have been used include designing the rotor so that high winds create thrust forces that push the rotor out of the wind, drag brakes on blade tips that are activated in high winds, or simple mechanical brakes. Yaw control has been accomplished by using tails for rotors upwind of the tower, wind-created drag forces for rotors downwind of the tower, and mechanical and/or electrical devices. Power transmission for propeller-type windmill designs is nearly always some form of gears, although pulley belts, crankshafts, and direct-drive systems have also be used.

High-speed, propeller-type windmills are low torque, high rotational speed wind machines that operate best in moderate to high wind speeds and can produce relatively large amounts of power. These windmills are almost always used for electricity generation by driving some type of generator or alternator. However, there are a few examples of the direct use of high-speed windmills to pump water.

Advanced vertical-axis designs

Advanced vertical-axis windmills include the Darrieus design and modifications of that design sometimes referred to as the giro-mill, cyclo-turbine, or straight-bladed Darrieus (See fig. 1.8). The basic Darrieus design has curved or a combination of straight and curved airfoil baldes attached to a central vertical shaft, whereas the modified designs have straight airfoil blades mounted parallel to the central shaft and attached by perpendicular support arms. Construction of these windmills requires at least the technical capability to form aerodynamically-shaped blades. Although there are examples of construction and testing of these designs in developing countries, opinions on the suitability of these designs to local manufacture differ.(27) For the foreseeable future, most advanced vertical-axis windmills will require sophisticated manufacturing and fabrication techniques not found at the village level.

Fig. 1.8 Schematic Diagram of a Darrieus-Type, Vertical-Axis Windmill.

Source: Reproduced from "ENERGY, Vol. II: Non-nuclear Energy Technology," 1975, written by S.S. Penner and L. Icerman, with permission of publishers, Addison-Wesley, Advanced Book Program, Reading, Massachusetts, U.S.A.

The advanced vertical-axis windmills have the advantage of requiring no yaw control, having low materials costs for a specified swept area, and not depending on auxiliary devices to transmit the shaft power from a horizontal to a vertical axis.(28) One disadvantage of the basic Darrieus design is the inability to be self-starting. Self-starting has been achieved in some designs by mounting small Savonius rotors on the central shaft and by incorporating mechanisms to alter the blade pitch.

Most Darrieus-type windmills have two to four blades usually constructed from fiberglass or aluminum. Overspeed control is achieved by alteration of blade orientation, disc brakes, or aerodynamic spoilers or drag brakes. In addition, the blades used in the basic Darrieus design will stall in high winds, thereby contributing to overspeed control. Power transmission mechanisms for these windmills typically include gears or flat or teethed belts. Darrieus-type windmills have been used primarily for electricity production. Although the basic Darrieus design was invented more than 50 years ago, advanced vertical-axis designs have yet to be fully explored and only a few Darrieus-type windmills are commercially available. Current installations using this design are extremely limited in developing countries and any installations in the foreseeable future will probably utilize imported designs and hardware.

LIGHT-CAPITAL WINDMILL DESIGN DEVELOPMENT EFFORTS

A number of windmill designs are inherently suitable for or may be modified to be suitable for construction at the village level. In this section, a sample of the windmill designs that have been used in wind energy experimental or implementation programs in developing countries is reviewed. The discussion is organized by categories of principal windmill designs.

Savonius Designs

Savonius windmills have been described frequently as attractive for use in developing areas because this design is inexpensive, simply constructed, and operates independent of the wind direction. However, this design is characterized by small wind collection areas, heavy rotors, low operating efficiencies, and tendencies to develop vibration problems which may outweigh the advantages of this design for many applications. (29)

Relatively successful efforts to implement Savonius windmills have been initiated at the Centro Internacional de Agricultura Tropical (CIAT) in Colombia and the International Rice Research Institute (IRRI) in the Philippines. The CIAT design, known as the Carimagua windmill, is an adaptation of the design published by Brace Research Institute in Canada, which uses rotors that are made from oil drums split in half lengthwise. Modifications made by CIAT include the replacement of a diaphragm pump with a positive displacement pump, the exchange of a bell crank for a pulley and V-belt power transmission system, and the offsetting of the oil drum sections by 90°. A manual mechanical brake is provided for overspeed control and the improved power transmission system keeps vibrations from being transferred to the pump from the rotor.(30) These windmills are located in open savannas on short towers, are used for applications with pumping heads of 5 to 15 m,(31) and are capable of providing water for approximately 100 cattle and domestic water for a family of five.(32) Experience with these windmills has been good when maintenance was adequate, and the maintenance costs for the oldest model (five years) have amounted to a total of only $25.(33) This windmill design appears to be attractive because the costs of cattle lost in a single season as a result of becoming mired in mud while attempting to reach stream water may easily outweigh the costs of one of these windmills.(34)

Experimental Savonius rotors have also been utilized by IRRI for rice irrigation. The IRRI design also utilizes oil drums for the windmill rotor which drives a tubular pump that uses the vertical shaft motion of the rotor directly and can be made out of standard plumbing pipe. The pump operates on suction developed by centrifugal force created by the turning of a T-shaped structure.(35) These windmills pump water from rivers, irrigation canals, or tubewells and are reported to be built easily by semi-skilled labor. However, the IRRI design has limited pumping capacity during dry seasons and may be too expensive for small farmers to purchase or build without financial assistance.

The Indian Institute of Science in Bangalore, India, has been experimenting with a stretched-fabric Savonius rotor that reduces rotor weight and allows the incorporation of a simple overspeed governing device.(36) The rotor is constructed from jute canvas coated with coal tar and is attached by a number of supporting wires to top and bottom wooden plates. Overspeed control is achieved by the use of twisted joints at the trailing edges of the rotor that release the edge of the canvas when the supporting wire tension reaches a preset level. The total rotor area of the prototype is 8 m^2.

Power transmission is accomplished by a bell crank attached to a variable-stroke arm crank which permits adjustments for changes in the wind regime and pumping head. A diaphragm pump made from rubber tire fitted with end plates is used in the design which includes a compression release to lower the starting torque required by the pump. The stated objectives of this development project are to produce a windmill design that will operate in a 2.8 to 5.6 m/s wind regime and that will be able to deliver about 0.1 kw for a first cost of about 1,000 to 5,000 Rs. (See Appendix C for currency equivalents.) which is about the cost of a pair of bullocks. Reported prototype costs, including labor, pump, plumbing, foundation, tower, and installation, are about 2,990 Rs (1977).(37) The prototype windmill appears to meet the design objectives.

Horizontal- and Vertical-Axis Sail Windmills

Extensive historical and current use of sail windmills, particularly for water pumping and salt production, has been reported in Crete, China, (38) and Thailand.(39) The Greek configuration sail rotor is considered by many to be the most appropriate for immediate water pumping applications in developing countries,(40) and this windmill type has been the most widely used horizontal-axis sail windmill design to date. Although this design may still have opportunities for improvement, the operation of Cretan-type windmills is well understood and has been thoroughly tested in the field. These designs are especially attractive for developing country applications in the near term because of the use of simple construction techniques and relatively inexpensive materials such as wood and cloth. However, Cretan-type windmills require frequent maintenance, have relatively low power outputs, and operate with relatively low efficiencies.

A number of designs for horizontal-axis sail windmills have been developed. Step-by-step instructions are given in the plans published by the Steering Committee for Wind Energy in Developing Countries (SWD) in the Netherlands for an eight-bladed, cloth sail Cretan design that has a 6 m diameter rotor and may be built primarily from wood and some steel parts. According to the introduction to the plans, prototypes were constructed using only manual tools.(41) Windworks in the United States has developed a 7.6 m (25 ft) diameter, six-bladed Cretan-type design capable of producing 5 kw in an 8.9 m/s (20 mph) wind. The design is constructed primarily from wood and steel plates and uses cloth sails and an automobile differential for power transmission.(42)

Another Cretan-type design has been developed by the National Aeronautical Laboratory (NAL) in India. This design is a 10 m, six-bladed sailmill that uses some steel parts and is expected to be able to irrigate about one hectare in low wind conditions. Because this windmill drives a swinging vane rotary pump that presents no stationary load, the windmill is able to have a cut-in velocity of only about 1.7 m/s. Costs (1978) of the windmill including pump, tower, and foundation are estimated to be Rs 7,000.(43)

A design for a Cretan-type windmill has been developed and implemented in an effort by the American Presbyterian Mission at the Omo River in Ethiopia. These sailmills are 10 to 14 ft diameter, eight-bladed machines that may be constructed using an electric arc-welding set, an electric drill, a wood lathe, and hand tools. The windmill design was developed by the mission personnel to provide an irrigation system for use by the local farmers to increase previously inadequate food supplies. A number of different types of windmills, including imported multivane models, Savonius rotors, and a variety of sail windmills were also tested and evaluated during the design development process. Because of the limited financial resources of local farmers, the mission absorbs most of the costs for the machines but charges a fee of Eth. $5.00/year for 20 years to ensure the interest of new windmill owners in the project. Installation packages include hand tools, information on windmill operation, seeds or seedlings of the farmer's choice, and initial help in erecting the windmill. The project has been judged to be generally successful in spite of some adaptation problems.(44)

Sail wing windmills with potential for use in developing countries have also been developed in the United States at the Princeton University Flight Concepts Laboratory and at the New Alchemy Institute - East. The Princeton windmill, termed the "Bishop's windmill," was developed for use in Haiti and is claimed to be appropriate for the production of either mechanical power or electricity.(45) The New Alchemy Institute windmill is a three-bladed sail wing design that uses an automobile crankshaft for power transmission.(46)

Plans for an eight-bladed, horizontal-axis sail windmill design which utilizes a helical-shaped rotor, is unidirectional, and requires welding, are available from VITA. Plans for a vertical-axis sailmill are distributed by Low Energy Systems of Ireland. Details of the design were not obtainable during the preparation of this review material.

Multivane Windmills

Multivane windmills have proven to be sturdy, capable water pumpers during the past century. A number of multivane designs are commercially available, but locally constructed designs may be less expensive than commercial systems in developing countries. Materials costs for this design may be high because of the large number of rotor blades, which do not, in contrast to blades for more sophisticated windmills, need to have airfoil shapes.

A number of designs for simple construction of these windmills have been developed and at least two sets of plans are widely available. Plans for a design that uses sheet steel blades and an automobile differential for power transmission are distributed by VITA. Construction of the VITA model requires the use of welding equipment, a drill, and a number of hand tools. Blade feathering is controlled by a spring mechanism. This design has an output of 2.8 kw in wind speeds of 10 m/s.(47) Another design, distributed by the World Council of Churches in Switzerland, uses canvas for the rotor blades, a steel crankshaft for power transmission, and has an automatic blade feathering mechanism. Costs of this windmill are reported to be a little more than US $400 (1976).(48)

A multivane windmill referred to as the WP-2 was developed in India in the late 1950s. By 1966, when the WP-2 phase of the India National Wind Energy Program ended, 68 WP-2 windmills were installed in various parts of India. This design has 12 blades, is 4.8 m in diameter, and is reported to be able to produce 0.15 kw in 6.7 m/s winds. Current (1978) costs of this windmill are approximately Rs 16,000.(49) Recently, the Organization of the Rural Poor (ORP) in India has been working with TOOL of the Netherlands to develop a low-cost, multivane design. This design has a 5-m rotor, 12 sheet metal blades with constant pitch, and an automatic overspeed control that must be manually reset. Costs (1978) of the windmill, including tower, pump, foundation, and labor, are estimated to be Rs 2,980.(50)

A design development project by the Intermediate Technology Development Group (ITDG) of the United Kingdom is currently in progress that has produced a basic windmill design with a hub which can be fitted with either 24 curved metal plate blades to produce a multivane design or with six airfoil blades to make a high-speed windmill. Blade materials other than steel are also being evaluated. The crank and radius-arm power transmission system is constructed entirely out of steel bars, plates, and tubes. The design objectives of this windmill are to reduce system weight, to create a design suitable for small-scale manufacture, and to utilize component machining techniques available in small- to medium-size machine shops. Prototypes of this design are currently being constructed in eight developing country locations to gain experience in design adaptation and to conduct performance and economic evaluations. The final phase of the project will involve the creation of a design package with construction information on the various forms of the windmill which will be made available to developing country organizations. Cost estimates of a manufactured windmill based on this design are estimated at about £1,000 sterling.(51)

Another multivane design has been developed by a small-scale manufacturer in Thailand which is 2.4 m in diameter and utilizes a crankshaft for power transmission.(52) The Peruvian National Institute for Development Technology (ITINTEC) has also developed a multivane design that is reported to combine simplicity and the use of locally-available materials with low cost and relatively high efficiency. This development effort is currently in the testing stage.(53)

Medium-Speed Windmills

Although the potential for durability, simplicity, efficiency, and low cost make medium-speed windmills attractive designs, limited work has been done on their application in developing countries. A six-bladed cambered steel plate design has been developed in Tanzania by the Ministry of Water, Power, and Minerals and the Arusha Appropriate Technology Project (AATP) in close connection with villagers of the Arusha area. The major goal was to produce a design capable of responding quickly to changing winds that could be constructed using techniques available to rural Tanzanians. An eccentric-wheel and rocker-arm arrangement is used for power transmission. The operating range of the windmill is approximately 2.7 to

8.9 m/s and an off-center tail pulls the rotor out of the wind in the event of high wind speeds. The Arusha windmill will produce about 2.2 kw in a 6.7 m/s wind, is easily transported, and can be erected in two days. Costs of constructing this windmill design in Tanzania, including labor, tower, and foundation, are estimated to be Sh 1,900 (1975). A manufacturing cooperative has been established to manufacture these windmills for a retail price of Sh 15,000 (1978).(54)

A windmill designed for water pumping at low wind speeds has been developed at Las Gaviotas, a development project in the Savannah region of Colombia. The Gaviotas windmill, shown in Fig. 1.9, appears to be a medium-speed design with five metal airfoil-shaped blades. The windmill is mounted on a wooden tripod frame with guys added for additional structural support. According to Chris Ahrens of VITA, who visited Las Gaviotas in July 1980, 25 windmills were being produced per day in accordance with a contract with the Colombian government to supply a total of 1,500 windmills for use in rural schools and public facilities. Ahrens reported a selling price of about 40,000 pesos per windmill, which is less than 25 percent of the cost of their nearest competitor.(55) A December 1980 communication from Jorge Zapp Glauser of Las Gaviotas indicated that 1,500 windmills had been produced during 1980.(56)

Fig. 1.9 The Gaviotas Windmill. Photograph courtesy of Jorge Zapp Glauser.

High-Speed Designs

High-speed windmills utilize aerodynamic blades and require construction precision to achieve high efficiencies. High-strength materials are also necessary to build a durable high-speed windmill.

One simple high-speed windmill that is commonly used for pumping water in Thailand is constructed commercially by a carpentry shop in Chacheongsao. This model may be fitted with two or four blades, depending on the wind regime and the required starting torque. Pairs of blades are formed from a single wood plank and shaped into a crude airfoil. Power transfer is achieved by a twisted cowhide rope or steel chain belt system with wooden pulleys. A cost estimate for ten years of operation of this windmill and pump, including replacement parts, is given as 4,850 baht (1976). Conflicting reports are given on the self-starting capability of this windmill design, but this design is known to require manual orienting in the direction of the wind.(57)

The Korea Advanced Institute of Science developed a three-blade high-speed windmill design in late 1974 to drive a generator for electricity generation. The windmill has a cut-in wind speed of 4.4 m/s and is rated at a maximum of 2.3 kw at 11.9 m/s. Modifications of the original design were made after blade failure occurred.(58)

The dual-purpose windmill currently being developed by ITDG can be fitted with six airfoil blades filled with polyurethane foam and covered with a light alloy skin to create a high-speed design.

Flapping-Vane Windmills

An unusual design for a flapping-vane wind pump has been developed at the Interdisziplinäre Projektgruppe für Angepasste Technologie (IPAT) in the Federal Republic of Germany. This machine uses a flapping sail at the end of a vertically oscillating boom arm which is allowed to rotate freely about a central shaft. The sail is allowed to pivot about a horizontal axis within a range limited by stops so that, when wind forces are encountered, the sail moves to one extreme position and the subsequent lift forces pull the boom up. At the limit of the range of motion of the boom, the force reverses and the sail drops to the other extreme position as the boom falls. The up-and-down motion of the boom can drive a pump directly. This design has only been tested as a model and is reported to have a power coefficient of 0.15.(59)

Windmill Manufacturers, Plans, and Kits

A list of manufacturers of windmills designed to produce mechanical power is given in table 1.4. Manufacturers of windmill designs for electricity generation are given in table 1.5. A summary of major sources of windmill construction plans and kits is given in table 1.6. These tables were compiled from the following principal sources:

(1) "A Guide to Commercially Available Wind Machines," Rockwell International Wind Systems Program, Rocky Flats Plant, Golden, Colorado, April 1978.

(2) "Commercially Available Wind Systems and Equipment," Rockwell International Wind Systems Program, Rocky Flats Plant, Golden, Colorado, April 1980.

(3) *Energy for Rural Development: Renewable Resources and Alternative Technologies for Developing Countries*, Report of an Ad Hoc Panel of the Advisory Committee on Technology Innovation, Board on Science and Technology for International Development, Washington, D.C., National Academy of Sciences, 1976.

(4) McGuigan, Dermot, *Harnessing the Wind for Home Energy*, Charlotte, Vermont, Garden Way Publishing, 1978.

(5) Merrill, Richard, and Gage, Thomas, editors, *Energy Primer: Solar, Water, Wind, and Biofuels*, Revised edition, New York, Dell Publishing, 1978.

(6) Wilson, Stuart S., "Wind Power in the Sudan," 1975.

Table 1.4. Manufacturers of Wind Machines for Mechanical Power Production

United States

Aermoter Division
Valley Industries
P.O. Box 1364
Conway, AR 73032

American Wind Turbine, Inc.
1016 East Airport Road
Stillwater, OK 74074

Bowjon
2829 Burton Avenue
Burbank, CA 91504

Dempster Industries, Inc.
P. O. Box 848
Beatrice, NB 68310

Dynergy Corporation
P. O. Box 428
1268 Union Avenue
Laconia, NH 03246

Heller-Aller Company
Perry & Oakwood Street
Napoleon, OH 43545
(Baker windmills)

KMP Parish Windmills
Box 441
Earth, TX 79031

Wadler Manufacturing Co., Inc.
Route 2, Box 76
Galena, KS 66739

Other Countries

H. J. Godwin, Ltd.
Quenington, Gloucester
GL7 5BX
UNITED KINGDOM

Maschinenfabrik
Ludwig Bening
2847 Barnstorf
Postfach 171
FEDERAL REPUBLIC OF GERMANY

Merritt Windmill, Inc.
P.O. Box 1374
Merrit, B.C.
CANADA VOK 2B0

Molinos De Viento, S.A.
Calle 53, No. 1
Chihuahua, State of Chihuahua
MEXICO

Sidney Williams and Co. (Pty) Ltd.
Williams Parade
P. O. Box 22
Dulwich Hill, N.S.W. 2203
AUSTRALIA
(Comet windmills)

Sparco
A/S Naesbjerg
Maskin Center
6800 Varde
DENMARK

Stewarts and Lloyds
Technical Products Division
P.O. 74
Vereeniging, 1930
SOUTH AFRICA

Toowoomba Foundry Pty. Ltd.
259 Ruthven Street
Toowoomba, Queensland, 4350
AUSTRALIA
(Southern Cross windmills)

Wind Energy Supply Co. Ltd.
Bolney Avenue
Peacehaven
Sussex BN9 8HQ
UNITED KINGDOM

Wyatt Bros. (Whitchurch) Ltd.
Wayland Works
Whitchurch, Salop. SY13 IRS
UNITED KINGDOM

Table 1.5. Manufacturers of Wind Machines for Electricity Production

Aerolectric
13517 Winter Lane
Cresaptown, MD 21502

Aero Power
2398 Fourth St.
Berkeley, CA 94710

Aerowatt, S.A.
c/o Automatic Power, Inc.
P. O. Box 18738
Houston, TX 70023

Alcoa
Alcoa Laboratory
Alcoa Center, PA 15069

Altos: The Alternate Current
P. O. Box 905
Boulder, CO 80302

American Wind Turbine, Inc.
1016 East Airport Rd.
Stillwater, OK 74074

Astral/Wilcon
P. O. Box 291
Milbury, MA 01527

Bergey Wind Power Co.
2001 Priestly Ave.
Norman, OK 73069

Bertoia Studio
644 Main St.
Bally, PA 19503

Jay Carter Enterprises
P. O. Box 684
Burkburnett, TX 76354

Chalk Wind Systems
P. O. Box 446
St. Cloud, FL 32769

Dakota Wind and Sun, Ltd.
P. O. Box 1781
811 First Avenue, N.W.
Aberdeen, SD 57401

Dragonfly Wind Electric
P. O. Box 57-A
Albion, CA 95410

Dynergy Corp.
P. O. Box 428
1269 Union Ave.
Laconia, NH 03246

Enertech Corp.
P. O. Box 420
Norwich, VT 05055

Environmental Energies, Inc.
Front Street
Copemish, MI 49625

Grumman Energy Systems
4175 Veterans Memorial Highway
Ronkonkoma, NY 11779

Hinton Research
417 Kensington
Salt Lake City, UT 84115

Independent Energy Systems, Inc.
6043 Sterrettania Rd.
Fairview, PA 16415

Jacobs Wind Electric Co., Inc.
Route 13, Box 722
Fort Myers, FL 33908

Kaman Aerospace
Old Windsor Rd.
Bloomfield, CT 06002

Kedco, Inc.
9016 Aviation Blvd.
Inglewood, CA 90301

(continued)

Table 1.5. (continued)

McDonnell Aircraft Co.
P.O. Box 516
St. Louis, MO 63166

Megatech Corp.
29 Cook St.
Billerica, MA 01866

Mehrkam Energy Development Co.
179 East Road 2
Hamburg, PA 19526

Millville Windmills and Solar Equipment Co.
P. O. Box 32
10335 Old 44 Drive
Millville, CA 96062

North Wind Power Company
P. O. Box 315
Warren, VT 05674

Pinson Energy Corp.
P. O. Box 7
Marstons Mills, MA 02648

Power Group International Corp.
Suite 106
13315 Stuebner-Airline Rd.
Houston, TX 77014

Product Development Institute
508 South Byrne Rd.
Toledo, OH 43609

Sencenbaugh Wind Electric
P. O. Box 11174
Palo Alto, CA 94306

Tumac Industries
650 Ford St.
Colorado Springs, CO 80915

TWR Enterprises
Sun-Wind-Home Concepts
72 West Meadow Lane
Sandy, UT 84070

Winco
Division of Dyna Technology
7850 Metro Parkway
Minneapolis, MN 55420

Windworks, Inc.
Box 329, Route 3
Mukwonago, WI 53149

Whirlwind Power Co.
2458 W. 29th Ave.
Denver, CO 80211

Wind Power Systems, Inc.
P. O. Box 17323
San Diego, CA 92117

W. T. G. Energy Systems, Inc.
P. O. Box 87
1 LaSalle St.
Angola, NY 14006

Zephyr Wind Dynamo Co.
P. O. Box 241
21 Stamwood St.
Brunswick, ME 04011

(continued)

Table 1.5. (continued)

Other Countries

Aerowatt S.A.
37, Rue Chanzy
75-Paris 11^e
FRANCE

Ampair Products
Aston House
Blackheat
Guildford
Surrey
UNITED KINGDOM

Domenico Sperandio & Ager
Via Cimarosa 13-21
58022 Folloncia (GR)
ITALY

Dominion Aluminum Fabricators
3570 Hawkestone Road
Mississauga, Ontario
CANADA L5C 2V8

Dornier-System GmbH
Postfach 1360
7990 Friederichshafen
FEDERAL REPUBLIC OF GERMANY

Dunlite Electrical Co.
Division of Pye Ind.
21 Frome St.
Adelaide 5000
AUSTRALIA

ELEKTRO GmbH
St. Gallerstrasse 27
Winterthur
SWITZERLAND

Eltecco Ltd.
Aldborough Manor
Boroughbridge
Yorkshire
UNITED KINGDOM

Enag S.A.
Rue de Pont-d'Abbe
Quimper (Finistere)
FRANCE

Industrial Inst., Ltd.
Stanley Road
Bromley BR2 9JF
Kent
UNITED KINGDOM

Lubing Maschinenfabrik
Ludwig Bening
2847 Barnsdorf
P. O. Box 171
FEDERAL REPUBLIC OF GERMANY

Natural Power, Ltd.
Yorkshire House
Greek Street
Leeds LS1 5SX
UNITED KINGDOM

Noah Energy Systems GmbH
Wippenhohner Str. 32
D-5202 Hennef 1
SWITZERLAND

USSR Institute for Farm
Electrification I-U
Veshniakouski
Prc. Dom Moscow J-456
USSR

Winflo Power, Ltd.
90 Esna Dr., Unit 15
Markham, Ontario
CANADA

Table 1.6. Suppliers of Wind Machine Construction Plans and Kits

Plans for Mechanical Power Machines

United States

Volunteers in Technical Assistance, Inc.
3706 Rhode Island Avenue
Mt. Rainier, MD 20822

Windworks
Route 3
Box 329
Mukwonago, WI 53149

Other Countries

Brace Research Institute
MacDonald College of McGill University
P. O. Box 900
St. Anne de Bellevue 800
Quebec
CANADA

Low Energy Systems
3 Larkfield Garden
Dublin 6
IRELAND

Steering Committee for Wind Energy in Developing Countries
P. O. Box 85
Amersfoort
THE NETHERLANDS

World Council of Churches
Commission of the Church's Participation in Development
150 Route de Ferney
P. O. Box 66
1211 Geneva 20
SWITZERLAND

Plans for Electrical Power Machines

United States

Budget Windcharger
MI Plans Service
Fawcett Building
Greenwich, CT 06830

Environmental Action Reprint Service
2239 East Colfax Avenue
Denver, CO 80206

Flanagan's Plans, Inc.
P. O. Box 891
Cathedral Station
New York, NY 10025

Forrestal Campus Library
Princeton University
Princeton, NJ 08540

Helion, Inc.
Box 445
Brownsville, CA 95919

Plans for Electrical Power Machines

United States (continued)

Homecraft
2350 West 47th Street
Denver, CO 80211

TWR Enterprises
355 South Riverside
Rialto, CA 92376

Volunteers in Technical Assistance, Inc.
3706 Rhode Island Avenue
Mt. Rainier, MD 20822

Windworks
Route 3
Box 329
Mukwonago, WI 53149

Windy Ten
P. O. Box 111
Shelby, MI 49455

Other Countries

Brace Research Institute
MacDonald College of McGill University
P. O. Box 900
Ste. Anne de Bellevue 800
Quebec
CANADA

Low Energy Systems
3 Larkfield Garden
Dublin 6
IRELAND

National Centre for Alternative Technology (NCAT)
Llwyngwern Quarry
Machnylleth
Dyfd
WALES

Kits for Electrical Power Machines

United States

Comcop
Box 1267
Redwood City, CA 94064

Topanga Power
Box 712
Topanga, CA 90290

TWR Enterprises
355 South Riverside
Rialto, CA 92376

WIND ENERGY APPLICATIONS SURVEY

Survey Design

A wind energy applications survey was initiated, during the summer of 1978, with the cooperation of VITA. Two questionnaires were prepared by E. Andrews while she was a joint Washington University (Center for Development Technology) - VITA intern. These questionnaires were mailed to a total of 194 individuals and organizations with potentially valuable information on and experience in windmill applications. Case files of requests received by VITA for information or advice related to wind power utilization were also analyzed.

Questionnaire sample

The sample was divided into two groups: (I) agencies, institutions, organizations, and individuals known to have been involved in some aspect of wind power development; and (II) VITA wind power information requestors. The second group was compiled from users of the VITA technical information and advice-by-mail service which was initiated by VITA in the early 1960s and includes a total of 214 cases on wind power subjects received during the period from 1967 to 1978. The final sample size for Group II of 78 was determined by eliminating cases in which (1) the requestor did not return the VITA questionnaire that is regularly sent upon completion of the case to request feedback on VITA services, (2) the requestor was a temporary volunteer (e.g., a Peace Corps Volunteer) who typically remains in an area for only a short period of time, (3) the case file was missing, and (4) two case numbers were assigned to the same case as a result of subsequent inquiries on the same project. Questionnaires were sent to both Group I, with a sample size of 116, and Group II, with a sample size of 78 members.

Additional data collection efforts and methodology

Data consistent with the objectives of the questionnaires were collected directly from the 123 case files for which questionnaires were not sent and from the case files of the 64 requestors who did not return the questionnaires. The information taken from the 187 VITA requestor cases was categorized into three subject areas: (1) requestor information; (2) windmill applications; and (3) aspects of implementation. Information in these three areas was available in most case files.

The VITA requestors may not be a random sample of the larger group of those interested in windmill utilization since, as a U.S. organization, VITA tends to be especially well-known and utilized by those persons and groups with some kind of U.S. affiliation. This situation is illustrated by the high percentage (26 percent) of wind power requests that were made by Peace Corps Volunteers. Possibly, such persons and groups may have a different perspective on and interest in developing country wind power applications than indigenous developing country elements. Another possible source of bias might arise from the VITA reputation for being a source of information on small-scale, build-it-yourself technology. This image may prompt a higher percentage of persons looking for such information to contact VITA than are actually present in the general pool of persons interested in the use of windmills.

Because of the inclusion of a large number of various wind-related groups and persons in the survey, in addition to the VITA requestors, any effects of bias introduced by the requestors alone are probably not serious. In addition, any bias toward small-scale, build-it-yourself technology may in fact be more appropriate to our focus on light-capital technology than would be the results of a more general survey of those interested in wind power utilization. Lastly, recognition of possible bias brings attention to the fact that this survey should not be considered as statistically pure because of the limitations of the data base, the reliance on sporadic correspondence between VITA and requestors which was carried on for another purpose than this survey, and the relatively small number of returned questionnaires. Even though the characteristics of the cases surveyed may not entirely reflect the general characteristics of windmill implementation attempts, these cases should contribute useful, and in many ways unique, information to the documentation of light-capital international windmill utilization experiences. Therefore, we have included this analysis in this discussion of wind energy utilization in developing countries.

Questionnaire design

The questionnaires were designed to: (1) gather information on past efforts to implement wind power technologies throughout the world, (2) provide insight into the kinds of applications that were being considered or attempted and the needs for which windmills were being utilized, (3) identify the types of windmills that were being used, and (4) document problem areas in past windmill application efforts. In addition, the questionnaires requested opinions on the most important applications for wind power, the potential of wind power technologies, and the kinds of inputs and activities that would support further wind power development.

Two different questionnaires were prepared. One type of questionnaire (questionnaire #1) was sent to the institutions, organizations, and individuals who were known to be involved in wind energy development, as well as to VITA requestors who had expressed an interest in a specific windmill design or application. VITA requestors who asked simply for general information on wind energy were sent a modified version of the first questionnaire which is referred to as questionnaire #2.

Questionnaire #1 was divided into seven major subject areas, namely:

(1) background information;
(2) environment;
(3) financial and material resources;
(4) management/human resources;
(5) cultural integration;
(6) dissemination; and
(7) additional questions.

This questionnaire was oriented to the case in which a windmill had actually been implemented or was considered seriously for a specific application.

Questionnaire #2 contained questions under four principal subject headings, namely:

(1) information use;
(2) information availability;
(3) windmill efforts; and
(4) additional questions.

Since we felt uneasy in asking for a detailed analysis of a windmill effort from a requestor who asked only for general windmill information, this questionnaire was made shorter than the first and dealt primarily with questions of information availability, familiarity with windmill development problems, potential, and requirements.

Questionnaire response

A summary of the survey samples and the questionnaire response rates is given in table 1.7. A breakdown of the cases surveyed and questionnaire respondents by type of requestor and location is given in table 1.8. A list of locations, a total of 69, that were included in the VITA requests for wind energy information (Group II) is given in table 1.9. The 24 locations from which questionnaires were returned (Groups I and II) are listed in table 1.10.

A total of 163 questionnaires #1 were sent and 32 were returned. Considering that overseas communication is often unreliable, some requestors had no doubt moved from the address given in their case file, a language barrier may have discouraged completion of the questionnaire, and many of those

Table 1.7. Summary of VITA Wind Energy Cases, Questionnaire Sample Size, and Response Rate

Item	Group I	Group II	Total
VITA case numbers	214		
less: missing files	11		
duplications	2		
rejected cases	123		
Total	78		
Sample Size			
Questionnaire #1	46	117	163
Questionnaire #2	32	-	32
Total	78	117	195
Returned Questionnaires			
Questionnaire #1	9	23	32
Questionnaire #2	5	-	5
Total	14	23	37
Cases Analyzed	187	-	187
Total Cases Surveyed	201	23	224

receiving the questionnaire may have never built a windmill and therefore felt their response to be of little value, the 20 percent return rate is felt to be satisfactory (see table 1.7).

A total of 32 questionnaires #2 were sent and five were returned. Even with the communication problems discussed with respect to the return of questionnaire #1, as well as the fact that because the original request to VITA was simply for general information and might therefore reflect a lesser interest in windmills, a 16 percent return rate was achieved (see table 1.7).

The case analyses and questionnaire responses revealed that institutions, agencies, governments, and organizations, dominated this sample of persons interested in wind power applications, the majority of which were concerned with wind energy applications in developing countries (see table 1.8).

Table 1.8. VITA Requestors and Locations of Wind Energy Applications

Case	Type of requestor and location of wind energy application						
	organizations[a]		individuals				
	DC[b]	NDC[b]	DC[b]	NDC[b]	volunteers	not identifiable	total
Questionnaire #1	19	9	1	2	1	-	32
Questionnaire #2	5	-	-	-	-	-	5
VITA files	77	17	51	6	51	4	187
Total	101	26	52	8	52	4	224

[a]Governments, universities, agencies, businesses, institutions, and groups are included in this category.

[b]The terms DC (developing country) and NDC (nondeveloping country) refer to the location of the wind energy application for which information was requested.

Table 1.9. Locations Included in the VITA Wind Energy Information Request Case Files (Group I)

Afghanistan	Guatemala*	Papua-New Guinea
Argentina	Haiti	Paraguay
Australia	Honduras	Peru*
Bangladesh	India*	Philippines*
Belgium	Indonesia	Portugal
Belize	Iran*	Puerto Rico
Benin	Ireland	Senegal
Bolivia	Jamaica	Solomon Islands
Botswana*	Japan	Somalia
Brazil	Kenya	Spain
Cambodia	Lesotho	Sri Lanka*
Canada	Liberia	Switzerland
Canary Islands	Malagasey Republic	Tanzania*
Caroline Islands*	Malawi	Thailand*
Chad	Malaysia	Togo
Chile*	Mexico	Uganda
Colombia	Nepal*	United States*
Costa Rica	Netherlands	Upper Volta
Ecuador	New Zealand	Venezuela
England	Niger	Vietnam
Ethiopia	Nigeria	Western Samoa
Fiji	Pakistan	Zambia
France	Panama	Zaire

*Locations from which at least one questionnaire was returned.

Table 1.10. Locations from Which Questionnaires Were Returned (Groups I and II)

Botswana	Haiti	Peru
Cape Verde	India	Philippines
Caroline Islands	Iran	Sri Lanka
Chile	Ireland	Tanzania
England	Mexico	Thailand
Federal Republic of Germany	Nepal	Turkey
Fiji	Pakistan	United States
Guatemala	Papua-New Guinea	Wales

Missionaries accounted for approximately 14 percent of the VITA cases. International service volunteers, primarily Peace Corps Volunteers, formed the next largest subgroup. The other primary subgroup was formed of individuals concerned with wind energy applications in developing countries.

Data Analysis

Windmill owners, operators, and beneficiaries

Windmill owners, operators, and beneficiaries were identified by many of the respondents. Nongovernment indigenous organizations (e.g., agencies, institutions, universities, groups) were more frequently described as owners, operators, or beneficiaries of windmill applications than either foreign or indigenous government organizations (see table 1.11).

Windmill applications

The results of the questionnaires and the case analyses indicate that water pumping (primarily for irrigation and domestic water supply, but also for such uses as salt production, stock watering, and water supplies for aquaculture) was the primary wind energy application under consideration. Electricity generation and mechanical power were also cited frequently as application areas (see table 1.12). When asked about the applications for which wind power has the largest potential in the local region represented by the respondent, all but two respondents included water pumping in the application list, one of which indicated that water supplies were available without pumping (see table 1.13). Nearly 45 percent of the respondents included electricity generation (for such uses as lighting, refrigeration, general domestic applications, and the electrification of remote villages) as an important application area. Mechanical power excluding water pumping applications was mentioned by only 23 percent of the respondents. Nearly all the respondents felt that there was considerable potential for windmill utilization, although a few people responded that they felt there was only some, little, or no potential. The reasons for the responses of less than considerable potential included economic barriers to the widespread implementation of windmills, the need for reliable designs, the difficulty in trying to disseminate technology at the village level, and problems encountered in finding an acceptable windmill site because of hilly terrain.

Windmill technologies

Nearly 85 percent of the respondents to questionnaire #1 were directly involved with a wind energy installation compared to only 20 percent of the questionnaire #2 respondents and 4 percent of the VITA case requestors. The value for VITA case requestors is probably unrealistically low because the requestor may neither have mentioned actual involvement with a windmill installation at the time of the request nor continued follow-up correspondence with VITA to the point at which a windmill may eventually have been built. A total of 38 (17 percent) of the sample of 224 questionnaire respondents and VITA case requestors were directly involved with wind energy installations.

A number of different windmill designs were included in the wind energy installations described by the questionnaire respondents and VITA case requestors. The most frequently mentioned designs were the Savonius rotor, horizontal-axis sail wing, and multivane (see table 1.14). In many cases, the exact numbers of specific types of windmills were not indicated. Windmills were much more often constructed locally (88 percent) than purchased elsewhere and installed locally. Wind energy projects typically consisted of one to three windmill installations, although one case involved the purchase of 25 imported water pumping multivane wind machines.

Approximately 66 percent of the VITA case requestors who expressed interest in acquiring a windmill (a sample of 112) were interested in information related to constructing windmills locally. Twenty cases indicated interests only in purchasing windmills, while 16 percent of the requestors stated interests in either purchasing or building a windmill.

Implementation problems

A number of problems (listed in the order of decreasing frequency) were encountered by windmill implementation efforts in developing countries:

(1) environmental (including wind-related difficulties);
(2) personnel;
(3) financial;
(4) information/knowledge;
(5) machine failure;
(6) post-implementation education; and
(7) political.

Environment-related problems cited by respondents to questionnaire #1 consisted primarily of wind variability, wind inadequacy, and terrain-caused problems (e.g., one wind

Table 1.11. Windmill Owners, Operators, and Beneficiaries

Role	Actor	Location of actor*			Total
		foreign	indigenous	not identifiable	
Owner	nongovernment organization	1	13	4	18 (56%)
	government	-	3	-	3 (10%)
	other/no response	-	8	3	11 (34%)
Operator	nongovernment organization	-	10	3	13 (41%)
	government	-	3	-	3 (9%)
	other/no response	-	11	5	16 (50%)
Beneficiary	nongovernment organization	-	9	3	12 (38%)
	government	-	2	-	2 (6%)
	other/no response	-	12	6	18 (56%)

*These data are based on 32 responses to questionnaire #1 and the values in parentheses indicate percentages of nongovernment, government, and other/no response organizations involved in each of the three categories.

Table 1.12. Principal Windmill Applications

Source	Number of persons indicating area of interest*		
	water pumping	electricity generation	mechanical power source
Questionnaire #1 respondents	21 (66%)	7 (22%)	2 (6%)
VITA information requestors	63 (34%)	30 (16%)	10 (5%)
Total	84 (38%)	37 (17%)	12 (5%)

*These data are based on 32 responses to questionnaire #1 and 187 VITA case files. The values in parentheses represent the percentage of respondents or requestors who indicated an interest in the specified windmill application.

Table 1.13. Applications of Wind Power Judged To Be The Most Important by Questionnaire Respondents

Application	Location[a]		
	DC	NDC	Total
Water pumping	23 (77%)[b]	5 (17%)	28 (93%)
Electricity generation	8 (27%)	5 (17%)	13 (43%)
Mechanical power	6 (20%)	1 (3%)	7 (23%)

[a]The terms DC (developing country) and NDC (nondeveloping country) refer to the location of the wind energy application for which information was requested.

[b]The values in parentheses are percentages of the total number (30) of responses to this part of questionnaires #1 and #2, although respondents frequently cited more than one application.

Table 1.14. Wind Energy Technologies Used by VITA Case Requestors

Windmill design	Number of windmill installations: constructed locally	purchased	total
Savonius rotor	10	1	11
Sail (horizontal axis)	9	-	9
Sail (vertical axis)	1	-	1
Multivane	7	1	8
Medium-speed turbine	3	-	3
High-speed turbine	4	3	7
Darrieus rotor	2	-	2
Flapping vane	1	-	1

pump was forced to stand in salt water which corroded the tower). Wind variability caused an owner of a sail windmill to have to be present much of the time in order to furl the sails with changes in wind speed. The handling of exceptionally high winds was also cited as a problem by some respondents. Corrosion by salty atmospheres was mentioned as a problem which was reported to be correctable with regular maintenance measures and painting.

Personnel problems consisted of those encountered because of a lack of general laborers and/or skilled workers, as well as problems in the coordination of operations both within the working group and between the group and overseeing organizations. Financial difficulties included lack of knowledge of funding sources, a lack of available funds, problems in obtaining credit, and the inability to acquire funds at the particular time when they were required. In many cases, a lack of sound advice or information was also given as a cause of difficulties. Problems with machine failures coupled with limited opportunities for repair or modification and problems with educating staff personnel as to proper windmill operation were also mentioned. For one case in which a windmill was not implemented, local political instability was cited as an important factor.

Persons responding to questionnaire #2 listed the following problems (in order of decreasing frequency) that may slow wind energy development:

(1) unfamiliarity with the technology;
(2) lack of technically skilled workers;
(3) lack of funds; and
(4) unsuitable wind conditions.

Review of the VITA case files indicated that wind inadequacy and lack of funds were the primary reasons for decisions not to implement windmills. However, specific data on problems encountered during windmill implementation efforts were not typically included in the case files. A summary of the problems reported by respondents to questionnaire #1 (Groups I and II) and questionnaire #2 (Group II) is given in table 1.15.

Table 1.15. Common Problems Encountered in Windmill Implementation Efforts

	Number of respondents	
Problem area	Group I	Group II
Environmental (includes wind-related issues)	17 (53%)[a]	1 (20%)
Personnel	9 (28%)	3 (60%)
Financial	8 (25%)	2 (40%)
Information/knowledge	7 (22%)	-
Machine failure	1 (3%)	-
Staff education	1 (3%)	-
Political	1 (3%)	-
Unfamiliarity with technology	-	4 (80%)
No response[b]	7 (22%)	-

[a]The values in parentheses indicate the percentage of returned questionnaires that reported the specified problem. More than one problem area was reported on most questionnaires.

[b]Three of the respondents who gave no response indicated that, at the time of the inquiry, a windmill implementation effort had not yet been undertaken.

Cultural integration

Responses describing the involvement in and acceptance of a windmill project by the local population were solicited by questionnaire #1. About two-thirds of the respondents addressed some portion of this section of the questionnaire. Many of these who did not respond were involved strictly in testing or R&D efforts or had not yet established an operating windmill. Approximately 88 percent of the responses indicated that a successful or partially successful effort was made to involve members of the local community in the construction of the windmill installation. More than 50 percent of the responses to this section of the questionnaire indicated that efforts to involve the local community in the operation, planning, and ownership phases of the project were felt to be at least partially successful (see table 1.16).

Table 1.16. Involvement of the Local Community in Windmill Implementation Programs

Project phase and location	Level of success of efforts to include members of the local community in windmill implementation and number of responses[a]			
	successful	partly successful	unsuccessful	Total
Planning				
DC	6	2	-	8
NDC	2	1	-	3
Construction				
DC	8	3	-	11
NDC	2	1	-	3
Operation				
DC	6	3	-	9
NDC	2	1	-	3
Ownership				
DC	4	3	-	7
NDC	1	-	-	1

[a] A total of 16 responses were received with 13 replies covering windmill applications in developing country (DC) locations and 3 replies describing windmill applications located in nondeveloping countries (NDC).

Based on the responses to the cultural integration section of the questionnaire only, approximately 75 percent of the respondents indicated that some or all of the windmill repair work would be or was being done by local people. Nearly 50 percent of the respondents felt that the windmill would probably remain in use by local people without outside intervention. Local initiative in arranging for windmill construction occurred in more than 50 percent of the applications (see table 1.17).

Table 1.17. Cultural Integration of Windmill Installations

	Number of positive replies[a]	
Area of integration	DC	NDC
Windmill fit in well when used directly by the local people	7	2
Windmill will probably remain in direct use by local people even without outside intervention	7	-
Local initiative was used in arranging for windmill construction	6	2
Local innovation occurred in windmill design	4	3
Some windmill repair/maintenance work carried out by local people	4	-
All windmill repair/maintenance work carried out by local people	5	2
Problems occurred in cultural integration	6	1

[a]A total of 15 responses were received with 12 replies covering windmill applications in developing country (DC) locations and 3 replies describing windmill applications located in nondeveloping countries (NDC).

A number of problems were reported with respect to the integration of windmills into the local community. These problems included:

(1) over expectation of machine capabilities;
(2) lack of understanding of maintenance requirements and, therefore, poor maintenance;
(3) need for the constant attention of skilled and/or experienced workers (e.g., persons to tend to furling and unfurling of sails on a modified Greek-type sail wing in a region with highly variable winds);
(4) skepticism of windmill potential in the early stages of the project;
(5) unwillingness or inability to accept the risk of possible losses through machine failure or vandalism;
(6) fear of the windmill or misunderstanding of windmill functions; and
(7) sound levels from the operation of high-speed electricity generating windmills.

Areas for additional development efforts

Respondents to questionnaires #1 and #2 supplied a wide variety of suggestions for and opinions on activities, materials, and information that were needed to encourage wind energy development. These responses may be categorized into three principal areas: (1) information dissemination and activity coordination; (2) research and development; and (3) implementation. The most frequently stated needs were related to the publishing and dissemination of information on wind power utilization and to further work on windmill design improvement and development. A summary of suggestions received is given in table 1.18. There is a wide variety of activities which, if carried out, could assist the development of wind energy. Although information dissemination and windmill design research and development received the most emphasis by respondents, the need for solid support of all aspects of implementation efforts should not be overlooked.

Conclusions

Several results of the wind energy survey have potential implications for U.S. initiatives in the area of wind energy development. There are a large number of various types of nongovernmental organizations that are interested in and/or active in wind energy activities in developing countries. The survey results suggest that these organizations are probably principal actors in wind energy activities. Much of the activity and interest in the application of wind power is related to water pumping applications which is an area that was emphasized by questionnaire respondents. Accordingly, this area may be an important target application for successful implementation of wind energy technologies in developing countries. The problems encountered by windmill implementation efforts indicate several areas in which activities could assist wind energy development, namely, creation of more wind energy information, generation of more funds to support wind energy implementation efforts, establishment of personnel training programs, and assistance to implementation efforts in overcoming difficult problems in matching or adapting a given windmill design to the local environment and wind regime. The survey shows that considerable potential for the integration of windmills into communities in developing countries exists; however, a number of problems commonly occur during this process, especially as the result of the misunderstanding of windmill operation. Finally, many of the needs for the further development of wind power utilization expressed by the questionnaire respondents are areas in which U.S. organizations may contribute effectively to the development of mechanisms to respond to these requests.

Table 1.18. Areas for Additional Development Efforts

Area	Number of times identified by respondents	
I. Information Dissemination and Activity Coordination		
publishing and dissemination:		16
windmill plans	7	
translation of available information and plans	3	
available technical information on wind system components	2	
addresses of contacts and places with information relevant to windmill utilization	1	
information on energy storage systems for windmills	1	
cooperation among and coordination of wind energy activities		3
publicity on wind energy activities		2
standardization of testing results		1
center for distributing information on wind energy programs and projects		1
II. Research and Development		
design improvement and development of new designs:		11
cheaper designs	3	
improved windmill-pump matching	2	
more reliable designs	1	
designs that can be more easily built locally	1	
water pumping windmills	1	
design evaluation		4
windmill project evaluations:		4
economic analyses of projects	3	
overall evaluations of projects for strengths and weaknesses	1	
wind resources information:		3
resource surveys and wind data	2	
methods for predicting missing wind data	1	
feasibility studies		1
III. Implementation		
windmill demonstration projects		5
skilled personnel to work in implementation efforts:		3
skilled personnel capable of long-term project involvement	2	
funding for implementation-related programs		4
government financing and encouragement of windmill utilization		2
courses and implementation work in developing country schools (e.g., agricultural training schools)		2
training efforts in windmill construction, operation, maintenance, and repair (possibly through workshops)		2
prototype building and field testing		2
windmill manufacturing in developing countries		1
coordinated follow-up of projects for technical adjustments and management of future production		1
advice for questions related to specific windmill implementation efforts		1

2 Direct Biomass Utilization for Cooking

NEEDS AND APPLICATIONS FOR IMPROVED COOKSTOVES

Wood is the primary energy source for more than one-third of the world's population who live primarily in rural areas of developing countries.(1) In many of these areas, 80 to 90 percent of the households depend on wood as the primary source of energy.(2) The use of wood for cooking fuel may account for 50 to 90 percent or more of the total energy consumption in rural areas in developing countries.(3) Perhaps as much as 80 percent of all wood cut in developing countries is used for cooking fuel and, in some areas, cooking is the only use for wood resources.(4)

Traditional Cooking Practices

Traditional methods of cooking in developing countries, which have been used for centuries with little modification, involve burning wood in an open fire, sometimes enclosed by a horseshoe-shaped alcove made of mud or bricks to act as a windshield.(5) In some cases, three stones are placed around the fire to serve as supports for cooking vessels, which is commonly known as the three stone cooking method.(6) Open-fire cooking is very inefficient because only 5 to 10 percent of the potential energy in the wood fuel is utilized in the cooking process.(7)

Traditional cooking over an open fire also has a number of undesirable characteristics, for example:

- control - intensity or rate of burn of open fires is difficult to control which may restrict the preparation of some kinds of food or limit the choice of cooking methods;

- unhealthy conditions - family members, especially the cook, are exposed to the constant emission of smoke and soot into the kitchen area and, in many cases, throughout the house;
- hazards - open fires subject family members and especially unattended children to possible burns and scalds from sparks from burning logs or from unstable cooking pots, and expose combustible structures to possible fire damage;
- unsanitary food preparation - build-up of soot in the kitchen work area and roaming animals, such as dogs, may bring filth in contact with food cooked low to the ground over an open fire;
- lack of cooking space - often only one food item may be prepared at a time and a supply of hot water cannot be kept heated while preparing food items;
- fatigue - open fires often require constant tending or fanning by the cook who is exposed to the heat and smoke emitted by the fire; and
- indirect costs - family members expend considerable time, labor, or money to obtain the large amounts of firewood needed for open-fire cooking because of the poor efficiency of wood use.

The main advantage of open-fire cooking is that the cooking device is essentially free.(8) In addition, the smoke from open fires often creates a welcome deterrent to insects and predatory animals, which could devour food stores and/or dwelling structures and harm family members. In fact, in some areas wood is burned, in part, specifically to provide protective smoke.(9) However, open-fire cooking methods leave considerable room for improvements to make the chores of food preparation and firewood collection and use easier, more efficient, and less time consuming.

Needs for Improved Cookstoves

Introduction of improved cookstoves in developing countries may fulfill two important needs: (1) reductions in firewood use for cooking, and (2) improvements in the cooking process and environment. Improved cookstove designs address the first need by seeking cooking efficiency improvements; and attempt to satisfy the second need by providing mechanisms for control of the fire intensity and, in many cases, smokeless operation. A device to enable firewood for cooking to be used efficiently, thereby reducing the overall firewood consumption to accomplish the same task, is urgently needed along with reforestation programs in the rural developing areas of the world because the pressures on forest lands from increasing fuel demands of growing populations have caused acute shortages of wood fuel in many areas. The ultimate goal to be achieved by the widespread effective dissemination of more efficient, wood-burning cookstoves is potentially a 50 percent reduction in the demand for firewood.

Improved cookstove designs suitable for dissemination to rural developing areas, with low costs and the ability to be made with local skills and materials, have produced efficiencies of up to 30 or 40 percent in laboratory tests. Under normal operation in an uncontrolled setting, actual efficiencies of 15 to 20 percent may be obtained which is about twice the efficiency of 5 to 10 percent for traditional, open-fire cooking devices.(10)

Deforestation issues

Firewood scarcities are most severe in areas such as the Sahelian zone of Africa, the Gangetic plain in India, in parts of Java, in the Andes, and perhaps in parts of China.(11) Rural village populations in many regions have grown rapidly which has often resulted in local deforestation of woodlands in an ever-spreading area surrounding a village as cooking fuel demands surpass the natural regenerative capabilities of nearby forests.(12) The immediate effects of local deforestation influence the daily lives of villagers who must first spend increasing amounts of time and energy searching for firewood to supply per capita demands that average about 0.75 tons or more per year.(13) However, according to Eckholm, the effects are much more widespread:

> The accelerating degradation of woodlands throughout Africa, Asia, and Latin America, caused in part by fuel gathering, lies at the heart of what will likely be the most profound ecological challenge of the late twentieth century. On a global basis, an ecological threat to human well-being . . . is the undermining of the productivity of the land itself through soil erosion, increasingly severe flooding, creeping deserts, and declining soil fertility. All these problems are accentuated by deforestation which is spreading as lands are cleared for agriculture and as rising populations continue their search for firewood.(14)

Firewood collection is second only to clearing land for agriculture as a cause of deforestation. However, these two actions are linked in a cycle that may result in accelerated damage to vanishing woodlands. When fuel wood availability becomes limited in certain areas as the result of resource scarcities and/or rising costs, considerable pressure exists to find substitute cooking fuels. The alternate fuels chosen will most often be animal dung and crop residues which diverts these natural fertilizers and stabilizers from agricultural use. Because each ton of cow dung contains the nutrients necessary to grow 50 kg of food grain, the burning of cow dung as cooking fuel may be considered analogous to burning food to cook food.(15) In India, the soil nutrients lost annually from the prevalent use of dung for fuel have been estimated to be the equivalent of more than one-third of the nutrients supplied by chemical fertilizers.(16)

Use of dung for fuel contributes to an accelerated soil degradation process because important nutrients are not returned to the soil. Without the compensating application of chemical fertilizers, crop production per hectare will drop as the soil suffers from reduced structure and fertility which creates a need to clear more agricultural land for growing crops in order to maintain the required food yields. In the clearing process, surrounding forests will be further destroyed and, thus, the already receding cooking fuel source will become even farther removed from the village. Consequently, a growing reliance on cow dung and crop residues as fuel substitutes will result, the land will be further deteriorated, new land will again be cleared, and so on.(17)

In addition to the clearing of land for agricultural use and firewood collection practices, improperly managed logging activities contribute to depletion of forest resources and can cause an overall decline in woodland area, especially in areas where delicate mountain slopes are clear cut which often leads to large topsoil losses from heavy rains making reforestation efforts almost futile.(18) Steep slopes and remote mountain regions often contain the only remaining vast woodland areas in nontropical developing countries. Deforestation in these areas of many developing countries, such as Colombia, India, Indonesia, Nepal, Nigeria, Pakistan, the Philippines, and Tanzania, is increasingly exposing lowland areas to potential flooding and siltation of rivers, reservoirs, and harbors; to decreased water quality and availability; to extinction of wildlife and plant species; to elimination of sources for building materials, paper products, and chemicals; and possibly to climatic change.(19)

A summary of the extent of deforestation in selected developing countries and projections of possible levels of future deforestation based on historical data and expected rates of population growth are given in table 2.1. For example, dramatic losses of forest area have occurred in the Philippine and Malaysian lowland forests, where remaining stands are predicted to be depleted in five to ten years. Prior to recent satellite surveys, the Philippines were thought to still have 35 to 50 percent of the forest cover remaining; however, the estimates have been revised downward to 20 percent as a result of the surveys. In Nepal, forests in the

Table 2.1. Deforestation Levels in Selected Developing Countries

Country	Estimated remaining forests, %	Estimated rate of deforestation, 10^3 ha/y	Estimated time to complete deforestation,[a] y
Brazil	67	740	30
Burundi	--	---	7
Colombia	74	1,000 to 1,300	30
Costa Rica	38	50 to 60	22
Ecuador	52	52	--
Ethiopia	--	200	--
Gambia	4	--	--
Ghana	--	50	--
Indonesia	--	400	--
Malaysia	--	150 to 280	12[b]
Panama	50	15	--
Papua New Guinea	>70	20	--
Philippines	<20	260	--
Thailand	35	300	9 to 15

[a]These estimates have been compiled from various sources, some of which account for anticipated population growth.

[b]This value refers to only timber resources.

Source: Adapted from Walter Parham, "LDC Deforestation Problem - A Preliminary Data Base," (Draft), Dept. of State, Office of Science and Technology, Washington, D.C., March 1978.

Terai and adjacent regions have suffered losses from 22 to 30 percent during the last decade and an area of about 26 square km has become so barren that desertification has begun. Northern Thailand may be losing 5 to 7 percent of its forests each year; Burundi may be totally deforested in seven years; Gambia may only have 4 percent of the original forested area remaining; Morocco, Tunisia, and Algeria may have less than 11 percent of the forested areas left; lowland forests of Indonesia and Papua New Guinea may be totally deforested in 15 to 20 years; and the Amazon Basin lowland forests are being diminished at a rate of 4 percent per year. In Haiti, 50 percent of the forests have been stripped in the last four years and all remaining forests are expected to be cut in the next 15 years.(20)

The widespread adoption of improved cookstove designs may be an important strategy to help slow these losses of forested areas because improved cookstoves should be able to demonstrate substantial, perhaps as high as a factor of two, efficiency improvements compared to traditional cooking practices. This potential increase in the efficiency of firewood use could result in significant reductions in the cutting of forests to supply cooking fuel needs. However, successful dissemination of improved cookstoves cannot reverse these deforestation trends and should be a component of a well coordinated program that would also include projects to specifically grow firewood supplies and reforestation efforts.

Improved cooking processes and environment

Nearly all improved cookstove designs incorporate features designed to alleviate many of the undesirable characteristics of open-fire cooking practices. Most improved cookstove designs have been equipped with internal dampers and a chimney to control the fire intensity and to eliminate smoke from the kitchen. Use of an improved stove with a higher efficiency and controlled fire may actually stimulate increases in nutritional intake by allowing the preparation of new foods and larger quantities of traditional foods. For example, in the highlands of Guatemala, the higher altitude alone may cause wood fuel to be burned inefficiently which makes the preparation of foods with long cooking times, such as beans which have high nutritional value, too expensive in terms of fuel consumption. The lack of fire control in traditional stoves also makes slow cooking extremely difficult. Introduction of an improved stove design may allow slow cooking of such foods overnight with a slow, constant fire.(21) In Nepal, Haiti, West Africa, and Latin America, curtailment of traditional cooking of certain foods or meals has been the result of firewood scarcities. Use of a stove with twice the efficiency of an open fire may again enable these foods to be included in the diet.

The height and stability of many improved cookstove designs remove the cooking area from the reach of animals and dirt, thereby improving the hygiene of cooking activities. These cookstoves usually have a large cooking area that allows at least two items to be cooked simultaneously. To cook or heat two items at once with an open fire requires the wasteful use of two fires or the precarious balancing of pots against each other over a single fire. Many improved cookstove designs have multiple cooking pot seats that receive various levels of heat intensity which allows frying, boiling, or slow cooking simultaneously.

Advantages and Disadvantages of Smokeless Cookstoves

One of the principal motivations behind the development of smokeless cookstoves was to relieve the burdens on women in developing countries associated with the use of cooking devices which are smoky, primitive, unhygienic, and wasteful of fuel resources.(22) Traditional cooking methods are simple and cheap but are highly wasteful of fuel, and the smoke nuisance often created during the cooking process is well known. These stoves typically are messy, unsanitary, and require constant attention and tending. According to Rao, any improvement in cookstove designs means bettering the lot of the housewife and lightening her drudgery.(23)

Another advantage of smokeless stoves is the potential improved stove efficiencies obtainable from the use of a chimney as part of the draught system. The chimney, duct system, and inset pot seats also serve to prevent the kinds of accidental burns that can occur from sparks and flames leaping from an open-fire arrangement and from scalds received when pots, precariously set on top of rock supports or other makeshift apparatus, are knocked over by small children or even uncautious cooks.(24)

Not all of the impacts of adopting smokeless cookstove designs to replace traditional cooking devices will necessarily be positive. An example of the conflicts involved is described by Ianto Evan's observation:

> In Guatemala the smoke given off by indoor fires gets out of the house by seeping out through the tiled roofs or any other opening. The effect of the smoke on the cook is terrible (it causes severe eye and lung problems which can lead to blindness, emphysema, bronchitis and lung cancer) yet smoke is a good preservative for roof beams. The acrid sooty smoke is a strong deterrent to most tropical wood-eating insects which would otherwise delight in eating the wood structure. The Lorena stove has

> tidily confined smoke to a chimney, and ended exposure of the cook to unhealthy smoke, but people using it must now find a new way to protect their wooden roofs.(25)

There may be important considerations in addition to maintenance of building structures. Designers typically assume, perhaps based on Western rather than local tastes, that stoves should be smokeless, although wood smoke often plays an essential role in African homes in discouraging bugs that would otherwise devour both people and dwellings.(26) In regions prone to high mosquito populations, the use of smoke in the rural dwelling as an insect deterrent should not be undervalued.

Useful functions of the smoke also include applications as a food preservative. In the Malabar and Konkan regions of India, the hot air from the fireplace often rises to an attic over the kitchen with a grid flooring of wooden rafters which serves as a dry storage place where provisions are stored during the monsoons when food stuffs are susceptible to spoilage by mold and mildew attacks.(27)

An additional disadvantage of the smokeless stove that is often overlooked is the potential fire hazard created by improperly cleaned chimneys. Although most of the stove designs assume that dry wood fuel will be used, green wood is often used which leads to tar deposition in the chimney. These deposits may lead to a marked rise in temperature which may ruin the chimney and may be ignited by a spark from the stove when a strong fire is going. Therefore, chimneys bearing signs of tarring should be cleaned or replaced.(28) Many of the descriptions of smokeless stoves acknowledge this potential hazard and include instructions to clean the chimney every six months with a long bamboo stick with rags or leaves tied to the end.

For some designs, daily cleaning of the chimney base opening is recommended to prevent soot build-up.(29) Many of the construction manuals include instructions on how to put out a chimney fire by closing off any air intake. In addition to causing rapid tar build-up, the use of green wood results in reduced stove efficiencies and, therefore, the need for increased quantities of fuel to perform a given cooking task. Another disadvantage of smokeless stoves is that many of the materials needed to construct the chimney may be expensive and/or hard to find in developing countries. In some cases, these materials are the only purchased components and, therefore, represent the only real money outlay needed to build a stove.

In conclusion, the main, and very important, positive impacts of smokeless stoves are the elimination of smoke as a health hazard in the kitchen and the potential increased fuel economy. A careful assessment of the balance of benefits and negative impacts of smokeless stoves in areas where the smoke has traditionally provided an indirect valuable health benefit by protecting humans, food stores, and houses from harmful pests is yet to be performed.

Hot Water Supplies

One provision of nearly all improved smokeless cookstove designs is a supply of hot water that is useful for both cleaning and cooking. The importance of a dependable hot water supply is stated by Rao:

> Kitchen craft does not begin and end with the preparation and serving of food. Cleaning and washing of utensils and vessels is an important activity and value of hot water in this important chore is not to be minimized. . . . Hot water also makes bathing and washing of clothes easier and more pleasant. It is invaluable in sickness, accidents and childbirth.(30)

A hot water supply may also serve to reduce the time spent in meal preparation. In some Latin American countries, women rise very early in the morning just to fix hot water for breakfast coffee. In any cooking process involving boiling, a continuous supply of hot water will reduce the normal time period required for cold water to heat to the boiling temperature. In regions where vegetables are traditionally heated to boiling by starting with cold water, much of the nutritive value may be lost. By starting the vegetables in preheated water at about 180°F, the total cooking time may be lessened considerably and many of the vitamins may be preserved.(31) In addition, the use of heated water for beverages and as a component of food preparation may improve health conditions since the preheating helps kill harmful organisms in the water.

There is considerable difference of opinion regarding the attractiveness of hot water supplies provided by improved cookstove designs. Some evaluations of cookstove designs have concluded that the production of hot water not only adds to the cost of the cookstove design but may not be all that useful to a majority of users whose main objective is cooking.(32) Other designers feel that a hot water supply is so important that more elaborate systems, including inserted pipes, old car radiators, water storage systems, supplemental solar heating equipment, and heated water pipes in the chimney systems are all ideas to explore further.(33)

Hot water supplies do seem to be an important additional function of improved cookstoves for applications in some regions, even though this capability may not contribute to the timely adoption of improved cookstove designs in other areas. In any case, waste heat from the stoves can also be used to effectively dry fuel wood before use, thereby increasing combustion efficiencies and decreasing the possibility of chimney fires. Some improved designs include mechanisms for both hot water production and dry storage for firewood.

Cultural Acceptability

Improved cookstoves must meet the needs of people in developing rural areas in ways that are acceptable to the users. First, the improved designs must be inexpensive to compete with the use of open fires. For example, the materials for the Lorena stove may cost as little as $4 to $5 (1979), which is a fairly low cost even for poor rural villagers. Secondly, the stoves must be durable to withstand heavy use without cracking for at least a year and preferrably even longer. Third, the designs should be able to be constructed with locally available tools, such as large kitchen spoons, knives, wood molds, and other simple hand tools, and from local materials such as mud and crop residues, sawdust, sand, or ant hill dirt, which may be molded to form the stove body or made into sun-dried or fired-clay bricks for stove construction. In most cases, the use of a chimney and, perhaps, dampers leads to the most difficult materials availability problems. In designs requiring the work of local skilled artisans such as brick and pottery makers, the design must take the typically available level of skill into account as well as the cost of procuring these services.

Fourth, improved cookstove designs should be simple to operate and understand, should utilize the types of fuel available, should be compatible with the preferred cooking methods, and should be consistent with the cultural or religious beliefs of the local people. For instance, the shape and height of improved designs should conform to the local cooking practices of women and should be able to be used with indigenous cooking pots and utensils. Finally, improved cookstove designs must be flexible enough so that minor alterations to meet local conditions and preferences may be made without markedly reducing the efficiency of the stove because, unless the new designs are capable of demonstrating a substantial improvement in performance during actual use compared to traditional cooking methods, an adequate incentive is unlikely to exist for women to adopt the improved cookstove designs.

COOKSTOVE DESIGN IMPROVEMENT EFFORTS

Efforts to improve upon traditional methods of cooking over open fires began in some countries (e.g., India, Egypt) about a quarter century ago. Improved cookstove designs are characterized by: (1) offering the potential for significant fuel efficiency improvements as compared to open fires; (2) having smokeless operation; and (3) providing a cooking environment that is more hygienic, aesthetic, and pleasant than that with typical open-fire cooking practices. Very early improvements to open-fire cooking in India consisted of small mud or brick horseshoe-shaped alcoves which served primarily as a wind break.(34) Recent cookstove improvement efforts have focused on two classes of cookstoves, namely, high-mass designs and low-mass designs and, to a lesser extent, on improved baking ovens.

High-Mass Designs

High-mass cookstoves are typically constructed from mud and/or measured clay and sand mixtures. These designs are often 1 m^2 or more in horizontal cross-section and are usually located inside the homes of rural villagers. Many of the high-mass cookstove designs provide controlled heat for cooking, limited amounts of hot water, and contributions to space heating requirements if needed.

HERL chula

The design for the HERL chula was developed in 1953 at the Hyderabad Engineering Research Laboratories (HERL) in India and was considered a major improvement over the cooking devices that were in common use at that time. The HERL chula (see fig. 2.1) was designed to be smokeless, to have a controlled fire to obtain a 20 to 40 percent efficiency improvement over the then current chulas, to provide hot water, to be constructed simply from local materials, and to have similar shapes and operation to the cooking devices already in use. This design also relieved women from the laborious task of blowing the fire and from exposure to the intense heat by incorporating an effective draught system to exhaust the combustion gases through a chimney vented above the roof.

Experimentation with the HERL chula spread to rural Egypt where traditional rural cooking devices were very similar to those used in India. However, direct application of the Indian design was unsuccessful because bulky fuels, such as corn and cotton stalks, were being used in Egypt and because the traditional location of the Egyptian stove (i.e., in the

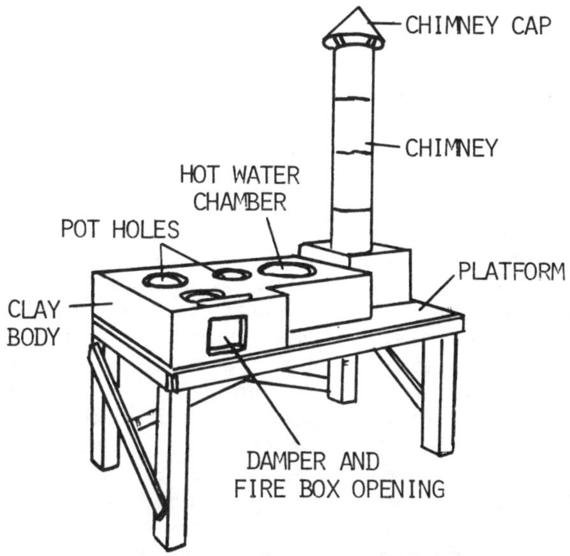

Fig. 2.1. Schematic Diagram of a HERL chula

Source: Volunteers in Technical Assistance, "Wood Conserving Cook Stoves," Energy Fact Sheet No. 3, Mt. Rainier, Maryland, April 1980.

inner court of the house and under the ceiling where fuel is stored) necessitated having a lower chimney as well as horizontal smoke outlets for fire prevention. Therefore, the Arab States Fundamental Education Centre developed a modified version of the HERL design, which had a lower and larger fireplace with shorter chimney (e.g., 2.0 to 2.5 meters in height) that was equipped at the top with horizontal gas outlets formed by inserting 10-cm drain pipes. Comparison of this modified HERL design with stoves commonly used in rural Egyptian areas showed a fuel savings of 16 to 28 percent.(35)

Additional design variations of the HERL smokeless chula subsequently appeared, all of which incorporated the basic improvements of adding a chimney, flues, and dampers to control the hot combustion gases to achieve better efficiencies. These designs utilized the waste heat remaining after the combustion gases passed through the flue channels beneath the main cooking vessels by placing a large vessel of water (e.g., five liters) near the chimney to absorb some of the waste heat or by including an area on the stove for storing fuel wood, thereby reducing the moisture content to provide a better quality fuel. Typically, design modifications changed the outward shape or size of the stoves, altered the internal course and/or cross-sectional size of the flue channels, modified the firebox size and/or shape, rearranged the location of dampers, and incorporated slightly different methods of, and materials for, construction of the stove, especially the chimney.

Magan Choola

The Magan Choola was developed in India as the result of a feeling that none of the Ceylon (rectangular-shaped), Hyderabad (L-shaped), or horseshoe (U-shaped) improved chulas in use were performing well enough as smokeless stoves.(36) The shape of this design accommodated the spacing of pot-seats on the stove top so that the centers formed an equilateral triangle and utilized an indirect flue channel path to the chimney. Beneath each pot a mound of the clay building material was built up in the flue channel to force the hot gases to pass across the vessel bottom in a more concentrated manner.(37) The Magan Choola provided a fuel savings over traditional cooking devices of about 33 percent and was capable of burning an assortment of fuels, including wood, charcoal, or agricultural wastes. In order to avoid back-draught problems associated with strong winds blowing around the chimney exit, special H-shaped cowls were designed for the chimney.(38) Back-draught problems have not been reported as a difficulty encountered by other smokeless chula designs. A place for wood to be stored and dried was incorporated into the stove design as a waste-heat utilization technique. The strength and integrity of the clay structure were best maintained by daily washing with a thin fluid of cow dung.(39)

Singer stove

The Singer cookstove design incorporates the HERL chula characteristics of being smokeless, with flues and a closed hearth, (see fig. 2.2). A major objective of this design was to improve upon the short (i.e., six months to a year) useful life of the indigenous Indonesian mud stoves.(40) Descriptions of this design emphasize the merits of building with burnt bricks rather than sun-dried bricks, either of which are preferred over solid dried-clay construction with respect to durability. The preferred material mixture for burnt and sun-dried brick components is 50 percent sawdust and 50 percent clay, which produces strong, fire-resistant burnt bricks and fairly strong sun-dried bricks. For clay stoves, a mixture of clay, rice straw, and ashes is preferred with no sawdust included because, in a large solid structure, the different drying rates of clay and sawdust cause cracking.(41) Clay building materials are recommended only as a last resort and, even then, only for models with heights less than 30 cm.

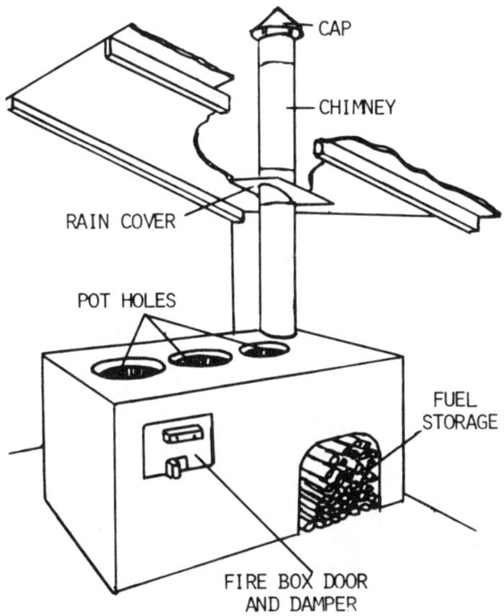

Fig. 2.2. Schematic Diagram of the Singer Stove Design

Source: Volunteers in Technical Assistance, "Wood Conserving Cook Stoves," Energy Fact Sheet No. 3, Mt. Rainier, Maryland, April 1980.

The Singer stove model which has performed best in cooking tests based on efficiency measurements and on uniformity of heat distribution over the potholes is a three-holed, rectangular, smokeless model with a straight, but slightly inclined, flue channel between the hearth and the chimney. A 70-cm high version of this "Model II" Singer stove performed with a 26.6 to 28.7 percent efficiency when using an assortment of wood fuels in tests conducted at the Regional Housing Centre of Bandung.(42)

Fijian designs

An improved cookstove design, smokeless and constructed from cast-in-place concrete, based on the Indian chula was introduced in Fiji. This design reportedly had a fuel efficiency below that of open fires but did offer the advantage of smoke-free cooking.(43) However, problems with the cracking of the stove structure resulted in unsuccessful dissemination attempts.

The Nasinu stove, a prototype design developed by the Department of Forestry in Fiji, was designed to be inexpensive, constructed from local materials, durable, efficient, and safe.(44) This stove design is smokeless and includes a removable oven for baking. Although this design has not been fully tested, the Nasinu model is expected to have a performance comparable to that of the Indian chula.(45)

Ghanaian stove

A Ghanaian smokeless stove is designed to be constructed from clay or laterite, to burn locally available fuel without recutting, and to incorporate a traditional clay oven large enough for family use. A single chimney serves the oven and stove units. The stove top can be used for low-temperature heating of soups and water while the oven is in operation. Dried or fired clay as well as waste components of sheet metal or asbestos water pipe are recommended materials for chimney construction. To strengthen the stove top, the use of a few pieces of steel reinforcing rod is proposed.(46)

Lorena cookstove

The Lorena stove (see fig. 2.3) was designed by workers at the Estacion Experimental Choqui (Choqui Experimental Station) following the 1976 earthquake in Guatemala. The design process was coupled to an extension effort focused on teaching stove construction methods to small groups of stove promoters.(47) The main design criteria were that the stove should conserve fuel, have smoke control, and utilize a broader range of fuels than is possible with an open fire. The stove also provides much desired hot water and may also be used as a kitchen table or work area.(48) The key to this design is the building material called "Lorena," which is a mixture of mud (lodo) and sand (arena). Since solid mud structures crack during drying, the Lorena stove is made mostly of sand which does not shrink as it dries. An additional design feature is that the tunnels forming the flue channels and the firebox are carved out of the Lorena block in domed shapes.(49) The stove is constructed on an adobe brick base, with thin layers of Lorena applied atop this structure and allowed to dry to form a solid block.

Lorena stoves tend to perform better the bigger they are and offer a potential fuel savings of about one-half over open-fire cooking. This stove design can be made any shape desired, but performs best when the flues are channeled to take as indirect a course as possible between the firebox and chimney so that the portion of heat imparted to the cooking area before the hot gases are exhausted is maximized. Indeed, an indicator that this stove is working efficiently is that the cook will be able to grasp the metal chimney without discomfort while the stove is fired.(50) In practice, some of the stoves

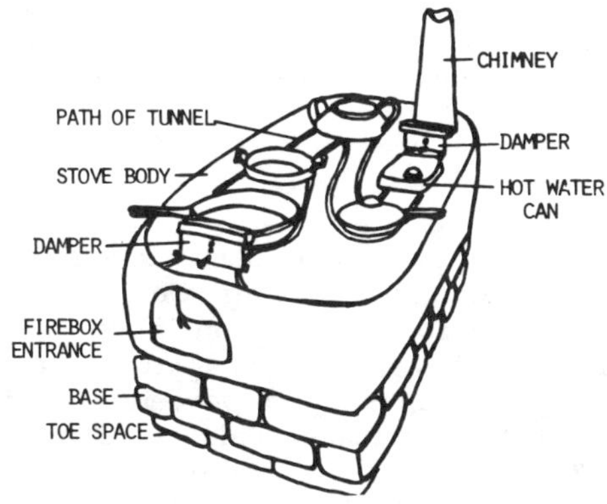

Fig. 2.3. Schematic Diagram of the Lorena Cookstove

Source: Volunteers in Technical Assistance, "Wood Conserving Cook Stoves," Energy Fact Sheet No. 3, Mt. Rainier, Maryland, April 1980.

have exhibited a tendency toward structural cracking and the potholes have tended to shrink too much while the Lorena is drying. However, this problem is attributed to using an improper mix of sand and clay and/or to improper drying procedures, both of which are correctable.(51)

Additional designs

Other adaptations of the smokeless chula stress: (1) the use of a mold for the stove top, a combination of cement and sand for the building mixture, sheets of zinc for the chimney, and inset tin cans to form the potholes;(52) (2) variations in chimney construction using tin cans fitted one atop the other, cement pipes, galvanized sheet metal, asbestos pipe, or a clay chimney formed around a central pipe which is withdrawn after the clay is dry;(53) and (3) designs which burn coal or coke and therefore need round-shaped gratings of mild steel below each pothole to hold the burning coals.(54)

Low-Mass Designs

Low-mass stoves usually have metal or fire-clay structures and are often built from waste metal products such as oil drums or paint cans. Single burners, simple construction, and portability are characteristics of these designs, some of which can burn sawdust, rice hulls, wood, or charcoal. Fuel savings of up to 50 percent compared to open fires are typically claimed. Low-mass, portable stoves are commonly used by urban dwellers in temperate climates where much of the cooking is performed outdoors.

Indonesian designs

Three single-burner designs have been developed and judged by workers at the Regional Housing Centre of Bandung to perform sufficiently well to warrant continued manufacture and dissemination. The first was a Y-shaped, single-walled fire-clay container, with a basin-shaped porcelain grate at the top and an air intake at the bottom. This model was reported to have a 27.2 to 31.2 percent efficiency. A second version, developed by the Ceramic Institute of Bandung, consisted of an upright cylinder made of a fire-clay sleeve casing, with a refractory basin grate and air intake at the side. This model incorporated an iron damper and, after some experimentation, a 30.5 to 36.0 percent efficiency was measured by eliminating air holes in the rim around the grate. A third model made of cement with an iron casing and iron-bar grate was found to have an efficiency of 29 to 31.6 percent after the slits in the initial grate design were narrowed.(55)

Komadu cooker

The Japanese designed Komadu cooker is a double-walled ceramic stove with an inner firebox, an upper and lower damper, a metal grill, and burns charcoal or wood (see fig. 2.4). Very efficient operation is claimed since the outer wall and inner firebox have very little contact with each other. In addition, baking can be done by placing a covered baking vessel on top of the burner. An adaptation of this cooker was constructed at the Georgia Institute of Technology using a five-gallon paint can as an outer shell surrounding a ceramic firebox liner. Cooking vessels are supported by the ceramic liner or a metal grill. Baking can be carried out as with the Komadu cooker. Special clay mixtures may be required to produce a ceramic with sufficient thermal shock resistance. This stove has a claimed potential fuel savings of 50 percent compared to open fires.(56)

Sawdust stoves

A simple sawdust burning stove for cooking and heating which is constructed of a five-gallon paint can with support rods for legs and a metal reducer duct pipe or chimney has been designed by Simon and Solis.(57) The can has four holes in the bottom into which broomsticks are placed, and dampened sawdust is packed in around them. The sticks are then

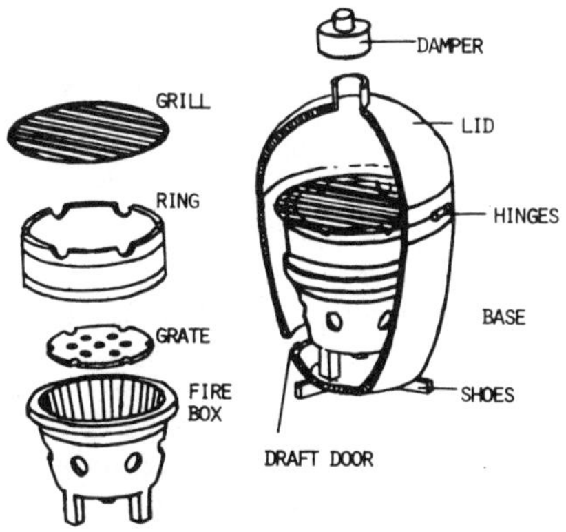

Fig. 2.4. Schematic Diagram of a Komadu Cooker

Source: Georgia Institute of Technology, "Recent Efforts to Develop Simple Wood Burning Stoves," A State of the Art Survey of Solar Powered Irrigation Pumps, Solar Cookers and Wood Burning Stoves for use in Sub-Sahara Africa, Small Industry Development Network, Atlanta, Ga., 1978.

removed and a fire is ignited with a small amount of fuel-oil in each hole. A four- or five-hour cooking fire can be obtained from 2 kg of sawdust. This single-burner stove is unique in that it is equipped with a chimney.

A variation of this stove has been designed for use in Honduras, where there are restrictions on firewood consumption in order to protect the forest industry. Although firewood is scarce, sawdust which could be utilized as fuel is regularly burned by the forest industry for waste disposal. The proposed stove is made of a five-gallon can, but has only one hole in the bottom for a broomstick to be inserted, and no cover or chimney. This stove is operated in the same manner as the first model by placing it on brick supports and maneuvering these so as to control air flow to the bottom of the can. One load of sawdust may last for six to seven hours.(58)

A slightly more elaborate stove was designed by Wartluft for space heating of rooms up to about 20 ft square. This stove is constructed using a 55-gallon oil drum for an outer shell, and a 30-gallon drum as an inner fuel container. A stove-pipe is included to function as a chimney, and the stove is fitted at the bottom with a side drawer opening that controls air intake, while the drawer serves as an ash pan. The stove is loaded leaving a single hole formed in the center of the packed-in sawdust, and may produce good quality heat for six to eight hours. Flare-up problems have been reported resulting from improper operation. These stoves, which have been used in Afghanistan, England, and Chile, can burn bark residues, and planer shavings, as well as sawdust.(59)

Rice hull stoves

Stoves burning rice hulls are in use in Bali. This design consists of a brick or metal cylinder placed on the ground, with a half-circle side opening at the bottom and metal grate bars across an open top for support of a cooking vessel. Two sticks are arranged to meet at right angles to each other, one inserted in the top at center and one at the bottom opening of the stove, to form a fire-hole when removed after rice hulls have been packed in around them. Another stick half as thick as the one used to form the bottom channel is ignited at the end and inserted into the stove to initiate the fire beneath the vertical channel. A two-hour cooking fire can be obtained by slowly pushing the stick in as it burns.(60)

Use of a rice-husk fuel cooking device is being revived in the Philippines where the use of firewood and natural gas has become too undependable and expensive. This design consists of a square cement block with a short pipe on top leading to a wooden funnel where the rice husks are poured and sift down into a firebox in the cement block. In front of the stove is an inset ash outlet connected to the firebox by a tilted grate. The firebox runs the length of the stove top, supplying heat to two burners made of cast-iron rings. The rice-husk fuel is often burned as waste and, hence, costs almost nothing to use as cooking fuel.(61)

Baking Ovens

Several designs for baking ovens have been developed. One design resembles a small, narrow igloo and is made of adobe blocks or brick, with a lower door opening and a smoke hole at the top. This oven operates on a hot-box principle. That is, the fuel is burned completely inside the oven and, after the ashes are swept out, the items to be baked are placed inside the heated structure. The door and smoke hole are then closed off with wooden or metal pieces so that the oven is airtight.(62)

A second design was developed to handle the large-scale baking requirements of a school in Kathmandu, Nepal. This design is a large brick and mortar structure with a 4 ft. square by 22 in. high inner baking space. This oven is equipped with a sheet-iron door, a metal stove pipe chimney, and operates on the hot-box principle. The oven also supplies heat to a storage room located behind the oven to keep flour and other staples dry.(63)

A third oven design consists of a 44-gallon oil drum, cut in half along its length to form two ovens, which are welded to rectangular steel plates and laid side by side on a hollow rectangular base made from cement, bricks, and two steel bars as cross supports. The ovens are then housed in a rectangular brick structure built up from the base, with door openings left in each open drum-end and a metal chimney pipe in the center. This oven operates when a wood fire is burned beneath the two ovens, with the hot gases circulating between the steel bars the oven rests on and then passing through the chimney.(64) This model is designed to supply baked goods for large numbers of people.

Comparison of Improved Cookstove Designs

A comparison of important design features of a number of basic improved cookstove designs is given in table 2.2. Representative materials choices, required construction skills and tools, fuel choices, labor needs, performance data, and costs are included.

COOKSTOVE PERFORMANCE CHARACTERISTICS

One of the primary design objectives of improved cookstove efforts has been the achievement of higher cooking efficiencies. As a result, improved cookstove designs typically have considerably higher efficiencies of fuel use under closely controlled laboratory construction and operation practices than open-fire cooking procedures. However, imperfect construction and/or individual operating preferences may easily offset the expected thermal efficiency improvements of a new design.

Efficiency Measurements

The procedures applied to measure the efficiency of operation of improved cookstoves typically consist of one or more of the following tests: (1) measuring the time required to boil a given amount of water with a specified amount of fuel; (2) measuring the number of times a given quantity of water can be brought to a boil for a specified amount of fuel; (3) measuring the amount of fuel required to bring a measured amount of water to a boil and/or to maintain boiling for a specified time; and (4) measuring the quantity of fuel and cooking time required to prepare a typical meal for a specified region. The most common efficiency test consists of measuring the amount of fuel required to bring a specified amount of

Table 2.2. Design Features of Improved Cookstove Designs

Design	Design features: major materials requirements	construction tools	labor needs	fuel	Performance estimates	Cost estimates	Source and date
High-Mass Stoves							
HERL chula	moist earth; cow dung; fired-clay attachments; tin	yard measure, boards, knife, string, trowel, tin template	local skills; potter	wood, charcoal, or coal	20 to 40% fuel savings[a]	--	1 (1953)
Egyptian stove	mud; mud bricks; fired-clay plates; drainage pipe	--	local skills; potter	corn or cotton stalks	16 to 20% fuel savings[a]	0.15 £ E	2 (1953)
Magan choola	prepared clay; straw or fiber; cow dung; iron rings and plates; fired-clay pipes	wooden rod	local skills; potter; blacksmith	wood, charcoal, and other combustible materials	33% fuel savings[a]	2 to 3 Rs	3 (1955)
Singer stove	clay; sawdust; rice husks or ashes; fired-clay bricks and plates	wood-forming blocks	local skills; potter	wood	27 to 29%[b]	--	4 (1961)
Nasinu stove	concrete; metal plates; oil drums	concrete molds, metal cutters	local skills	wood	--	- $30	5 (1978)
Ghanaian stove	clay or laterite; sheet metal; steel rods	--	local skills	wood or charcoal	--	--	6 (1976)
Lorena stove	sand and clay; sheet metal	shovel, spoon, sifting screen, machete	local skills	wood, charcoal, or other combustible materials	50% fuel savings[a]	$15 to $35	7 (1977)
Low-Mass Stoves							
Indonesian stove	fired clay or ceramic materials; iron grate	--	local skills; potter	charcoal	27 to 36%[b]	--	4 (1961)
Komadu cooker	ceramic materials; metal grill	--	local skills; potter	wood or charcoal	--	--	8 (1978)
Honduras stove	metal paint can; broomstick; metal pipe	welding, metal cutters	metalworking	sawdust	--	--	9 (1977)
Philippine stove	concrete; metal pipe; wooden funnel; metal grate	concrete molds, metal cutters	metalworking; potter	rice husks	--	$80 to $135	10 (1978)

No entry indicates that no information is reported in the literature.

[a]These values refer to fuel savings compared to traditional cooking practices.

[b]These values refer to percentage of the heating value of the fuel used for cooking.

Sources:

1. S. P. Raju, Smokeless Kitchens for the Millions (Madras, India; The Christian Literature Society, 1953).
2. B. Theodorovic, "Experiments with the Improved Egyptian Rural Stove," Arab States Fundamental Education Centre, Sirs-el-Layyan, Menoufia, Egypt, 1953.
3. T. Kalluppatti, "Magan Choola," Gandhiniketan Ashram, Maduri District, India, 1955.
4. H. Singer, "Report to the Government of Indonesia on Improvements of Fuelwood Cooking Stoves and Economy in Fuelwood Consumption," FAO Report #1315, Rome, 1961.
5. Institute of Natural Resources, University of the South Pacific, and Department of Forestry, Fiji Government, "Wood as an Alternative Energy Resource," Report of the seminar held at the University of Bure, Laucala Bay Campus, University of the South Pacific, Suva, Fiji, July 3-4, 1978.
6. Canadian Freedom from Hunger Foundation and Brace Research Institute, "The Smokeless Stove - Ghana," Case Study #7, Appropriate Technology Handbook, Ottawa, Canada, 1976.
7. Estación Experimental Choqui, "The Lorena Mudstove," Quezaltenango, Guatemala, 1977.
8. Georgia Institute of Technology, "Recent Efforts to Develop Simple Wood Burning Stoves," A State of the Art Survey of Solar Powered Irrigation Pumps, Solar Cookers and Wood Burning Stoves for Use in Sub-Sahara Africa, Small Industry Development Network, Atlanta, Georgia, 1978.
9. Eric Simon and Pedro Solis, "Economic Stove that Burns Sawdust as Fuel," Appropriate Technology 4 (May 1977): 23-24.
10. Mario P. Chanco, "The Rice Husk Stove," Appropriate Technology 5 (November 1978): 18-19.

water to a boil, in which the cooking efficiency is defined as the ratio of the theoretical quantity of energy required to boil the water to the energy content of the fuel consumed. In addition, the effect of using differing kinds of wood and other fuels is often examined, as well as testing for performance sensitivity to alterations in stove designs, cooking vessels, and fuel choice.

The objective of these measurements is to provide comparisons of the performance of different stove designs under actual use conditions, which is very important in evaluating the relative usefulness and impacts of different stove designs. However, this goal has not always been achieved in the testing programs conducted to date because the procedure usually adopted for measuring the efficiency of a cooking stove only involves heating a pot of water to a temperature just below boiling.(65) This test is not representative of actual cooking practices which vary greatly from region to region.

A number of studies have been conducted to evaluate the operation of a single stove design and/or to compare the operational efficiency of different designs.(66) The accuracy and reliability of many of the reported results leave much to be desired.(67) Nevertheless, the principal determinants of cookstove performance have been fairly well identified for operation under closely-controlled laboratory conditions. However, further work is needed to validate cookstove performance under actual use conditions, to determine the effects on efficiency of different cooking requirements and practices, and to evaluate the performance sensitivity to variable fuel inputs.

In addition, cookstove efficiency measurements, based on heating water trials, may be heavily influenced by the rate of energy delivery to the cooking vessel and the temperature of the delivered heat. Cooking efficiency rises rapidly with increases in both the input power and temperature.(68) For example, a simple water heating test favors stoves with high heat output and stoves which perform well on this test may put out much more heat than is needed for operations such as simmering, resulting in much of the stove heat output being wasted. However, for many actual cooking sequences (e.g., rapid boiling followed by long simmering periods), high input power levels and temperatures are desirable for only relatively short intervals during the cooking process, and often lead to low efficiencies for food preparation because of the still insufficient fire control capability of many improved cookstove designs.

In order to improve upon the past methods of evaluating stove performance, Dutt suggests that an easily reproducible test, such as boiling water, should be followed by actual cooking tests consisting of preparation of typical meals or by simulations of actual cooking activities with scientifically-constructed, water-heating sequences.(69) Although these schemes may be preferred from a theoretical viewpoint, in practice such tests are likely to be difficult to perform adequately because tests based on preparation of typical meals may be difficult to control for accuracy, and the determination of proper water heating cycles requires the tabulation of heat capacity values for standard food items (i.e, information which apparently has not yet been compiled in a systematic way).

Performance Sensitivity

A number of design parameters have been altered in cookstove testing procedures (i.e., boiling water tests) to evaluate performance sensitivity. For high-mass stoves, the results of these studies indicate that the following factors may have significant effects on cookstove performance: (1) combustion chamber design; (2) distance from the central pot seat to the chimney; (3) width and shape of the flue ducts and chimney; (4) slope of the flue ducts; (5) location of the dampers; (6) size(s) of the damper opening(s); (7) dimensions of the pot seats; (8) penetration distance of the pots into the flue channels; and (9) dimensions of the cooking vessels.(70) Perhaps the two most important variables affecting cookstove performance are the design (e.g., size) of the combustion chamber (firebox) and the control of the draft in the flue-chimney system.(71)

Most high-mass cookstove design efforts have implicitly assumed that installation of a flue-chimney system will automatically produce a more efficient design; however, some performance test results and cookstove analysts question whether this assumption is correct.(72) For example, in some cases, high-mass smokeless stoves with chimneys have not performed as well as low-mass, single-burner, portable designs in laboratory test sequences.(73) In fact, an improperly designed flue-chimney system may produce an intense fire but, in doing so, may pull the combustion gases out of the stove so quickly that heat is not effectively transferred to the cooking vessels which results in lower cooking efficiencies than might otherwise be achieved. Thus, the flue-chimney design must be sensitive to the other stove design parameter choices to ensure the development of a functional cookstove design.(74)

Heat losses from cookstoves generally occur as the result of (1) incomplete combustion of the fuel, (2) inadequate heat transfer mechanisms from the fire to the cooking vessels, (3) convection and radiation losses from the stove and cooking vessels, and (4) transient processes that do not contribute effectively to the cooking process.(75) These losses are best controlled by careful design and operation of the stove. For example, accurate control of the combustion air intake and damping systems will improve combustion efficiency. Convection heat losses may be reduced by careful stove construction to minimize cracks and leaks around cooking vessels. Fire-to-cooking vessel heat transfer losses can be reduced by proper design of the cooking vessels (i.e., utilizing large surface areas and materials with high thermal conductivities) and flue channels (e.g., installing raised mounds below pot seats to force the flue gases to squeeze past the bottom of the cooking vessels). Transient processes can be best controlled by proper operation of the cookstove, for example, by selecting the correct quantity of fuel to perform the cooking task to avoid fuel being consumed unnecessarily after cooking has been completed.

Fuel Selection and Typical Cookstove Efficiencies

Most of the high-mass improved cookstoves are designed to burn firewood, while the low-mass portable designs typically utilize charcoal or soft coke. Cooking efficiencies, based on water boiling tests, generally range from 20 to 40 percent for both high- and low-mass improved cookstove designs, compared to open-fire efficiencies of 5 to 10 percent. An example of measured cookstove efficiencies as a function of stove design and fuel choice is given in table 2.3. Kerosene stoves are included in this tabulation for comparative purposes.

Although these data support the often claimed performance advantages of improved cookstove designs, conflicting data are not uncommon.(76) In fact, laboratory cookstove efficiency measurements and comparisons have little utility without accounting for operational practices and, hence, are not particularly useful measures of the efficiency of day-to-day cooking schedules. Furthermore, villagers are more likely to evaluate cookstove design comparisons by daily firewood requirements than by relative efficiency data.

In summary, significant shortcomings and inconsistencies are commonly found in the design, performance, and/or documentation of many of the past cookstove performance testing efforts. For example:

- laboratory tests tend to use cookstoves which are specially constructed for the tests under closely-controlled building practices;
- laboratory tests typically utilize specially prepared firewood or other fuels;
- laboratory tests are based on water heating trials which do not accurately reflect rural cooking practices;
- laboratory test data are not directly comparable in most cases because of the use of inconsistent testing procedures and data collection methods; and

Table 2.3. Typical Cookstove Efficiencies

Fuel and stove design	Average measured efficiency, %
SOFT COKE	
Mud-lined sigri	26.65
Brick chula	25.3
Iron sigri	12.8
CHARCOAL	
Brick chula	25.7
Iron sigri	27.5
FIREWOOD: Logs	
Brick chula	14.0
FIREWOOD: SMALL PIECES	
Brick chula	19.1
Mud-lined sigri	18.7
COW DUNG	
Special mud-lined chula	10.4
KEROSENE	
Valve-fed capillary stove	41.6
Wick-fed capillary stove	38.2
Pressure stove	54.6

Source: Adapted from Z.T. Bandukwala, Domestic Fuels in India (Bombay: Asia Publishing House, 1959), Appendix A, pp. 53-60.

- laboratory test data which are currently available appear to be unreliable, not easily validated, and inconsistently and/or incorrectly reported in many cases.

Important Cookstove Design Characteristics

The results of studies on cookstove performance may be used to identify important design choices for the development of efficient cookstove designs.(77) Thus, desirable basic design specifications appear to be as follows:

- the firebox should not be insulated;
- the combustion chamber should be closed with some type of air intake control;
- the chimney should be as straight as possible and should not be tapered at either end;
- the chimney should be the downcast type (i.e., each successively higher segment should be fitted outside of the lower one) if the chimney is segmented;
- the chimney and flue duct system should be designed to operate so that the exhaust gases are hot enough to avoid condensation;
- the pot seats on smokeless stoves should be designed so that the cooking vessel fits into the seat with no gaps and sits as low as possible in the flue duct;
- the flue-duct system should contain raised areas below each cooking vessel and should be flat rather than sloped;
- the cooking vessels should be curved-bottom vessels with the area exposed to the heat source blackened; and
- the materials used in the stove construction should be chosen to improve durability as should the stove dimensions and construction methods.

FUEL USE FOR COOKING

Factors Influencing Fuel Choices

The primary source of cooking fuel in rural areas of developing countries is firewood, but other sources of cooking fuel are also commonly used. Typical fuel substitutes for firewood are: animal manure, agricultural wastes, wastes from lumbering activities (e.g., sawdust, wood chips, and shavings), charcoal (a derivative of wood fuel), soft coke, coal, and kerosene. The reasons for substitution of these fuels for firewood in a particular region are dependent on a number of factors:

- availability of firewood;
- availability of substitute fuels;
- immediate relative costs in terms of time, labor, and money of procuring the alternative fuels;
- relative costs of acquiring the cooking apparatus needed to burn the substitute fuels;
- efficiencies with which the alternative fuels can be utilized;
- income level or social status of the user;
- legal restrictions on the use of firewood;
- cultural or religious preference; and

- preferred cooking methods (e.g., frying, boiling, steaming, baking).

Fuel Collection Practices

In most rural areas of developing countries, firewood collected from the region immediately surrounding villages is the preferred cooking fuel. For example, in East Africa, firewood is collected for village use over traveling distances up to about 80 kilometers away before charcoal becomes an attractive replacement.(78) In Indian villages, fuel requirements met by forests and associated vegetation diminish quickly as the forest edge lies farther away from the village point of use. Villages surrounded by a forest typically use firewood to supply 100 percent of their fuel needs; however, if the forest is 10 km away, firewood supplies only 70 percent of the fuel requirements and, for distances greater than 15 km, the fuel needs fulfilled by forest growth decline to nearly zero.(79) In other words, wood fuel is apparently preferred as long as the source is within a reasonable walking distance of Indian villages.

In Nepal, migrants moving from the sparsely forested hill regions down to the tree-covered Terai areas often use twice as much wood for fuel as do the remaining hill residents. In Thailand, people living in regions of scarce wood supply use less than half the amount of wood that is used in more heavily wooded areas. In Tanzania, the household use of firewood in well-forested areas is over three times that used in other regions.(80) These examples highlight the important role played by wood availability in determining the level of use; that is, as long as forested areas are within easy access, firewood is used extensively, sometimes to fulfill all fuel needs.

In addition, firewood is often the preferred fuel in rural developing areas because wood is usually freely gathered for use by villagers, with only the associated cost of the time and labor spent retrieving the wood. Since most rural firewood use does not entail a monetary transaction, wood is often classified as a noncommercial energy source as are, for example, agricultural wastes and animal manure. As long as the wood source is close by, even the nonmonetary costs of firewood collection are relatively low and, thus, the total cost of cooking using the traditional open fire or simple mud stove is very small. This situation changes rapidly, however, as local wood supplies are diminished. For instance, as forested areas retreat from points of energy use, wood tends to be sold commercially rather than being gathered by final users. This results in disproportionate rises in energy costs for rural and urban poor alike.(81)

Even when firewood, though scarce, is still collected for free, the cost is becoming great in terms of invested time and labor. In sub-Saharan regions of Africa, distances of 50 km may be traveled by family members to collect firewood.(82) In some rural Indian areas, the search for the weekly firewood supply for a family can entail up to two days labor.(83) Moreover, as cities and towns in developing countries experience increased population growth, landholders in villages are enticed to transport firewood to densely populated areas rather than to permit gathering by local families. This situation has been observed in India and Pakistan where, as fuel wood prices rose dramatically in urban areas, families in rural villages were prevented from continuing the traditional practice of collecting wood from village land holdings.(84) Thus, as demand places a monetary value on fuel wood, rural supplies become inaccessible to rural populations dependent on subsistence collection.(85) In areas where firewood is purchased, the cost may absorb 25 to 30 percent of the income of an average family. This situation is common, for example, in the Sahelian regions of West Africa,(86) and also in the highlands of Guatemala where 1 m^3 of firewood (i.e., a one-month supply for an average family) cut and stacked may cost \$6 to \$15 (1978).(87)

Use of Green Wood

An additional cost incurred as local firewood scarcities increase in regions surrounding villages is the loss in burning efficiency of wood when it is not allowed to dry properly before use. Estimates for burning efficiency reductions resulting from burning green instead of air-dried wood range from 10 to 20 percent or more, depending on the type of wood burned and the cooking device used. As firewood becomes scarce in rural areas, the opportunities for stockpiling wood to provide dry fuel supplies are diminished greatly.(88) Consequently, as acquiring wood supplies becomes more of a hardship, an increased amount of the heat potential of the gathered wood is likely to be wasted. The heating value of a solid cord of wood varies substantially depending on whether the wood is green (i.e., a 70 percent moisture content or more on an oven-dry basis, which is the ratio of the weight of moisture removed in oven drying to the weight of the oven-dry wood) or air dried (i.e., 20 percent moisture content on an oven-dry basis). The inefficiencies associated with the necessity of burning freshly cut, green wood as opposed to properly air-dried wood are encountered precisely when firewood is in short supply and should be conserved. For example, on a volume basis, typical increases in heating values range from about 10 percent to over 44 percent as the result of air drying green wood.

For the same weight, air-dried wood shows an average increase in heating value of 63 percent over green wood(89) (see table 2.4).

Wood availability is certainly a major determining factor in the propensity of rural villagers to use alternate fuels, since the costs of firewood increase rapidly as the supply becomes limited. Typically under these conditions either fuel substitutes are found or, with unfortunate curtailments in basic consumption, fuel use is reduced. Fuel use reduction may, in some cases, result in villagers giving up one of the traditional daily hot meals reducing the consumption of vegetables or other foods that require cooking. These nutritional impacts of firewood scarcities have not been adequately investigated.(90)

Wood Fuel Substitutes

Common substitute fuels for cooking needs are animal manure, agricultural wastes, and, when affordable, charcoal transported to the village from forested areas, or kerosene. Animal dung and crop residues are the most frequently used substitutes in rural areas; however, these fuels are estimated to provide, respectively, about the equivalent of 13 percent and 2 percent of the energy provided by firewood in rural developing areas.(91) Nonetheless, 60 to 80 million tons of fried dung cakes* are estimated to be burned annually in India for fuel.(92)

In rural areas of India and Bangladesh, animal dung represents 14 percent and 18 percent, respectively, of the aggregate energy consumption in the forms of wood fuel, dung, and crop residues; crop residues comprise 25 percent and 52 percent of the total, and wood fuels account for the remainder of the noncommercial consumption of organic fuels.(93) Statistics reported for noncommercial energy use surveyed for towns

---

* Cow dung patties, a mixture of cow dung and crop residues or straw, have heating value estimates ranging from 2,130 kcal/kg to 2,609 kcal/kg, which for the average value of 2,370 kcal/kg yields a caloric value of 2.15 million kcal/ton. A significantly higher value has been given for pure dried cow dung by Makhijani and Poole, equal to 3.53 million kcal/ton; dried crop residues were estimated to have a heating value of 3.78 million kcal/ton, so that any combination of the two fuels to form "dung cakes" would produce a per ton caloric value greater than that for dung alone. This conclusion differs from the earlier estimates by over 1 million kcal/ton or about 65%. The discrepancies here may indicate how little is known about the real caloric values of these noncommercial fuels.

Table 2.4. Comparison of the Weights and Heating Values for Solid Cord (2.24 m^3 or 80 ft^3) Selected Woods in a Green or Air-Dried State

	Type of wood										
	green wood				air-dried wood			air-dried vs. green wood			
Tree type	moisture content* %	weight, kg/cord	heating value, 10^6 kcal/cord	heating value, kcal/kg	moisture content* %	weight, kg/cord	heating value, 10^6 kcal/cord	heating value, kcal/kg	Percent more heating value/cord	Percent more heating value/kg	Percent less weight/cord
Ash, White	45	1,743	4.16	2,387	20	1,562	5.04	3,227	21	35	10
Aspen	100+	1,562	2.60	1,665	20	981	3.15	3,211	21	93	37
Beech, American	60	1,961	4.36	2,223	20	1,707	5.49	3,216	26	45	13
Birch, Yellow	73	2,070	4.36	2,106	20	1,671	5.37	3,214	23	53	19
Douglas Fir (heartwood)	37	1,453	3.28	2,257	20	1,090	4.54	4,165	38	85	25
Elm, American	93	1,961	3.60	1,836	20	1,317	4.34	3,295	21	80	33
Hickory, Shagbark	60	2,288	5.22	2,282	20	1,925	6.20	3,221	19	41	16
Maple, Red	75	1,816	3.78	2,082	20	1,453	4.69	3,228	24	55	20
Maple, Sugar	68	2,034	4.64	2,281	20	1,671	5.37	3,214	16	41	18
Oak, Red	65	2,325	4.51	1,940	20	1,671	5.37	3,214	19	66	28
Oak, White	68	2,288	4.84	2,115	20	1,780	5.72	3,214	18	52	22
Pine, Eastern White	68	1,308	3.08	2,355	20	944	3.35	3,549	9	51	28
Pine, Southern Yellow	70	1,816	3.58	1,971	20	1,180	5.17	4,381	44	122	35
AVERAGE VALUES	68	1,894	4.00	2,115	20	1,458	4.91	3,412	23	63	23

*Moisture Content on an oven-dry wood basis = $\frac{\text{Weight of moisture removed in oven drying}}{\text{Weight of oven-dry wood}}$.

Source: Based on data from Synthetic Fuels Office, Development Division, California Energy Commission, "Woodburning Stoves," Staff Draft Report, November 1978.

and villages in India differ slightly from those given here, showing wood fuel, dung, and crop residues comprising 57, 26 and 17 percent of total organic fuel consumption, respectively, based on conversion factors of 3.53 million kcal/ton for dung and 3.78 million kcal/ton for crop residues.(94) Using an estimate of 2.15 million kcal/ton for cow dung, the percentages are 63.6 percent for wood, 17.7 percent for dung, and 18.7 percent for crop residues. It is interesting that the amount of dried cow dung used was reported to be 50 million tons annually, whereas Eckholm reported an estimate of 60 to 80 million tons per year for India alone.(95) Dung is also a source of fuel in parts of Africa's Sahelian zone, in Ethiopia and Iraq, and in the sparsely forested Andean valleys and slopes of Bolivia and Peru where, in some areas, the dung of llamas has served as the primary fuel since the time of the Incan empire.(96)

Commonly burned crop residues include cotton stalks, corn stalks and husks, sorghum stalks, sugarcane bagasse, and rice hulls. For example, in Egypt, a traditional source of fuel in winter has been corn stalks and husks.(97) In rural areas of Upper Volta, millet stalks serve as a replacement fuel for wood from the end of the harvest in November until May.(98) In the Punjab of India, cotton and sorghum stalks, pigeon pea, and sugarcane bagasse are widely used.(99) The use of animal manure and agricultural wastes in place of firewood does not entail an immediate monetary outlay; however, the delayed costs in terms of lost soil structure and fertility from diverting these natural fertilizers from the tilled land may be high. Reduced food production from loss of topsoil containing important nutrients or the necessity of revitalizing the land with chemical fertilizers are examples of the indirect costs of using these organic wood fuel substitutes for cooking.

Commercial Fuels

Villagers who can afford to pay for commercial fuels may turn to using charcoal, soft coke, coal, or kerosene. The preferred fuel choice will depend on availability and cost. The attractiveness of a particular fuel choice will also depend on the availability and cost of any special cooking device needed to utilize the fuel. For coal or kerosene, the supply is dependent on having a workable network of roads or railways for transport.(100) In cases where wood and charcoal may be used interchangeably in the traditional cooking apparatus, fuels such as kerosene, coal, or soft coke may require relatively expensive stoves to be utilized. In Fiji, pressurized and wick-type kerosene stoves may cost as much as $40 (1978).(101) The ease of use with which stoves are turned off and on and the control of flame intensity are the major attractions of kerosene stoves.(102)

Fuel prices have increased substantially during the last six years which may have discouraged the substitution of kerosene for wood fuels by rural and urban poor populations in developing countries. In Malawi, for example, use of kerosene declined by 24 percent between 1973 and 1976, allegedly due to higher prices.(103) The high cost of commercially distributing fuels such as coal or kerosene to rural areas may result in prices too high for any but the wealthiest villagers.(104)

In areas where desertification has set in as a result of deforestation, a pattern has been observed where first wood and charcoal, and later kerosene or other imported fuels, are brought in commercially to villages that have depleted the local forests.(105) In rural Fiji, increases in cash income are usually followed by switches to kerosene stoves because the people dislike the chore of collecting firewood.(106) In Guatemala, as incomes rise, people often switch to a propane stove costing $150 to $200 (1978).(107)

The use of kerosene or other commercial imported fuels is ordinarily infeasible for poor rural villagers in developing countries. An increase in kerosene use must be preceded by a rise in income or an input of government support providing, or subsidizing the purchase of, stoves or fuel. In some urban areas of the developing world, the sale of wood fuels (including charcoal) has been declared illegal and/or the use of fuels such as kerosene has been subsidized by the government to help reduce the consumption of firewood. For example, this situation occurred in the Republic of Korea where urban sale of fuel wood is now illegal, in India where kerosene and soft coke use has been subsidized through price supports and development of more efficient stoves, and in Senegal where tax incentives and design of better stoves have encouraged the use of butane.(108)

In areas fortunate enough to have an indigenous stockpile of burnable refuse from agriculture or lumbering activities, a switch from firewood may be a monetary and waste saving proposition. In the rural Philippines, where a cubic meter of firewood may cost $5, rice hulls may be purchased for fuel at 2.5¢ for a 30-kilogram bag, which is enough to supply cooking needs of an average family for a month. Even though an initial outlay of about $140 (1978) is required for the rice hull stove, many people are able and willing to pay for the stove in return for fuel savings and a more reliable fuel supply.(109) In Honduras, where tree cutting for wood fuel is illegal, the export forest industry produces substantial quantities of sawdust. Lumbering activities in other parts of the world also produce large amounts of waste in the form of sawdust, wood planings, and unused tree parts. The challenge in many of these cases is to find an economic way of transporting wood wastes to the point of use.

Charcoal Utilization

A convenient way to transport wood fuel to rural villages is in the form of charcoal. Converting wood to charcoal produces a cooking fuel with approximately twice the heating value per unit weight as wood (see table 2.5). Average heating values for charcoal are about 7,400 kcal/kg, while average values for air-dried wood are approximately 3,700 kcal/kg. Charcoal has nearly the same heating value as soft coke or coal which are nonrenewable and usually imported. Charcoal may also be easily substituted for firewood in most cookstoves and burns with a nearly smokeless flame. The cooking efficiencies of charcoal, coal, and soft coke are essentially equivalent, but coal and soft coke require special and more costly stoves. Kerosene, with an 11,100 kcal/kg heating value and a utilization efficiency of 35 to 50 percent, is a far superior source of useful heat than an equivalent weight of charcoal (see table 2.5).

Charcoal has traditionally been the preferred fuel in many regions. In Thailand and Senegal, for example, almost 50 percent of all wood consumed is first converted to charcoal in kilns. Charcoal consumption in rural areas of Thailand accounts for about 40 percent of the total rural wood fuel use. In Gambia, charcoal is estimated to represent 26 percent of total wood fuel consumption, but that figure decreases to about 9 percent in rural areas.(110) As forested areas recede from points of use, charcoal will most likely comprise a greater portion of fuel consumption as a result of the relative ease and low cost of transporting it. The greatest concentration of charcoal use is now, and will probably continue to be, in urban areas. However, charcoal is used to some extent now in rural areas and increased utilization may occur where fossil fuel use is infeasible, where use of agricultural residues and animal manures is discouraged, and where wood fuel supplies are a large distance away from the point of use.

The processing of wood to produce charcoal normally results in substantial energy losses. In rural areas, 50 to 84 percent of the energy in wood may be lost in conversion to charcoal.(111) For example, traditional charcoal kilns in Ghana operate with only 5 to 15 percent efficiency in converting wood to charcoal by weight. A more efficient, Tranchant kiln, which is portable and made of steel, yields about 20 to 25 percent charcoal by weight.(112) If air-dried wood is used to make charcoal in a traditional kiln (i.e., 10 percent efficiency by weight) and the end use efficiencies of using the fuels is taken into account (e.g., 10 percent for wood fuel and

Table 2.5. Comparison of the Heating Values and Utilization Efficiencies of Alternate Cooking Fuels

Fuel Source	Sources and estimates of heating values, 10^6 kcal/ton						Average heating value		Utilization efficiency, %	Useful heat, kcal/kg
	1	2	3	4	5	6	10^6 kcal/ton	kcal/kg		
Oven-dry wood		4.26	4.34		4.28	4.09	4.24	4,670	5 to 20	235 to 935
Air-dry wood	3.5		3.21				3.35	3,690	5 to 20[b]	185 to 740
Green wood							2.08[a]	2,290	5 to 20[b]	115 to 460
Cow dung			(cakes) 2.37	3.53	(cakes) 1.93		2.61 2.61	2,875 2,875	5 to 20 5 to 20	145 to 575 145 to 575
Bagasse		4.28	2.02	3.78			3.36	3,700	5 to 20	185 to 740
Cotton sticks			2.72	3.78			3.25	3,580	5 to 20	180 to 715
Rice hulls				3.78			3.78	4,160	5 to 20	
Saw dust										
Charcoal	7.1		6.51		6.30	7.08	6.75	7,430	20 to 35[c]	1,490 to 2,600
Soft coke					5.72		5.72	6,300	20 to 35	1,260 to 2,205
Indigenous coal			4.28 5.54				4.91	5,410	20 to 35	1,080 to 1,890
Imported coal		5.90	6.30				6.10	6,720	20 to 35	1,345 to 2,350
Kerosene	10.4	9.90	9.88 9.53		9.81	10.90	10.07	11,100 (9,100/liter)	35 to 50[c]	3,880 to 5,545

[a]Green wood is estimated to have a 38% lower heating value than 25 to 30% moisture-content air-dry wood on a weight basis.

[b]Burning tests using oven-dry wood yield efficiencies of 5 to 10% for open fires or traditional stoves and 10 to 20% for improved cookstoves. Burning air-dry wood containing moisture to be evaporated will utilize the heat potential less efficiently than if the wood was oven dry. Burning green wood with a high moisture content will further reduce heat-potential utilization by as much as 20% below that for air-dry wood. Thus the heat-potential utilization efficiencies for air-dry and green-wood fuels will probably fall in the lower portion of the 5 to 20% range for the kinds of traditional or improved stoves discussed in this report.

[c]The efficiency ranges for charcoal and kerosene are taken from J. E. M. Arnold, Source 1 below.

Sources:

1. J.E.M. Arnold, "Wood Energy and Rural Communities," Position paper for the Eighth World Forestry Congress, Jakarta, October 16-28, 1978 (Rome: FAO, October 1978), Table 5, p. 11.
2. Institute of Natural Resources, University of the South Pacific, and Department of Forestry, Fiji Government, "Wood as an Alternative Energy Resource," Report of seminar held at the University of Bure, Laucala Bay Campus, University of the South Pacific, Suva, Fiji, July 3-4, 1978.
3. S.A. Draper, "Wood Processing and Utilization at the Village Level," Third FAO/SIDA Expert Consultation on Forestry for Local Community Development held in Semarang, Indonesia, December 1977, Appendix 3, p. 1, table for "Comparison of Data on Heat Equivalent of Different Fuels."
4. Arjun Makhijani and Alan Poole, _Energy and Agriculture in the Third World_ (Cambridge, Mass.: Ballinger, 1975), Table 4-6, p. 109.
5. Z.T. Bandukwala, _Domestic Fuels in India_ (Bombay: Asia Publishing House, 1959), p. 58, Appendix A, Table 2.
6. From Source 1, Table 6, p. 15. Taken from W.M. Floor, "The Energy Sector of the Sahelian Countries," Policy Planning Section, Ministry of Foreign Affairs, The Netherlands, April 1977.

35 percent for charcoal) about 30 percent of the potentially useful energy in the wood is lost by conversion to charcoal. Even using a 25 percent efficient kiln on a weight basis to produce the charcoal, the efficiency of direct wood utilization for cooking would only have to be about 18 percent before the potential useful energy of the wood fuel would be exceeded by the potential useful energy contained in the charcoal if the charcoal is burned with a 35 percent efficiency. For the same assumptions for charcoal kiln and consumption efficiencies and a 20 percent efficiency for wood burning, charcoal use is about 12 percent less efficient than direct wood combustion. Thus, on a net energy basis alone, if air-dried wood were burned in an improved stove with an efficiency of 18 to 20 percent, charcoal production and consumption would result in an accelerated utilization of renewable energy resources.

Charcoal may become economically attractive when the wood source is a great distance from the point of use. Approximately 50 percent of the weight of a unit of wood energy would have to be transported as charcoal to deliver the same amount of heat potential, which may then be burned twice or more as efficiently as the original wood source. If transportation costs are very high, producing charcoal may be the only economical way to supply remote areas with fuel. In fact, if transportation distances are extremely long, the energy losses in the wood-to-charcoal conversion process may conceivably be returned in the form of transportation energy savings. Because charcoal is susceptible to losses during transport and handling as the result of pulverization, the use of medium to high density wood for charcoal production may be required to ensure that the charcoal product will withstand transport over long distances. On the other hand, wood is heavy and bulky to transport, must be seasoned before being burned with any reasonable heat utilization efficiency, has to be chopped to a usable size, requires a more complicated stove than is traditionally used to be burned as efficiently as possible, and is subject to deterioration in storage from rotting and damage by pests. Charcoal is not subject to deterioration in storage. In addition, charcoal may have some cultural attractiveness because when used for cooking over an open fire, a desirable flavor is imparted to cooked foods(113) and because some people believe charcoal to have beneficial magical powers.(114)

Concluding Remarks

A switch to nonrenewable fossil-based fuels such as coal, soft coke, or kerosene appears not to be a feasible alternative for the widespread relief of the cooking fuel shortage in rural areas of developing countries. Switches away from the high dependence on fuel wood may occur in urban centers of developing countries where a preference may already exist for smaller, portable stoves which could utilize coal, soft coke, kerosene, or butane.(115) The potential for effectively transporting these fossil-based fuels to urban centers is a much more real possibility than to remote rural villages. This action could bring relief to poor rural dwellers who have increasingly had to forego fuel wood supplies in the form of firewood and charcoal which have been diverted for sale at a high price to city dwellers. However, this approach might only provide a temporary solution because of the rising costs of fossil fuels and may be difficult to achieve on a wide scale in areas where the traditional cooking methods (e.g., slow simmering) may be better accomplished by the kind of heat produced by a wood or charcoal fire.(116) The cultural biases in some urban regions favoring wood and charcoal use may thus make large-scale diversion of these fuels to rural areas a difficult undertaking.

An additional possibility exists where, as in Ghana and Fiji, lumbering activities may leave behind a quantity of felled but unusable waste wood that may be comparable to the amount of lumber removed for processing.(117) If these wastes were suitable to produce high-density, transportable charcoal in efficient kilns, conversion of waste wood to charcoal for commercial sale could truly extend the fuel wood energy base. Similarly, where land clearing for agriculture is practiced, which is often cited as the primary cause of deforestation, large expanses of forest area are burned. If an efficient method of converting wood from cleared areas to charcoal could be developed, a substantial source of wood-based fuel would result and revenue could be received from the sale of an otherwise wasted energy source.

Production of charcoal in rural areas for sale in local villages and towns has provided a livelihood for rural entrepreneurs. Although traditional earth kilns are very inefficient from an energy viewpoint, in Ghana, for example, attempts to encourage use of the more efficient, portable, Tranchant kiln have failed because of the relatively high financial investment required for this kiln as compared to the inexpensive earth kilns.(118) Although the loss from inefficient conversion of wood to charcoal may be high on an aggregated level, individual producers may have little incentive to purchase efficient kilns because a comparable increase in charcoal output could be achieved from the traditional earth kilns by merely increasing wood-gathering efforts for little or no cost. The most promising application of more efficient kilns probably lies with commercial lumbering operations which may be able to afford to invest in modern charcoal production technology to utilize the lumbering wastes.

Since wood fuel currently is, and will probably remain, the most practical fuel option in rural areas, all possibilities for increasing the supply to meet the cooking needs of villagers should be explored. Increasing the supply through an efficient conversion of wood wastes to charcoal is one possibility. Another way to augment wood fuel supplies, while also increasing the utilization efficiency, is to develop village fuel wood plantations. This option, combined with the dissemination of more efficient wood stoves, could greatly relieve the severe localized wood shortages being experienced in the rural areas of many developing countries.

FIREWOOD PRODUCTION PROGRAMS

Benefits of Fuel Wood Plantations

In rural areas where nearby forest growth is no longer sufficient to meet village needs for cooking fuel, local fuel wood plantations (i.e., managed timber cultivation programs) may provide necessary firewood supplies within reach of village consumers. A secondary benefit may accrue by allowing fuel wood to be properly stored and seasoned to an air-dried state before consumption. This practice may produce substantial energy savings in regions where wood fuel has been in such short supply that stockpiling has been impractical and where wood has been burned when still green. Fuel wood plantations may also offer substantial ecological benefits where unchecked deforestation and the use of animal manure and crop residues for additional fuel sources have created a destructive cycle in which increased soil exposure from tree cutting is followed by accelerated soil deterioration resulting from burning animal dung and crop residues for fuel. This loss of soil fertility, in turn, increases demands for land which, along with increasing fuel demands, cause further clearing of forest land, and so on. Establishing fuel wood plantations could possibly halt this degenerative process and perhaps even reverse the cycle.

Additional benefits from establishing fuel wood plantations may be the reduction in time and labor now spent searching for firewood in fuel-short areas and an effective increase in rural incomes by eliminating the need to purchase high-cost commercial fuels. Since children commonly participate in the time-consuming job of firewood collection, reductions in collection times may allow children to spend more time attending school.(119)

Conditions for Implementation

Successful implementation of a fuel wood plantation program to fulfill most or all of the firewood requirements of a rural village will depend on a number of local conditions. Some of these conditions are: amount of land available for planting trees; climate and soil conditions; existence of effective links between the local community and government forestry services; traditional land-use and ownership patterns; and level of community participation.

Candidate sites for fuel wood plantation projects may be best chosen by first determining the types and levels of local fuel consumption through village surveys, and by secondly employing techniques such as aerial surveys and satellite imagery to classify land areas into those that are currently unattractive for plantation efforts, those that are still heavily forested enough to receive secondary attention, and those where an immediate effort could make the difference between rapid deforestation and indefinitely sustained yields of firewood.(120) In these borderline regions, such as at the edge of an advancing desert which is expanding as a result of deforestation for fuel, fuel wood plantations are likely to provide the most immediate benefit by supplying additional fuel wood supplies and by staving off an increase in deforested land area. Fuel wood plantation programs coupled with the dissemination of more efficient, locally-constructed cookstoves may make large impacts over one- to five-year period in regions of growing fuel wood scarcity by, first, reducing fuel demands by utilization of more efficient stoves and, then, by producing increased fuel wood supplies after woodlots have matured.

The primary physical barrier to extensive fuel wood plantation programs is the amount and quality of land available for planting. In areas of subsistence farming, land areas able to support a heavy growth of trees will probably already be fully utilized for grazing or crop growing.(121) In most cases, land available for tree plantations will be that which is not only marginally suitable for agriculture, but which is also marginal for growing trees.(122) Establishing beneficial combinations of farming or grazing and tree growing, and making use of marginally productive land, may be prerequisites for the successful implementation of fuel wood plantations.

One approach to the land use problem is to use intercropping techniques (i.e., crops planted between rows of trees). Fuel wood species in intercropping arrangements may provide beneficial windbreaks, shade, watershed characteristics, and, in some cases, soil enriching mulch from leaf residues. Modifications of this scheme include fuel wood trees planted in combination with fruit and/or fodder trees and grasses for grazing.(123) An example of grazing and intercropping is the planting of elephant grass with trees in areas of Central Java where the grass is used for fodder for stall-fed cattle and the trees provide watershed protection as well as fuel wood.(124) In Indonesia, fuel wood needs have been met in local areas by planting Caliandra calothyrus, which is a vigorously coppicing (i.e., the ability to regenerate new growth from cut stumps), branching tree shrub that can, with proper trimming practices, produce usable fuel wood in the first year. This species is also easily established on poor soils, competes with undesirable lalang grass, is a legume with soil improving qualities, and has leaves that can be used as fodder.(125) Another example is Leucaena leucocephala which is a vigorously coppicing tree that has been traditionally intercropped for shade with coffee, cocoa, cinchona (which is a genus of evergreen trees or shrubs widely cultivated in the East Indies, India, and tropical America), pepper, and vanilla, as well as being used for fuel wood in the Philippines and Guam.(126)

In areas where fuel wood trees must be relegated to unused and/or wasteland areas, land plots may be artificially upgraded by application of irrigation and fertilizers. Trees may also be planted along roadsides, field boundaries, waterways, and around houses or on other small unused land areas. This technique has apparently been used successfully in China in recent reforestation efforts. A third approach to planting on unused, low-grade land areas is to select tree species that are considered to be weed trees but may still be suitable for use as firewood.(127)

Candidate Trees for Fuel Wood Plantations

A list of some species grown for fuel wood is given in table 2.6. One of the most intriguing species is the Leucaena leucocephala, sometimes called the giant ipil ipil. In the Philippines, the giant Hawaiian variety is now grown to provide fuel wood and building materials. This tree can potentially reach a height of 55 ft and a diameter of one foot in six years. Five to six years are typically required for this species to mature before being harvested, while other fast growing species with similar annual yields typically have rotation periods of six to ten years.(128)

Although 30 to 40 m^3/hectare is the average annual biomass yield for Leucaena trees, annual production may exceed 100 m^3/hectare.(129) Leucaena is a coppicing species that grows well in semi-arid and savannah regions of the subtropics and in humid, low-land areas. It thrives on steep slopes, in marginal soils, and in areas with extended dry seasons, but may exhibit retarded growth at elevations above 500 m.(130) Leucaena can improve the soil by increasing nitrogen and organic matter content, rebuilding tilth and surface texture, breaking up compacted soil layers, improving

Table 2.6. Selected Potential Fuel Wood Species and Regions of Application

Tree species	Potential application region
Acacia senegal	Africa (arid areas)
Albizia falcataria	Indonesia, Philippines
Anthocephalus chinensis	Philippines
Caliandra calothyrus	Indonesia
D. Sissoo (Sissoo)	West Bengal, India (north)
Dalbergia latifolia (Shisham)	West Bengal, India (north)
Eucalyptus deglupta	Philippines
Gliricidia muaculata	Indonesia
Gmelina arborea	Philippines
Indian neem	India
Leucaena leucocephala	Australia, Fiji, Hawaii, India, Indonesia, Java, Mexico (west & central), Papua New Guinea, Philippines
Lespedeza	Republic of Korea
Sebana grandifolia	Indonesia

Sources:

1. J.E.M. Arnold, "Wood Energy and Rural Communities," Position paper for the Eighth World Forestry Congress, Jakarta, Indonesia, October 16-28, 1978 (Rome: FAO, October 1978).
2. National Academy of Sciences, "Leucaena: Promising Forage and Tree Crop for the Tropics," Report of a study conducted by the Philippine Council for Agriculture and Resource Research and the United States National Academy of Sciences, Washington, D.C., July 1977.
3. Institute of Natural Resources, University of the South Pacific, and Department of Forestry, Fiji Government, "Wood as an Alternative Energy Resource," Report of the seminar held at the University of Bure, Laucala Bay Campus, University of the South Pacific, Suva, Fiji, July 3-4, 1978.
4. Keith Openshaw, "Wood Fuels the Developing World," New Scientist 61 (January 1974): 271-72.

water absorption, reducing moisture evaporation, providing a forest cover, and reducing soil erosion.(131) Although this species has not been adapted to arid mountain areas where firewood consumption is often high, potential exists for its use as a fuel wood plantation tree in denuded regions of Africa, India, and Central and South America.(132)

The planting of special, fast-growing, fuel wood species is especially attractive in areas with acute land shortages because the rates of wood production on a plantation can be five or more times the production of an indigenous forest of an equivalent land area. In Fiji, the natural growth rate for indigenous forests is estimated to be 4 to 6 m^3/hectare-year, compared to expected fuel wood plantation outputs of 20 to 30 m^3/ hectare-year. In addition, the rotation period for the fuel wood plantation may be eight years or less, whereas the natural forest may require a 25-year rotation for fuel wood production.(133) However, demand for fuel wood may be so large that the land requirements for fuel wood production are prohibitive in some land-poor areas, even if special fast-growing species and plantation techniques are used.

Land Requirements

The quantity of land needed to supply firewood for a family of five, consuming the annual average of 0.5 to 1.0 m^3/person for cooking with a traditional (i.e., 5 to 10 percent efficient) stove,(134) is 0.125 to 0.25 hectare if planted with a tree species producing 20 m^3/hectare-year. For villagers who need a cheap, reliable fuel source, the land needed for private fuel wood production would consume from 8 to 50 percent of an average family farm of 0.5 to 3 hectares.(135) In areas characterized by subsistence farming, even 8 percent of the family land may be too much to dedicate to tree growing. Accordingly, viable fuel wood plantations may require the use of community plots contributed, perhaps, by large landholders in the village.

For a typical Indian village of 500 people or less, the land requirements for fuel wood plantations yielding 20 m^3/ hectare-year are about 25 hectares to supply a village wood fuel demand of around 500 m^3/ year (1.0 m^3 per capita).(136) The feasibility of committing land to tree growing alone will depend on village resources and whether or not wealthy landholders can be persuaded either to grow trees for the village on their land or to dedicate the required land area to a village plantation project. In most cases, either of these options will require large changes in traditional land ownership rights along with strong administrative involvement by the government in the local fuel wood plantation effort.

Production Costs

Even if the land is donated, the expenses of planting and maintaining the fuel wood plantation must be covered. For the case in which both land and labor are free inputs to a village plantation effort with an annual yield of 15 tons/hectare (sufficient to supply annual firewood needs of 40 to 50 people), the cost to provide seedlings and to teach villagers to manage the plot is estimated to be about $20/hectare (1974).(137)

Makhijani and Poole estimated that the cost of providing cooking fuel from fuel wood plantations is half that required to produce the fuel requirements with kerosene. In this analysis, kerosene is assumed to require no land investment, and land is taken as a free input to a plantation because the land is used to provide the same amount of fuel that would have been gathered by cutting wood over an area five times as large in an already receding forest.(138)

Assignment of values to plantation schemes in terms of the degree to which further destruction of natural forests is prevented has been suggested by French.(139) For example, a large-scale fuel wood plantation in the African Sahel was estimated in 1978 to cost from $565 to $725/hectare to establish and maintain during the first year and to produce firewood for $10/ton in a region where firewood is sold for $6/ton. French argues that the price of firewood is only $6/ton because the wood has been collected by careless destruction of already thinned woodlands and that the real price for firewood should, therefore, be at least $6/ton plus the cost of replanting.(140)

Estimates for costs of fuel wood plantations vary depending on whether the projects are large-scale or small, local woodlots and whether the projects are located on already improved land or on poor, marginal lands. For large-scale plantations located on irrigated land in the Sudan of Africa, the costs (1978) may be $100/hectare as compared to $690/hectare near urban centers in the Sahel.(141) Revelle estimated in 1978 that a 30-hectare fuel wood plantation in the Nepalese hills, which could provide 250 villagers each with 0.75 tons of fuel wood per year, would cost $20,000 including installation of protective fencing, or about $670/hectare.(142) This estimate implies a production rate of 6.2 tons/hectare-year, while the estimate by Makhijani and Poole(143) is based on a yield of 15 tons/hectare-year. The smaller estimate appears to correspond to a natural forest regeneration rate rather than a growth rate for a plantation of fast-growing trees.

Cultural Adaptation

The creation of a successful fuel wood plantation program is unlikely unless the cooperation of the local community is obtained. In areas where wood scarcities have prevailed for long periods of time and firewood prices are high, local residents must be convinced that tree planting efforts will not be thwarted by illegal removal of immature trees by profiteers. In addition, in areas of scarce fodder for livestock, unprotected new trees may be destroyed quickly by roaming, underfed livestock.(144) Incentives to help plant or to abstain from prematurely harvesting new saplings are not easily established or maintained without an organized cooperative effort by villagers to participate in the management of a fuel wood plantation.

Traditional feudal land tenure systems can be a barrier to obtaining the willingness of villagers to supply labor to a fuel wood plantation. For example, in Ethiopia, local poor people, who were hired as laborers to plant trees as part of a public works scheme, deliberately planted the seedlings upside down because the benefits of the program were known to accrue to the wealthy members of the village, rather than to the poor.(145) The importance of community support is further demonstrated by plantation failures in both Niger and Colombia. In Niger, villagers pulled out newly planted trees and allowed animals to feed on the trees uncontrollably, which destroyed efforts to establish 500 hectares of village woodlots. Because the villagers had not been involved in the project formation, the woodlot areas were perceived to be traditional grazing grounds.(146) Similarly, efforts to preserve natural forest areas within a resettlement region of Colombia were unsuccessful because the new settlers felt the land could be better used for agriculture.(147)

In regions of the world where planting trees for fuel wood has been successful, an important factor has been a strong link between the central government and the local residents. In some cases, the image of the forester as being someone who guards the forest from use by local people in need of firewood must be changed dramatically before cooperation between forestry experts and local villagers can be obtained. Likewise, governmental attitudes toward forestry as an industry to produce timber for only export has had to change to include production of timber primarily to fulfill local firewood needs.

Changes of these types have apparently occurred in China and in South Korea. Successful plantation programs in China have been a result of almost 30 years of working to reforest denuded landscape for both environmental protection and production of fuel wood. Forested areas are reported to have increased from 5 to 12.7 percent of the land between 1949 and 1978 as a result of integrating replanting efforts with other, broader social reforms.(148)

In South Korea, the countryside was so devoid of fuel wood supplies by 1970 that residents were forced to further degrade the sparsely covered land by gathering leaves, grass, and forest litter to supply fuel wood needs.(149) Community involvement in a 1973 government forestry campaign was achieved by the creation of Village Forestry Associations (VFA) in most villages that required mandatory membership by a representative from each household. In areas where large landholders in the villages had by tradition allowed free firewood collection on the land by other villagers, little incentive existed for reforestation. As a condition of the government program, large landholders were given the choice of reforesting the land or relinquishing control to the local VFA for a tree planting program in return for 10 percent of the proceeds and most chose the latter. Because of the massive replanting of 643,000 hectares of village woodlots by local villagers, the government estimated the fuel wood supplies were nearly adequate for local needs by late 1977. The members of each VFA tend and harvest the woodlots without pay and, through each VFA membership in the nongovernmental Korea National Federation of Forest Association Unions, funds, technical assistance, and policy direction from the government are received.(150) In addition to the growing of trees, the program has been combined with an effort to improve the efficiencies of cookstoves so that the newly grown fuel supplies will be used more efficiently.(151)

The successes achieved by tree planting efforts in South Korea and China are based on the strong political commitment by the government, massive public participation, and the sharing of benefits by the local community members. Eckholm suggests that the community involvement would not have been obtainable without the self-generated desires of local people to make the tree planting programs work.(152)

COOKSTOVE DISSEMINATION EFFORTS

Recently, there has been increased interest in encouraging the widespread use of improved wood burning cookstove designs on the part of institutions supporting the advancement of living conditions in developing countries. This interest comes nearly thirty years after efforts to improve upon the indigenous, inefficient cooking methods in rural developing areas were initiated in the early 1950s. A main impetus behind efforts to introduce an improved wood burning cooking technology continues to be concern about the effects of firewood shortages upon poor rural dwellers and the local environment. For example, in 1953, Raju noted the primary advantage of the new HERL smokeless chula as being higher thermal efficiency and fuel savings.(153) Raju pointed out that traditional Indian cooking devices constituted a world problem which simultaneously affected health, housing, fuel economy, and forest economy; and reported that use of improved stoves was spreading gradually in developing areas such as India, Pakistan, Burma, Ceylon, Africa, the West Indies, Iraq, and Lebanon.(154)

Early Dissemination Efforts

Although there is some evidence that improved cookstove designs have been disseminated successfully in parts of Asia, Africa, and Central America, universal acceptance of improved cookstove technology has not occurred as a result of the early implementation efforts.(155) Apparently, many of the improved designs that have been placed in use have efficiencies only slightly above that of traditional cooking techniques, have not been very durable because the clay materials used crack very easily, and have not been well designed because the recommended methods of construction yield a lightly built stove with low thermal mass with a tendency for heat to be lost up the chimney.(156) Thus, many successful cookstove dissemination efforts are needed before more efficient, durable, clean, and simple cooking devices will be in widespread use throughout rural villages in the developing world that are highly dependent on wood fuel.

Eckholm notes that, because those suffering from wood fuel shortages are the illiterate poor of rural developing areas, the affected villagers cannot speak out to call attention to the localized crises and, therefore, problems of wood fuel shortages have not drawn much attention.(157) Furthermore, discussions of the basic needs of the rural poor have often ignored the fundamental importance of forests to human well-being, and few programs have been established to reduce the forest deprivation now suffered by many.(158) According to Eckholm, programs to conserve wood through such efforts as widespread dissemination of more efficient cookstoves have not received the priority attention the situation demands.(159)

Examples of Unsuccessful Cookstove Implementation Efforts

Improved cookstove extension efforts may fail for a number of reasons, including poor design, construction problems, insufficient performance improvements (either real or perceived), and cultural incompatibility. For example, a case has been recorded in the village of Karimpur, India, where women who were given improved smokeless chulas continued to use open, horseshoe-shaped mud fireplaces in spite of the benefits of the new stoves. An important reason for this situation was that women kept the traditional mud chulas ceremonially clean by applying coatings of a clay paste each morning, and this process was perceived by women to be difficult to perform and of questionable value for the more complicated smokeless cookstoves. Consequently, the new stoves were left unused and the tin chimneys rusted in the rain, which quickly made the stoves inoperable.(160)

In southern India, an improved patented cookstove design was introduced that required a special effort from woodcutters to cut firewood into small pieces to fit in the stove. The woodcutters refused to do the extra work and so the stove was not accepted.(161) In some areas (e.g., Fiji), where traditional open-fire cooking involves wrapping the meat or vegetables to be cooked with leaves or putting the food into bamboo tubes sealed with mud and placing them in the fire,(162) there are no pots or utensils to be washed after a meal is prepared. The better efficiency of an improved cookstove may not bring immediate benefits sufficient to compensate for the need to clean cooking utensils with each meal. In any case, there must be obvious and large benefits to be gained by the users from spending the time and money necessary to procure and learn to use new stove designs for dissemination programs to be successful.

Lorena Stove Dissemination Strategy

Lorena stoves have been widely extended to areas in Guatemala, Honduras, and Mexico. This stove design was developed at the Estación Experimental Choqui (Choqui Experimental Station) in Guatemala to help mitigate fuel wood shortages following the 1976 earthquake. Extension efforts have been aided by the capability of the design to be modified to respond to local requirements involving stove size, shape, fuel choice, and selection of cooking utensils, without significantly jeopardizing the performance. This stove was designed specifically to be built with locally-available tools, materials, and skills. However, dissemination efforts for Lorena stoves have been inhibited by problems in finding the required mix of sand and clay to make a strong stove structure that will not crack and by difficulties in developing an understanding by users of the proper operation of the dampers and chimney. Actual extension of the stove design to villages has been accomplished by having prospective stovebuilders, who were initially community leaders, persons with adobe or brick-making skills, or representatives from other agencies involved in extension work (e.g., CARE, the Peace Corps, national forestry programs, and farming or health agencies), come to a three- to five-day workshop to learn how to build and use the stove. Individuals from established organizations are encouraged to attend these workshops because these people are felt to be better prepared to actively promote and continually evaluate the cookstove technology. Church groups and missionaries are examples of organizational resources that have contributed to the stove dissemination efforts.(163)

Because the dissemination strategy relies on demonstrating the stove technology to the local people, workshop participants are encouraged to build a second stove for themselves so that, by the time a third stove has been constructed, it will be good enough to use as an example for other villagers.(164) In areas where the stove has been disseminated most successfully, there has been a good local promoter, who is usually already a community leader and who may have had experience working with a national or international development institution. Promoters may simply teach others to build the stoves or may set up a private business to build stoves for others at a price of $15 to $35 (1979), including materials. Typical profits for these entrepreneurs are $8 to $18 per stove.(165)

Workshops were at first held at the Estación but are now also held in the villages where the stoves are to be extended. Approximately 500 to 1,000 stoves have been built in Guatemala as a result of these workshops. Average workshop attendance is about nine to ten people, of which one or more generally become promoters of the technology. The course costs $4 to attend. By February 1979, approximately 650 people in Guatemala had been trained by the Estación to build Lorena cookstoves.(166)

The goal of the Estación Experimental Choqui program is to have Lorena stove promotion carried along by its own reputation in one to two years so that there will be no need to continue to train promoters. Perhaps one unique, self-sustaining success with Lorena stove dissemination to date is in Quezaltenango, Guatemala, where Lorena stoves were promoted as part of a local reforestation program conducted by a Peace Corps Volunteer and a volunteer from the government forestry department. This effort has resulted in this region having the largest number of stoves in Guatemala.(167)

The Centro Mesomericano de Estudios Sobre Tecnologia Apropiado (CEMAT) is also working in Guatemala to promote Lorena stoves by running courses for 12 people that last one week with the goal of training artisans who intend to build stoves for a fee. After successfully constructing five stoves, a participant receives a diploma as a qualified multiplier of the technology. Out of the 12 attendees, six typically build stoves in their own homes and three go on to become multipliers. The goals of CEMAT are to construct 400 to 500 stoves over a two-year period with funding from Appropriate Technology International (ATI). In addition to instruction on the way

Lorena stoves operate, CEMAT emphasizes ways to improve nutritional quality of food prepared on these stoves.(168)

Save the Children Federation (SCF) in Honduras expects to promote the Lorena stove through groups which are already successfully involved in development activities.(169) The programs of CEMAT, SCF, and the Estación Experimental Choqui teach both men and women how to build stoves.

Lorena Stove User Acceptance Surveys

A recent evaluation of the success of extension efforts in Guatemala by the Estación Experimental Choqui was conducted by surveying 64 families during a six-month period to determine how the Lorena stoves were working.(170) The survey found that 91 percent of the stoves were used daily and the remainder were either inoperable or not being used. Twenty-six percent of the stoves had been made by Station personnel and the rest were constructed by trainees. One-half of the stoves that were working used only 50 percent of the fuel that would have been used with traditional cooking devices. Although 73 percent of the stoves had cracked in some way, the performance of most of these was not measurably affected; most of the stoves had nearly smokeless operation. The survey concluded that 95 percent of the villagers with working stoves were well satisfied with this new way of cooking.(171)

Social acceptance problems were first encountered by the Lorena stove design in the developmental stages because the stove was initially designed low to the ground to imitate open-fire cooking and was made small in surface area because the designers felt that women would want the stove to require as little space as possible. Both of these perceptions of the desires of Guatemalan women were objected to strongly by the village women who came to see the first stoves because they preferred cookstoves at waist-height, similar to modern western stoves, and large, top surface areas for use as work space and even as a kitchen table, since the stove served, in many cases, as the only piece of furniture in the kitchen.(172) Even though the Lorena stove is adaptable to desired shape, size, and locally-available fuel types, some users want to change the design in ways that indicate a misunderstanding of the functioning of either the dampers or the chimney.(173) In one community, the sample stoves cracked and were abandoned, which led to rejection of the whole idea of a Lorena stove by the people, because the promoters did not have sufficient organizational backing to encourage the villagers to rebuild the stoves.(174)

Another evaluation of the success of Lorena cookstove dissemination efforts surveyed 40 families with Lorena stoves in two different communities in Guatemala.(175) This study revealed a high overall level of acceptance of the Lorena stove; however, the nature of this acceptance was influenced, in many cases, by the degree to which the stove was used to replicate traditional functions of an open fire, a practice which may seriously compromise the fuel conserving characteristics of the stove design. The principal advantages of the Lorena stove according to the users are elimination of the smoke from the kitchen area, fuel savings, increased ease and safety of cooking in a standing position, availability of hot water supplies, and pride in the attractive appearance of the stove. Although nearly 70 percent of the users observed some fuel savings, only 6 percent claimed a fuel use reduction of 50 percent or more.(176) Nevertheless, improvements made by the stove owners included the use of a reinforced firebox and door attachments, the application of a cement coating to prevent surface cracking, and the replacement of the metal water can with a more durable clay vessel.

Cited disadvantages of the Lorena stove included inadequate space heating from the stove, difficulty in cooking large quantities of food during harvest or fiesta days because of the fixed pot seat size, cracking or crumbling of the stove structure, difficulty in locating the proper type of clay material nearby the village, and the leakage of smoke into the kitchen and the buildup of creosote in the flue and on cooking vessels when green wood was burned. To overcome these perceived disadvantages, users commonly operated the stoves in ways which resulted in the stove yielding little or no fuel savings, or the users would simply rely on an open fire in addition to operating the Lorena stove.(177) These deviations from the recommended stove operation have the effect of making dissemination of the stoves less attractive from a fuel conservation viewpoint but more attractive from the users perspective because the stove may be adopted without requiring major life-style changes.

Shaller concludes that the lack of understanding of the function of dampers and pot seats on improved cookstoves comes from the overemphasis on the stove construction process compared to the discussions of proper stove operation during the extension programs. In addition, men dominated the stove construction courses and nearly always were responsible for construction of stoves in village homes. Consequently, women were forced to learn about stove construction, operation, and cooking practices from their husbands who often had little interest in or understanding of the proper use of the stove.(178)

Conditions for Successful Cookstove Dissemination

Current activity

Drastic forecasts of potential total deforestation in some regions before the turn of the century have at last promoted both high- and low-level government organizations, as well as private organizations, to consider the active dissemination of improved cookstove designs as an important tool for relieving some of the ecological stresses resulting from deforestation. Included in this group of institutions are large and influential organizations such as the World Bank and Appropriate Technology International (ATI) who are involved either directly or indirectly in financing efforts to expand the use of more efficient cookstoves. The World Bank includes introduction of more efficient wood burning stoves as a possible component of financed forestry projects for developing countries(179) and ATI is currently funding efforts by CEMAT in Guatemala and SCF in Honduras to encourage the widespread use of Lorena stoves in these Central American countries.(180) VITA has initiated an improved cookstove project in the African Sahel and additional cookstove coordinating activities are being planned in collaboration with the Intermediate Technology Development Group (ITDG), and a number of African organizations.(181) Cookstove design manuals have been published by VITA in collaboration with ITDG for four improved cookstove designs(182) and by Volunteers in Asia to aid in the dissemination evaluation and improvement of the Lorena cookstove design.(183).

In addition, preparations for the 1981 United Nations Conference on New and Renewable Sources of Energy are aiming, through the FAO, to stimulate participation by individual governments in assessing wood fuel supplies and demands. They are also helping to further endorse and coordinate interest in extending improved cookstoves to firewood deficient areas throughout the developing world.(184) These kinds of high-level support of, and institutional commitments on an international and national level to, more efficient use of firewood for cooking has been largely missing until recently, which could be a major reason why there has been only moderate success in dissemination of more efficient cookstoves since the improved designs were first developed in the 1950s.

Important design considerations

The main factor required for the success of cookstove dissemination efforts is to ensure that the improved cookstove not only has advantages from the viewpoint of the promoters but also has benefits that can be perceived by rural villagers as being worth the time and investment to change from a traditional form of cooking which, while inefficient, may be simple to understand and culturally acceptable. The new designs must be capable of demonstrating a substantial improvement in performance, which is readily translated into noticeable improvements in village life, because the introduction of improved cookstoves will inherently mean making changes in cooking practices which are often constrained by traditional, religious, and economic factors. For example, in parts of Africa, where women collect the firewood as well as cook, men may have little desire to spend time or money to help obtain a more efficient stove which might cut firewood collection time in half. In some regions, women have control or responsibility for the household budget covering firewood and cooking utensils.(185) In such a case, women must spend money if they want to acquire a better stove. All of these considerations illustrate the unique problems that will have to be dealt with in each local region to successfully extend the use of more efficient stoves.

The success of a cookstove dissemination program may be enhanced by the involvement of an influential and knowledgeable person or group in the local village that understands the needs of the local people, including skills, materials limitations, and cultural requirements, as well as understanding the basic principles behind the workings of a more efficient cookstove. Careful consideration of this type of information may produce an adapted cookstove design which is attractive to the potential users while still performing with a better efficiency and having the necessary durability. For example, one adaptation which might be useful to aid the acceptance of improved cookstoves in areas where villagers value the smoke from traditional cooking devices would be to develop an efficient stove without a chimney.(186) Similarly, substitute insect repellants and/or protective coatings for roofing and other building materials could be developed and disseminated as an integral part of smokeless cookstove extension efforts in regions where the smoke from traditional cooking practices plays an essential role in insect control. In Kaya, Upper Volta, a cement brick cookstove design was modified to eliminate the use of flue dampers by employing a smaller bore chimney to limit the draft automatically because the local women found the use of dampers too difficult.(187) The importance of the adaptability of improved cookstoves to local conditions and/or limitations in materials supply without completely compromising design principles essential to achieving a better stove performance should not be underemphasized. Questions surrounding the adaptability issue should provide a fertile area for future cookstove research efforts.

Mechanisms for implementation

Perhaps an appropriate vehicle for cookstove extension efforts is an organized program operating through or in connection with persons already involved in other development projects in the region, particularly fuel wood plantation or reforestation projects. However, once an improved cookstove design has been selected and presented, emphasis must be directed toward assisting villagers in thoroughly understanding how to build the stove and how the stove operates so that the technology will find a permanent place in village life. The extension effort in Guatemala where the input of the local reforestation personnel stimulated Lorena stove dissemination is a good example of incorporating these considerations into the extension program. Implementation mechanisms incorporating a community-based approach to promote the adoption of improved cookstoves, to increase local participation in design adaptation, to assist in the control of stove construction practices, and to support the development of user proficiency are apparently an important element in successful cookstove extension efforts.

In contrast to disseminating specific designs, an alternative extension strategy that would make available to target areas descriptions of several promising cookstove designs, together with the training and simple equipment required to evaluate cookstove performance, has been suggested by Dutt.(188) This approach encourages talented individuals to adapt stove designs or invent new ones that are compatible with local conditions, and introduces the concept of performance measurement into rural villages. Dissemination efforts of this type may be strengthened by establishing cookstove evaluation efforts in laboratories and by developing communication channels between extension centers and centralized research facilities.

In areas where villagers may not have the time or resources to build their own stoves, entrepreneurs may construct stoves for a fee which may create, if the demand is large enough, a cottage industry based on stove-building skills and, thus, provide a source of rural employment. Because this dissemination strategy may result in improved cookstoves becoming out of the reach financially of many poor rural villagers, organizations such as Volunteers in Asia (VIA) and Save the Children Federation have been investigating mechanisms to establish low-interest, guaranteed, revolving loan funds in developing areas so that rural poor could obtain loans to pay for professionally built Lorena stoves.

Concluding Remarks

Although active cookstove dissemination efforts are only in the early stages of development, a number of elements appear to be necessary to achieve the goal of widespread use of improved cookstoves in developing countries. Some of these elements are:

- communication channels should be established between cookstove users, village representatives, and extension workers to identify necessary adaptations for cookstove designs to be compatible with local conditions;
- involvement of village community leaders and/or development institution representatives stationed in the village should be encouraged in the promotion efforts;
- financing opportunities for villagers who could not otherwise pay to have a stove built should be made available as part of the extension effort;
- extension efforts should be tied in with other development activities occurring in the village that have community involvement whenever possible;
- communication links between extension efforts and technical support groups should be established to ensure that adaptations of an improved design to local conditions do not result in cookstove safety, durability, or performance problems;
- feedback from users of improved cookstoves should be used to make improvements in the stove design or dissemination technique and to record the impacts of cookstove dissemination on village life; and
- commitments should be sought from high-level agencies in governments to efforts to ease firewood shortages, through combined efforts of forestry extension workers and cookstove promoters at the village level.

3 Solar Crop and Timber Drying

INTRODUCTION

In this chapter, we focus on two applications of direct solar energy which we believe to be of importance to developing countries, namely, solar drying of crops and timber. These technologies clearly were not the only applications we could have considered. However, they do appear to be applications which, among others, have priority in the developing countries themselves and illustrate some of the problems and opportunities with direct solar energy use. These applications were also chosen based upon some of the criteria discussed in the Introduction.

In the draft report of the UNIDO Working Group on Energy for Rural Requirements at the 1978 New Delhi Forum on Appropriate Industrial Technology, priorities for flat-plate collectors were listed as (1) hot water, particularly for process heat; (2) provision of potable water; (3) crop drying; and (4) cold storage for agricultural products. Concentrators as a source of heat for cooking were singled out, as was mechanical shaft power from solar powered engines for several applications of which pumping was probably most important.(1) Discussions at the same forum with an Indian energy planner indicated that cooking, water pumping, and crop drying were three high priority needs for energy in rural areas in India.(2) The Workshop on Direct Solar Energy at the Symposium on Energy and Development in the Americas held in Brazil in 1978 came to a number of conclusions that are summarized in table 3.1.(3) A summary of the rationales for using solar energy for drying three classes of products is given in table 3.2. In this chapter we concentrate on drying of grain crops and timber; relatively little attention is given to the drying of fruits, vegetables, or other uses of solar energy.

SOLAR GRAIN CROP DRYING: NEEDS AND APPLICATIONS

Introduction

Food losses which occur between harvesting and consumption, commonly called postharvest food losses, are a problem which has become increasingly important to development agencies and developing country governments. For instance, the Seventh Special Session of the United Nations General Assembly in 1975 adopted a resolution to take as a goal the reduction of postharvest food losses worldwide by 50 percent by 1985.(4) These losses have important economic impacts and hamper developing country attempts to achieve self-sufficiency in food. Although the importance of the postharvest food loss problem is well known, the magnitude of the actual losses is not. However, a commonly quoted estimate is not less than a 10 percent overall loss for cereal grains and legumes.(5) In any case, these losses represent a vast potential supply of food that is worth billions of dollars and would be capable of feeding tens of millions of people.

The reasons for losses vary widely and are a function of crop variety, pests, climate, harvesting technique, processing, storage, handling and marketing, and the social and cultural setting. However, surveys show that storage and drying are perhaps the most critical areas for reducing postharvest food losses. According to the National Academy of Sciences:

> Drying is a particularly vital operation in the chain of food handling since moisture may be the most important factor determining whether, and to what extent, grain will be liable to deteriorate during storage. Drying is used to inhibit germination of seeds and to reduce the moisture content to a level that prevents the growth of fungi and bacteria; it can also retard attacks on the grain by insects and mites.(6)

Many farmers in developing countries realize that proper drying techniques are a very important factor in successful storage programs and devote much attention and care to achieving good drying and storage conditions. For example, a survey conducted in the Philippines on the nature of farm paddy storage found that all the small farmers surveyed felt drying of the grain was a very important factor in preventing storage losses; however, the same farmers did not seem to be aware of the importance of proper storage.(7)

Table 3.1. Observations Concerning Utilization of Direct Solar Energy at the Symposium on Energy and Development in the Americas

Applications	Observations
RURAL SECTOR	
Solar Drying	Technology at a level which deserves commercialization.
Water Pumping	Use for pumping potable water has greatest immediate potential.
Distillation	Some successful tests and some failures in several rural areas.
Water Heating	Lack of local industries has limited implementation.
Solar Refrigeration	Small-capacity units could strongly influence productivity of rural industries.
Passive Applications	Potential great for ventilation, cooling, dehumidification, and heating.
Solar Cells	Even at current costs, can be considered for use in providing cultural, medical, and communications services to isolated areas.
Solar Cookers	Technology permits implementation but local customs have prevented use.
URBAN SECTORS	
Hot Water for Domestic Use	In most places, this technology is the only solar application in commercial production. Cost reductions are needed.
Helioarchitecture	Passive systems have potential in urban areas and need to be included in building manuals and regulations.
Hot water for Industry	Industrial use of flat plate collectors could be and, in some cases, is already economically feasible.

Source: Compiled from "Reports and Recommendations of Workshop on Direct Solar Energy, Symposium on Energy and Development in the Americas," Gustavo Best Brown, Chairman, Jaime Moraques, Secretary. Interciencia 3 (May-June 1978).

Table 3.2. Rationale for Solar Drying

Product	Rationale
Grain	Improve product quality Improve storage capability Reduce time and space requirements for drying Facilitate quick preparation of fields for next cropping Improve marketing control of farmer Facilitate wet season harvesting and storage Improve drying hygiene
Timber	Improve product quality Reduce period capital is tied up in drying stock Maintain competitiveness of small operators Improve exportability of finished wood products Provide low-expertise, low-capital, improved drying options Expand range of utilizable timber species Improve attainable drying level
Fruits, Vegetables & Fish	Reduce product seasonality Improve marketing control of farmer Reduce spoilage Improve product transportability Improve drying hygiene Improve storage capability Reduce nutritional fluctuations

Most farmers in developing countries have limited opportunities to dry harvested grain. The most common and traditional drying method is to spread the grain out on the ground, on mats or plastic sheets, or on concrete or asphalt roads to dry in the sun. These methods are unsatisfactory for a number of reasons. First, limited surface area is available in most developing countries for drying grain, which limits the quantity of grain that can be dried and thus stored. Secondly, this type of drying requires prolonged sunshine which may not be always available. Traditionally, crop harvesting takes place once a year during the dry season which in most developing countries is characterized by a high chance of strong, prolonged sunshine. However, the introduction of new high-yield varieties of grain that mature and must be harvested during the wet season and the shift toward double cropping practices have made successful drying programs much more difficult. Reliance on unacceptable drying techniques may contribute to the reluctance of farmers to plant wet season crops and may hamper attempts to increase farm productivity.(8)

Thirdly, traditional sun drying techniques make the crop susceptible to damage from insects and foraging animals and require the crop to be brought under cover during rainy periods. In addition, drying is often uneven in quality, with some grain being overdried and some underdried. Drying rates vary greatly, which makes obtaining consistent quality difficult. This problem, in turn, hampers export sales or the growing of grains for cash rather than for local consumption.

Grain Drying Practices

Grain may be damaged by excessive heat which causes splitting of the grain, bleaching, discoloration, loss of germination capability, and nutritional changes (see table 3.3). Too rapid drying may cause bursting or case hardening; on the other hand, inadequate or slow drying which is often a problem in humid regions may result in infestations by insects and attack by fungi and bacteria (see table 3.4). Drying damage occurs not only as quantitative losses (i.e., losses in the amount of grain) but also as qualitative losses (e.g., losses of nutritional quality, germination capability, preferred structure, and market value). Successful drying is a complex process requiring skill and effort on the part of the farmer, who normally appreciates the importance of quantitative losses but may not be so perceptive of the nature and impact of qualitative losses.

Some grain dryers are available in developing countries. However, the systems are usually imported and consist of large-scale, fossil-fuel fired air dryers.(9) In many cases, the hot combustion gases are passed directly through the grain, which often becomes contaminated by unburned fuel, fumes, and soot.(10) Electric heating of the drying air is preferred but is very expensive and is not applicable in rural villages without electricity. Solar drying technologies have become of greater interest as the result of the continued energy price rises in world markets. Unfortunately, solar dryers operate least effectively during periods when drying is most needed, that is, during inclement weather.

Despite the advantages of low operating costs, solar dryers are rarely found in developing countries, which may be partially explained by the immature status of the development of appropriate technical designs, the lack of proven perfor-

Table 3.3. Maximum Recommended Drying Temperatures for Crops

Crop and Use	Maximum drying temperature, °C
Grain used for:	
livestock feed	74
milling and manufacturing	60
human consumption	57
seeds	43
brewery processes	43
Rice used for human consumption	43
Beans used for human consumption	35

Source: D.W. Hall, "Handling and Storage of Food Grains in Tropical and Subtropical Areas." FAO, Rome, 1970, p. 70.

mance data, and the unavailability of economically attractive systems.(11) Because the majority of developing country farmers are poor, solar dryers may not even be affordable if there is a guaranteed market for improved quality grain. Solar dryers are subject to the same obstacles that impede any efforts to improve farming productivity in developing countries, i.e., lack of financial credit and market outlets for improved products. In addition, the scale of most solar crop dryer designs is not well matched to the typical product output of individual farmers in developing countries. On the other hand, groups of farmers, or cooperatives, and large industrial and government marketing organizations may be able to use forced-convection, indirect solar dryers because of the volume of grain that is handled.

Currently, most crop dryers found in developing countries are owned and operated by large entrepreneurs or government organizations and are of the fossil-fuel fired type.(12) These designs are typically introduced by government subsidy programs with the dominant cost being an internal combustion engine. Continuous dryers of this type

Table 3.4. Maximum Recommended Moisture Content for Crops to be Stored for One Year

Crop	Maximum moisture content,* %
wheat	13.5
maize	13.5
paddy rice	15.0
milled rice	13.0
sorghum	13.5
millet	16.0
beans	15.0
cow peas	15.0

*An average climate condition of 70% relative humidity and a temperature of 27°C is assumed.

Source: "Small Farm Grain Storage, Vol. 1: Preparing Grain for Storage," VITA Publication No. 35E, Volunteers in Technical Assistance, Mt. Rainier, Md., 1977.

are usually imported whereas batch models could be built locally by experienced craftsmen. However, unless heat exchangers are used, the crop can become contaminated by combustion products, and often the operators try to accelerate the drying process which can result in damaged grain.(13) The point was made at the VITA/CDT Solar Drying Panel (see Appendix A) that crop dryers must have temperature control mechanisms if significant improvements over open-air sun drying are to be made.

Solar Grain Crop Drying Experience

Preliminary experimental programs for development of solar dryers in the United States and Canada indicate that large quantities of grain can be successfully dried to produce high-quality products at nearly competitive costs.(14) These systems employ a high air-flow rate at a low temperature over a long drying period, as compared to fossil-fuel fired systems which use high temperatures and low air-flow rates to give rapid drying times. Some crops that are commonly grown in developing countries, for example, rice, may actually be more adapted to solar drying technologies than to fossil-fuel fired drying systems because case hardening and other damage is less likely to occur at the low temperature conditions of solar dryers. However, solar dryers have yet to win acceptance in Asia, India, and Africa because of the high capital costs of both imported and locally-constructed dryers which have given unsatisfactory performance due to design flaws and lack of experienced operators.(15)

Solar drying technologies are still largely unproven in developing countries, and there is a need for demonstration programs to gain design and operational experience.(16) The most immediate equipment need at the village or small farm level appears to be for low-cost, batch-type dryers which may be fabricated from locally-available materials.(17) The dryers must produce dried grain of good milling quality in reasonable drying times. Systems can be envisaged in which preliminary drying is done in small batch dryers on the farm before being transported to the large central drying plants for final drying, storage, and milling.(18)

SOLAR GRAIN CROP DRYING: TECHNOLOGIES

Classification of Designs

Solar crop dryers may be categorized by the choice of three principal design parameters: (1) type of air convection (i.e., natural or forced); (2) type of crop exposure to solar radiation (i.e., direct or indirect); and (3) temperature of circulated air (i.e., ambient or preheated). Only three of the eight possible designs are commonly found (only two in developing countries): direct-exposure, natural-convection dryers using either ambient or preheated air; and indirect-exposure, forced-convection dryers employing preheated air supplies (see table 3.5).

The terms natural or forced convection refer to whether the air passing over the crop is drawn by natural convection currents driven by temperature-pressure differentials or by an artificial pressure differential generated by a motor-driven or wind-driven fan. The terms direct or indirect drying correspond to designs in which the crop is directly exposed to solar radiation or in which the crop is placed in an enclosed bin and thereby shielded from solar radiation. In the former case, the crop container, or chamber, acts as a solar collector

Table 3.5. Classification of Solar Dryers

Convection	Drying Mode	Heating Type	Examples	Occurrence	Crops	Distribution	Supplemental Power
Natural	Direct	Ambient	Racks, chambers	Common	Fruit, vegetables, coffee	Developing Country tropics	No
		Preheated by solar	Chamber	Common	Paddy, corn	Developing Country tropics	Rarely
	Indirect	Preheated	Bins	Rare	-	-	-
Forced	Direct	Preheated	Chamber	Rare	-	-	-
	Indirect	Ambient	Bins (can be preheated by electricity or fuels)	Rare	Corn, grains	USA, Canada temperate zones	Common
		Preheated by solar	Bins (can be preheated by electricity or fuels)	Common	Corn, grains	USA, Canada temperate zones	Common

by directly trapping solar radiation. The terms ambient or preheated air signify whether the air entering the drying chamber has been heated above the outside, surrounding (i.e., ambient) air temperature by a solar collector, or enters at the prevailing ambient temperature. Preheated air has a higher drying potential and is normally used in indirect solar crop dryers.

Examples of direct-exposure solar crop dryers are racks and chambers which are directly exposed to the sun. These designs are most commonly found in developing countries and/or tropical areas where the high air temperatures (e.g., 30 to 40°C) combined with a moderate air-flow rate provided by natural convection currents may be used with small batch loads to lead to successful drying. The design and operation of these dryers is elementary, but this design may be an effective, inexpensive, and easy to maintain solution to crop drying needs in developing countries. Crops dried in this fashion include coffee, paddy, grains, fruits, and vegetables.(19) Supplementary power supplies are not typically included in these designs so the drying depends entirely on the availability of solar radiation.

Bins and silos are examples of indirect-exposure solar dryers that normally include systems for forced convection. These designs are larger in scale than direct drying chambers and require high air-flow rates. Generally, the roof of the structure containing the crop serves as a solar collector and the preheated air is drawn through the crop by a fan. Supplemental power supplies are often available and may be used to further heat the preheated air. The fan is usually driven by an electric or gasoline motor, although a wind-powered fan has been used in at least one case.(20) Although the drying bin could serve as a storage container as well, in most cases in developing countries, the dryers are operated with small batch loads with drying times of a few days; storage is provided elsewhere. The construction procedures for indirect dryer designs are more complex and costly than for direct dryers and, therefore, indirect dryers are not usually suitable for individual farmers in developing countries. However, indirect dryer designs may be applicable to co-operative farming projects and would be suitable for regional or national grain storage centers in developing countries.

The majority of research activities we have information on focused on direct-exposure solar crop dryer technologies and is centered in developing countries in southeast Asia and in India where rice is the primary crop to be dried, although some research is ongoing in Latin America on coffee drying applications. Research on indirect-exposure designs is centered in the United States, often with support from the Departments of Energy and Agriculture.

Table 3.6 summarizes information available to us as of Spring 1979, on solar crop dryers. The various solar crop drying technologies and individual projects will now be described.

Natural-Convection Dryers

Direct-exposure rack dryers

The most widely used and traditional method of drying crops in developing countries is to lay them out on the ground to dry in the sun. This method requires a large area, offers no protection against insect and animal damage, and is very labor intensive because the crops are gathered up as rainstorms approach. A simple solution is to dry the crop on racks raised above the ground and covered with a roof thus taking advantage of the availability of solar radiation. This type of direct-exposure drying is used extensively for the drying of fruit, particularly grapes, in Australia,(21) where 100,000 tons are estimated to be dried in 8 to 14 days. One 50 m^2 drying unit is generally considered to provide enough rack space to dry the fruit harvested from three acres of vines.

In Colombia, similar rack dryers are used for drying cassava and coffee. A vertical, wire-mesh dryer for drying cassava designed at the University del Valle, Cali, consists of two wire-mesh panels attached to two wooden uprights which serve as end walls. Openings at the top and bottom of the structure provide for easy loading and unloading. A wooden roof covers the dryer to protect the product from rain and to permit the drying process to occur during periods of no sun. This dryer design is reported to produce a high-quality cassava which is easy to handle and store.(22)

Coffee farmers in Colombia traditionally dry coffee on trays and wooden platforms exposed to the sun. These drying practices have remained nearly unchanged with time because these designs are more economical than most other techniques which have been developed as replacements. One design, known as a tray dryer or pasera, is usually composed of wooden trays which are supported on elevated frames in order to avoid direct contact with the ground.(23) Since these dryers do not have covers, the trays must be carried inside at night and during rainy periods. Some designs mount the trays on wheels which move on tracks of wood, cement, or iron so that the trays may be moved under corrugated metal roofs during rainstorms or at night.

A similar coffee dryer, called a marquee, has a raised platform covered by transparent plastic to increase the drying rate and to protect the crop.(24) The coffee beans, which have a moisture content of about 50 percent before being

Table 3.6. Comparisons of Solar Crop Dryers

	Drying Mode and Design						
	DIRECT: NATURAL CONVECTION						
Performance Criteria	Rack	Chamber	Preheated Chamber				
Crop	coffee	paddy	paddy	corn	paddy	paddy	paddy
Dryer capacity, kgs.	46	-	500	260	550	50	-
Drying bin area, m^2	1.9	6	5.7	6	-	-	-
Solar Collector Area, m^2	-	-	17	27	-	6.75	6.75
Moisture Reduction. From: To	50: 11	-	22: 14	30: 14	23: 12	30: 14	30: 14
Batch time	8-9 days	2 hours	1-2 days	8 hours	6 hours	1 hour	1 hour
Temperature in dryer, C°	ambient	42-54	<45	43	+8 ambient	70	80
Airflow, $m^3/min/m^3$ crop	-	-	1-5	520 ft^2/min	3.3-4.3	393kg/hour	3
Capital cost of dryer, $	2.50-4	15	120	300-500	-	125	-
Cost/m^2 drying area, $	1.25-2.10	2.50	12	50-80	-	-	-
Cost/m^2 collector, $	-	-	+	11-18	-	19	-
Cost/kg dried/day, $	0.23 0.39	-	0.24	1.15-1.92	-	0.42	-
Assumed batches/day	0.18	3	0.5	1	-	6	-
Field Tested in	Colombia	India	Thailand	none	India	India	India
Source	1	2	3	4	5	6	7

(continued)

Table 3.6. (continued)

	Drying Mode and Design					
	INDIRECT: FORCED CONVECTION					
Performance Criteria	Preheated Bin					
Crop	paddy	grain	corn	corn	coffee	grain
Dryer capacity, kgs.	3000	7250	-	770	250	900 bu.
Drying bin area, m^2	12.5	8.4*	Ratio* $2m^3$	6.18*	56	240 1125*
Solar Collector Area, m^2	-	9.3	to $1m^2$	-	-	-
Moisture Reduction. From: To	35: 15	20: 13	-	30: 13	50: 12	18: 14
Batch time	8 hours	3+ weeks	-	1 day	1 day	2+ weeks
Temperature in dryer, C°	-	+8 ambient	+5-12 ambient	-	+11 ambient	+10 ambient
Airflow, $m^3/min/m^3$ crop	-	5	2	-	$50ft^3$/ min	2
Capital cost of dryer, $	150	-	-	2400	100	-
Cost/m^2 drying area, $	12	-	-	-	2	-
Cost/m^2 collector, $	-	-	-	-	-	-
Cost/kg dried/ day, $	0.05	-	-	0.03	0.40	-
Assumed batches/day	1	-	-	1	1	-
Field Tested in	India	USA	USA	Barbados W. Indies	Puerto Rico	USA
Source	8	9	10	10	11	12

*Bin volume m3

(continued)

Table 3.6. (continued)

Sources:

1. "A Handbook on Appropriate Technology," Section B5, Canadian Hunger Foundation/Brace Research Institute, Canada, 1976.
2. Algal Division, "Solar Dryers: Periodical Technical Notes No. 3," Shri A.M.M. Murugappa Chettiar Research Centre, 11/12 North Beach Rd., Madras 600-001 India, March 1978, mimeo.
3. Robert Exell and Sommai Kornsakoo, "Solar Rice Dryer," Asian Institute of Technology, Bangkok, Thailand, 1978.
4. "Report of Design Projects in the Department of Mechanical Engineering: Appropriate Technology," Professor Clifford Sayre, Coordinator, College of Engineering, University of Maryland, College Park, Maryland, December 1978.
5. O.P. Singhal and C.P. Gupta, "Solar Drying," Seminar Proceedings: Industrial Applications of Solar Energy, Madras, June 4-6, 1975, National Productivity Council, New Delhi, 1976.
6. V.R. Muthuveerappan et al., "Utilization of Solar Energy for Paddy Drying," Sixth Meeting, All Indian Solar Energy Working Group, Allahabad, December 23-24, 1974.
7. S.K. Tandon and D.K. Biswas, "Studies on Drying of Paddy in Solar Drier," Journal of IASAE (Indian Agricultural Research Institute, New Delhi), February 1974.
8. D.K. Biswas, "A Solar Dryer for Small Farmers," Indian Farming 24 (August 1974): 21-22.
9. Fredrick H. Buelow, "Drying Grain with Solar Heated Air," Quarterly Bulletin of the Michigan Agricultural Experiment Station 41 (November 1958): 421-29.
10. T.A. Lawand, "The Potential of Solar Agricultural Dryers in Developing Areas," presented to UNIDO Conference, Vienna, February 14-18, 1977.
11. Allan L. Phillips, "Drying Coffee with Solar Heated Air," Solar Energy 9 (1965): 213-16.
12. Greg Schoenau, "Solar Grain Drying," unpublished, undated.

placed in the dryer, are stirred frequently by hand with wooden tools during the drying process. A typical loading density for freshly harvested coffee is about 24 kg/m^2 which is reduced to about 12 kg/m^2 when parchment coffee is produced with an 11 percent moisture content. The average drying time is eight to nine days, or about 100 hours of daylight with about 40 hours of direct sunlight.(25)

The costs of these solar coffee dryers vary with the type of dryer and the materials used in construction. An average pasera tray dryer (e.g., 2.4 m x 0.8 m x 0.15 m) costs from \$2.40 to \$4.00 (1975) or about \$1.25 to \$2.10/m^2 of drying surface. The models mounted on wheels cost about \$9 to \$10/m^2, including the trays, roof structure, and rails. Typical lifetimes of tray dryers are five to ten years. A very simply constructed marquee dryer, covered with polyethylene sheeting which must be replaced annually, may cost as little as \$2.00/m^2. If a metal frame structure, covered with glass and mounted on a concrete floor, is used the cost rises to about \$12/m^2.(26)

The construction of these dryer designs poses no technical problems which could not be easily handled by villagers in developing countries. Perhaps the only limiting factor to the use of these dryers is the size or level of coffee production because, when coffee output is greater than about 12 ton/year, prohibitively large drying surfaces are required. Inclement weather conditions may also discourage the use of these dryers when the rainy season coincides with the harvest period. However, solar coffee dryers have been used very successfully in Columbia, where as much as 70 percent (400,000 ton/year) of the national coffee production is solar dried in spite of the fact that the climate is not ideal for solar drying. Colombian coffee farmers are reported to use solar dryers because no other methods meet the small- to medium-scale drying requirements as well.(27) Solar dryers of similar design are also reported to be in use in other areas throughout the world where the climate and crops to be dried are different than those in Colombia.(28)

Direct-exposure chamber dryers

Direct-exposure chamber dryers are the most commonly used design in developing countries. In this design the crop is placed in an enclosure covered by a transparent cover and/or side panels. Solar radiation is absorbed directly by the crop and by the internal surfaces of the chamber to provide heat and convection currents to evaporate the moisture in the harvested crop. Workers at the Algal Division of the Shri A.M.M. Murugappa Chettiar Research Centre in India have developed a simple design consisting of a structure constructed from casuarina poles, bamboo, coconut thatch, and polyethylene sheeting.(29) This design is typically 4 m x 1.5 m with the roof sloped upward from 0.2 m to 0.7 m. Mud-coated cow dung is used for the floor; and the sides, which contain two doors for loading and unloading the crops, are made of thatch. Construction costs (1978) are about $15, with labor and the polyethylene accounting for approximately 40 percent and 25 percent of the costs, respectively. These dryers are used primarily for drying rice paddy and may reach internal temperatures of 42 to 54°C.

Another simple dryer design has been proposed by McDowell.(30) This design consists of an outer and an inner box with the space between the boxes packed with an insulating material (e.g., wood wool, dried grass, coir fiber). Holes in the bottom and slots in the top of the boxes serve as inlets and outlets for air circulation, and the top of the structure is covered with polyethylene film. Typical box dimensions are 2 m x 1 m x 30 cm with a capacity of 18 to 24 kg of grain which can be dried from a moisture content of about 25 percent to less than 12 percent in one day. A possible problem with this design is that internal temperatures in the dryer may reach 65 to 80°C, which can damage rice grains used for food by cracking ($>$50°C) or rice grains used for seed by overheating ($<$40 to 50°C). An adaptation of this design is to install a firebox at one end and run an 11 cm diameter metal flue pipe through the length of the dryer to provide heated air during periods when solar heat is unavailable because of inclement weather.

Although we do not have firm data, it seems likely to us that there is considerable experimentation going on in developing countries with natural-convection solar dryers of the types just described. For example, at the UNIDO Forum on Appropriate Industrial Technology in New Delhi (1978), one of us (Morgan) met an Indian entrepreneur who had developed a collapsible, tent-like structure using plastic sheeting which could be taken by truck from village to village and placed over stacks of grain to form a simple solar dryer. A villager was hired to fan accumulated moisture out of the structure which also provided shelter for the villager. It seems to us that inexpensive dryers of this type could be very important to expedite drying during the rainy season. A key to the economics might be the development of plastic sheeting which lasts longer than those currently available.

Direct-exposure, preheated air dryers

In direct-exposure, preheated air dryers, preheated air from an adjacent solar collector is introduced into the drying chamber. One of the best known designs of this type was developed at the Asian Institute of Technology in Bangkok, Thailand(31) (see fig. 3.1). The solar collector consists of a

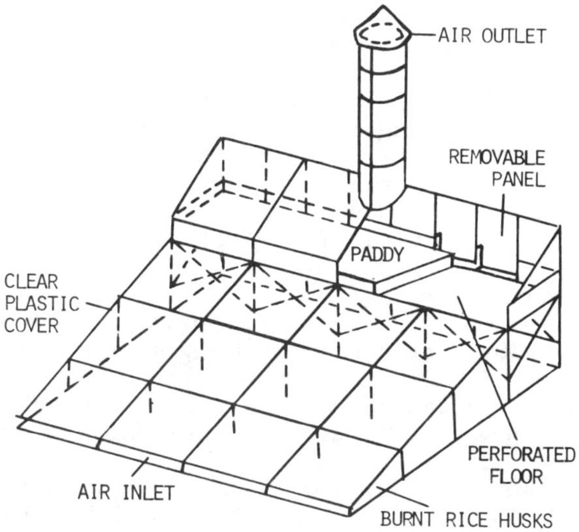

Fig. 3.1. Schematic Diagram of the Paddy Dryer Developed at the Asian Institute of Technology

Source: Robert Exell and Sommai Kornsakoo, "Solar Rice Dryer," Asian Institute of Technology, Bangkok, Thailand, 1978.

layer of burned rice husks, which form a black mat to absorb solar radiation, and a 0.15-mm clear plastic cover stretched over a simple wooden framework to form an air duct 15-m wide by 3.6-m long. The heated air passes into the 4.8-m x 1.2-m drying chamber from below through a 0.8-mm thick steel sheet perforated with 1.5-mm diameter holes. The air exits through a covered chimney consisting of a bamboo frame covered with dark-colored plastic sheeting to absorb heat from the sun. Removable panels at the back of the drying chamber provide access to crop loading and unloading. Capital expenditures (1977) for the dryer were $120, with the perforated steel sheet and the plastic sheeting, which will require annual replacement, accounting for about 40 percent and 15 percent of the costs, respectively. Perhaps a cheaper floor material could be fabricated from bamboo matting or wire-mesh mosquito netting supported by a wooden lattice, but application of these materials would probably be restricted to chambers with less than 2 m^{-2} floor areas.

Performance tests for paddy drying indicated that the preferred paddy depth in the drying chamber is about 15 cm and that the air flow for this configuration is sufficient to dry paddy from a moisture content of 22 percent to 14 percent in two to three days, even in inclement weather.(32) The recommended operating capacity is about 500 kg (80 kg/m^2). Paddy dried in this experimental dryer produced a head yield (i.e., the percentage weight of unbroken kernels to the paddy weight) after milling of 58 percent, which is equal to the head yield from mechanically dried paddy and superior to the maximum head yield of 47 percent obtainable from paddy sun-dried in the open air. These improved yields are important because higher yields lead to higher incomes for village farmers. In these experiments, cracking of kernels resulting from excessive drying temperatures did not occur; however, if the air temperature inside the dryer becomes too hot (i.e., 45°C), the cover should be shaded. One of the drawbacks of this dryer design is its limited capacity of 0.5 metric tons which may be insufficient to dry a full harvest before some of the paddy waiting to be dried spoils.

A similar dryer design has been tested by Muthuveerappan, et al.,(33) and Tandon and Biswas,(34) which uses a 4.5-m x 1.5-m corrugated, galvanized steel solar collector. This design can dry 50 kg of parboiled paddy from a moisture content of 30 percent down to 14 percent in one hour and costs about $125 (1974) to build.

A more sophisticated solar collector, described by Singhal and Gupta, consists of a glass-covered collector with a corrugated steel absorber plate.(35) This design can dry 550 kg of paddy in five hours from a moisture content of 23 percent. Internal chamber temperatures are about 8°C above ambient which prevents the paddy kernels from cracking.

The enthusiasm for applications of natural convection solar dryers implied by these examples is not shared by all investigators. For example, Macedo and Altemani conclude that, in spite of being attractive conceptually, natural-convection dryers are currently being built without basic knowledge of natural-convection air heaters, including information on air-flow rates, collector temperatures, and overall efficiencies.(36) However, if this lack of knowledge impedes the otherwise attractive application of this technology, a modest coordinated research and development program should be sufficient to establish any missing fundamental information necessary for design purposes.

Forced-Convection Dryers

Combined Direct and Indirect Heating Models. Combined direct and indirect, forced-convection dryers utilize a drying chamber with transparent covers to supplement the heat that is provided for the drying process from an adjacent solar collector. One design which appears very promising to some investigators, for example T. A. Lawand,(37) employs a wind-powered ventilation system and dampers for temperature and air-flow rate control. This model avoids the limitations of small batch sizes as the result of low air-flow rates which characterize nearly all natural-convection dryers without being dependent on a nonrenewable energy source for fan power. In this design, the solar collector consists of a blackened hardboard sheet which is insulated and covered by plastic or glass. The preheated air from this collector is convected into the drying chamber which contains 12 trays placed in two adjacent six-tier stacks. Additional heating is supplied by direct heating of the chamber by solar radiation through the transparent walls of the chamber. Air convection is provided by a rotary wind ventilator mounted on top of a chimney above the drying chamber which also induces a natural convection current.(38)

Indirect heating designs

Workers at the Indian Agricultural Research Institute (IARI) in New Delhi have designed an indirect, forced-convection dryer that is reported to dry and store grain in the same structure at nominal costs.(39) The main components of this design are the roof which serves as the solar collector and is fabricated from corrugated steel sheets blackened with bituminous paint, a plywood air duct, a blower, and the grain bin. Estimated construction costs are about $150 (1974) for a dryer with a capacity of three metric tons with about 60 percent of the cost attributed to the blower-motor assembly. Cost reductions of up to 30 percent are expected if the solar collector is integrated into the roof of farm houses, livestock shelters, or other adjacent buildings.(40) Another cost evaluation concludes that dryers of this design should be able to dry grain for significantly lower costs than comparable models which use oil as the heat source.(41)

Performance tests of the IARI design indicate that grain may be dried from a moisture content of 35 percent down to 15 percent in about eight hours, even in periods of intermittent sunshine and in regions with high relative humidity. Covering the metallic roof-collector system with glass or polyethylene sheets produced significantly higher drying temperatures.(42)

Another forced-convection dryer developed for use in developing countries is designed to dry feed, primarily corn, for consumption by livestock.(43) A two-stage drying process is used that includes drying in a mobile solar-heated cart and then in an 18-m^3 solar air-heated storage bin. The drying procedure is designed to process 770 kg (0.9 m^3) of freshly shelled corn per day by first reducing the moisture content from 30 to 18 percent in one day in the mobile cart and then by slower drying to an equilibrium moisture content of 13 percent in the storage bin. The collector is fabricated from three 29-m long plastic sheets, with a transparent upper sheet, a black mesh center sheet, and a bottom sheet consisting of an insulator sandwiched between a black and an aluminum film. The sheets are sealed along the long edges and inflated. During drying operations, the end of the collector is attached to the mobile drying cart which is insulated to reduce heat losses and is covered with a double-layered plastic roof.(44)

Many of the forced-convection solar dryer designs use supplemental heat which is often provided by electricity. One example of this type of configuration is a solar drying system incorporated into the roof of a coffee processing building in Puerto Rico.(45) The system dries 250 kg of coffee per day from a moisture content of 50 to 12 percent. A 56-m^2 solar collector is built into the roof which preheats the drying air by about 10°C from an average ambient air temperature of 29° C prior to being heated to 49° C by electric heaters. The costs of collector construction were recovered in the form of reduced expenditures for electricity in less than two years.

In the United States, indirect, forced-convection solar drying has been the subject of limited research efforts and is currently supported by the Departments of Energy and Agriculture.(46) These systems are typically of a larger scale than is suitable for use in developing countries and are often designed to be retrofitted to existing bin-dryer facilities. Inflated tubes of black polyethylene are also being developed as solar collectors for forced-convection dryers but the (1977) costs (about $2,000) may be too high for many applications in developing countries.(47)

An example of the bin-wall collector design has been developed by Peterson at South Dakota State University.(48) In this design, the roof and the south facing wall are painted black and covered with a transparent plastic film. This collector provides preheated air at temperatures from 5 to 12°C above the ambient conditions. Drying times for corn range from one to two weeks depending on the detailed design. Collector area requirements are approximately 0.5 to 1 m^2 per m^3 of grain to be dried and the fan should be capable of supplying 2 m^3/minute or more per m^3 of grain.

In general, forced-convection dryer designs must balance parameters such as batch size, air-flow rate, depth of crop to be dried, maximum permissible moisture content of the harvested crop, temperature rise, and pressure drop of the air passing through the crop.(49) All attempts to develop solar drying systems in the United States have focused on low-temperature, high air-flow rate designs that have relatively long drying times. In developing countries, shorter drying times would appear to be required.

SOLAR CROP DRYING: NON-TECHNICAL ASPECTS

Crop drying is only one part of the food production chain, and is thus heavily influenced by what occurs before and after it. Also crop drying is not solely a technical issue but is influenced by farmers' perceptions of the importance to crop storage and the role of crop storage in a particular society. Attempts to introduce crop dryers to reduce crop losses and increase food production will be strongly affected by cultural, economic, and social factors. As the National Academy of Sciences report warns:

> Past experiences with agrarian reform have demonstrated that programs must be sensitive to the cultural, socioeconomic and political characteristics of a society and that technical and scientific components of change cannot be divorced from the social context within which they are applied.(50)

Thus, before any new technical change is introduced it is important to understand the many complex beliefs and attitudes that underlie traditional ways of growing, harvesting, and storing food. Fortunately for the advocates of improved crop drying technologies, small farmers are usually aware of the importance of crop drying to successful storage. However, small farmers are hampered in their ability to dry their crops because of the lack of suitable drying surfaces and limitations on time and labor to supervise drying, which can be a very laborious task if the grain has to be taken up frequently at the onset of rain. Traditional crop drying has been complicated by the introduction of high-yield varieties that mature and must be harvested during wet seasons or by the production of a second, irrigated crop (i.e., double cropping) that must also be harvested during the rains. In these cases, the grain requires artificial drying and, thus, taxes the farmer's ability to handle grain by traditional methods. Consequently, a new drying and storage procedure must be adopted or the crop must be sold undried for a lower price. The alternative may be to forego the new variety and the associated increased productivity.

In many societies, there is much reluctance to relinquish individual control over grain to newly formed cooperative or government organizations; thus, unless larger-scale sharing arrangements can be worked out, drying equipment and storage facilities will prove ineffective. Traditional practices, often rooted in notions of security and status, will not be abandoned unless it can be demonstrated that the new technologies and methods will be a real improvement on the old ways, and that these technologies will not result in intolerable strains on social structures, income levels, and distribution.

Past experiences with agrarian reform through technical improvements suggest that unforeseen effects can turn people against any future attempts to "improve" their lives. The lessons of past failures include:

- price depression resulting from increased availability of food;
- taxation, especially tithes on amounts of food stored;
- fixed quotas for commodities to be purchased after harvest;
- inadequate means of storage or marketing surplus production; and
- lack of transport facilities to markets.

Improved methods of crop drying must be part of a national policy for food production and conservation. Past experience indicates that the effectiveness of intervention depends on adequate communication between central governments and local communities. The committee responsible for the NAS Report on Postharvest Food Losses places special emphasis on the need for national policy bodies which can take into account the broad range of local interests involved and can examine the technical and scientific considerations in the light of local conditions and attitudes.(51)

While the loss to the individual farmer by crop damage due to ineffective drying can be measured in monetary terms, the impact on the government of national crop losses involves not only economic consequences but also social responsibilities and national development plans. One technique to evaluate crop losses and costs of prevention is social cost-benefit analysis. This technique may be useful for evaluating the cost of reducing a unit of food loss and the quantity of units that can be effected within a fixed budget. However, this approach requires good information on costs and anticipated results and is only one of the factors that policymakers and program planners should consider. Other factors are the sociocultural acceptability of possible programs, overall national development priorities, and secondary impacts. Important secondary impacts include effects on employment, income distribution, nutrition, social stability, and balance of payments.

For instance, importing crop dryers could have adverse effects on the balance of payments and on local employment of craftsmen, while designing and fabricating crop dryers locally from local materials could reduce rural unemployment and income inequalities. Thus, poorly conceived programs, as has been found by past experience, may impose social costs that negate the benefits derived from increased grain storage and production. The introduction of crop dryers bears many similarities to other types of intervention into rural practices such as rural health delivery, agricultural extension, and other community development projects conducted at the village level.

Because the traditional agricultural sector has, in many countries, the largest population and the highest food production, crop loss reduction can mean substantial savings of food even when the percentage losses are relatively low. As the National Academy of Sciences report comments:(52)

> The producers and consumers in this sector comprise the largest poverty-level group in many countries, and therefore can provide a focus on the poor in national development activities. Because loss reduction activities may require some capital investment and must be perceived to provide practical benefits, governments may need to demonstrate the value of these efforts and to subsidize them. At the village level, good opportunities exist for using local raw material and labor and artisan skills in loss reduction activities; these elements of a country's resources are relatively cheap and abundant, yet the opportunity cost - their value in alternative investments - is relatively low. Use of these resources, moreover, provides direct social benefits by generating employment and distributing income, in addition to reducing food losses for the benefit of poor families and consumers. Increased food availability in this sector can also provide the poorest farmers and farm women with access to, and perhaps integration into, the urban and export markets.

There are many other, nontechnological needs which should also be addressed, including:

- the need to develop a better understanding of traditional farm practices and the conditions that facilitate or hinder corrective action, including more research into the links among economic and cultural practices and food losses;
- the need for more understanding of the effects of government financial policies, subsidies, and price control on crop losses and the incentive to reduce losses;
- the need to perform specific case studies to illuminate village-level problems, such as the impacts of new innovations and the effect of financial subsidies to introduce them;
- the need for data on the costs of increasing the availability of food through crop reduction losses, in order to be able to compare the effectiveness of measures to reduce crop losses with other types of intervention, such as government fiscal intervention, which will make for

better decision making and planning of development programs; and

- the need to give crop drying a more prominent place in agricultural development plans and in the work of international agricultural research centers.

SOLAR TIMBER DRYING: NEEDS AND TECHNOLOGY APPLICATIONS

The majority of lumber in the developing nations is, by industrialized country standards, inadequately dried. Open-air drying, the most prevalent method, often is incapable of producing a low wood moisture content on an oven-dry basis (i.e., the percentage weight of moisture removed in oven drying to the weight of the oven-dry wood) after long periods of seasoning. The equilibrium moisture content (EMC) is the moisture level of a product when it is in balance with the moisture content of the atmosphere. In many developing countries, the values of the EMC under ambient conditions are quite high. Tschernitz and Simpson reported average monthly EMC values of 14 to 17 percent for different regions in the Philippines.(53) The mean monthly EMC at Rio Piedras, Puerto Rico, was reported to vary from 13 to 15 percent.(54) However, because the drying rate of a given product slows greatly as the moisture content (MC) approaches the EMC, the practically feasible level of drying by open-air methods is greater than the EMC value. In the Philippines, 17 to 18 percent MC is the lowest practical level of drying.(55) Drying curves for air-dried timber in Uganda tend to flatten around 20 percent MC.(56)

The moisture content of wood products is particularly important in regard to high-quality applications such as furniture making, joinery, and wood carvings. In the temperate climates, wood with 10 percent, or even less, moisture content is considered necessary for such uses in order to prevent shrinkage related problems such as splitting, warping, and rupturing seams which result when drying occurs in the finished product. Although conditions within buildings in the developing nations are, on the average, less controlled than in the developing nations, the interior EMC may still be lower than that attainable through open-air drying. For example, the average EMC within buildings in Uganda was reported to be 12 percent.(57) Although storage of air-seasoned timber in buildings can reduce the moisture content to this value, the time required for this to occur is quite long, at least several months. Small operators unable to afford conventionally-fueled kiln technologies face difficulties because of both the large space requirement of air drying and the extended, uncertain period capital must be tied up in drying stock. As a result, much of the wood products manufacturing in developing countries occurs without regard to the moisture content of the lumber.(58)

Historically, this problem has not been as great as might be expected in developing countries. The most commonly used woods for high-grade purposes have been those species with low shrinkage coefficients such as mahogany, teak, or deodar.(59) Regions of fairly low climatic variation, where internal heating is not a necessity, may have fairly high, consistent interior EMCs, minimizing the likelihood of further moisture loss in finished wood products. Finally, consumer expectations of the performance quality of wood products have not been high in many developing countries.

For several reasons, however, proper seasoning of timber is, and will increasingly become, more of a recognized problem for developing countries. Central to most development plans are increased efforts for involving indigenous businesses in value-added processes associated with export commodities. For example, the Philippines has considered restrictions on log exports in order to boost the indigenous secondary wood products manufacturing capacity.(60) For such programs to succeed, product reliability in temperate and controlled environments must be established, which requires drying capabilities beyond that of open-air drying techniques. As Tschernitz and Simpson have written:(61)

> The chances are great that a significant percentage of wood products made from lumber above 15 percent moisture content, and then put into use in an environment with an EMC of 10 percent or less, will develop splits, distortions, ruptured gluelines, and other excessive-shrinkage related problems.

The domestic markets of Third World countries for high-grade wood products should expand concomitantly with their development. Improved wood seasoning may be necessary to improve housing construction and furnishings. For example, a self-help project for introducing wood frame housing in Central Java encountered difficulties in obtaining dry timber supplies locally. As a result, simple village drying plants were proposed to support the program.(62) With increasing income levels in developing countries, consumer demand for higher quality wood products may be expected to increase.

A third incentive for improving drying facilities is related to forest management and conservation efforts. More thorough drying would help increase product life which could contribute toward a more efficient use of a potentially limited resource. Improved drying techniques could also help relieve extraction pressure on preferred timber species by allowing the utilization of presently unacceptable ones. For example, in regard to India, Sharma, Nath, and Bali have stated: "The prevailing practice of using insufficiently seasoned timber has been a major stumbling block in the replacement of conventional low-shrinkage coefficient timbers . . . with other equally suitable but lesser known timbers. . . ."(63)

The introduction of improved timber drying technologies may contribute toward the implementation of the utilization of a wide range of species (so-called "any species" utilization) in tropical forests, a concept which is receiving special attention in recent forestry forums.(64) Some investigators feel that the opportunity to utilize more timber species can improve the effectiveness of management for more valuable species by permitting the economic removal and marketing of less valuable species.(65) Although the effect of improved drying on national forestry practices in the near future is probably not very great, improvements could have significant local impact, particularly in those countries with poorly developed transportation networks. If pressures on domestic forest reserves begin to affect the availability of low-shrinkage coefficient timbers, utilization of other species requiring more careful drying may become unavoidable.

Two important drying improvements are lower attainable final EMCs and reductions in the required time for drying. The former relates to product quality and market penetration and the latter affects capital turnover. Tschernitz and Simpson have estimated the effects of temperature on the drying times and the EMCs of several timber species for the climatological conditions at Quezon City, Philippines.(66) If the drying temperature is increased only 10°F (5.6°C), a given MC level can be attained in 45 to 75 percent of the time required for ambient temperature drying. If the temperature is increased 20°F (11.1°C), only 30 to 50 percent of the open-air drying time will be required. The EMC is also very responsive to temperature. At ambient temperatures, the EMC varies between 13.2 and 17.8 percent; however, a 10°F rise in drying temperature reduces the EMC range to 9.0 to 10.9 percent, whereas a 20°F increase reduces the EMC range to 6.7 to 7.7 percent. Apparently, relatively modest increases in drying temperature are sufficient to attain both high-quality moisture contents and to reduce greatly the time involved in the drying process.

The required temperature increases are relatively easy to obtain with fairly simple, nonfocusing solar collectors. While conventionally-fueled kilns dry timber more quickly, these designs are dependent on fossil-fuel or waste fuel resources which may be subject to increased scarcity and competitive uses. Furthermore, solar kilns are less costly to operate and perhaps to construct than fossil-fuel fired kilns as well as being more flexible with respect to scale, which is important for many developing country applications.

A significant portion of the developing world's wood handicrafts and furniture are manufactured by small operators. Tschernitz and Simpson have estimated the lumber consumption of small to medium operators in the Philippines to be on the order of 1,000 to 3,000 bd. ft./week (2.4 to 7.1 m^3/week).(67) Many of these operators supply export markets, and product rejections as a result of improper drying are a significant problem. The small operators are, however, disadvantaged relative to large-scale operations in obtaining improved drying technologies. The smallest commercial steam-heated or direct-fired dry kiln designs have at least an order of magnitude larger capacity than the weekly timber processing capacity of small operators.(68) These technologies cannot be scaled down easily because severe diseconomies of scale are encountered. Tschernitz and Simpson estimated that the capital costs per board foot (bd. ft.) holding capacity of a given hardwood kiln design in 1976 would double from \$2 to \$4 per bd. ft. when the capacity was reduced from 30,000 bd. ft. (71 m^3) to 10,000 bd. ft. (24 m^3).(69)

Solar kilns, on the other hand, at present levels of development are best suited for small-scale uses. The most common experimental designs have utilized integral wall and roof solar collectors. The size of these dryers is limited by the diminishing ratio of surface collecting area to volume holding capacity. Oversized dryers of this type could not trap sufficient heat energy to appreciably speed the drying rate compared to open-air drying. Solar kilns have, however, been successfully constructed and operated up to 8,000 bd. ft. (18.9 m^3) holding capacity. Using four weeks drying time as a rough approximation of solar kiln performance, the holding capacity of small and medium operators in the Philippines would fall in a range of 4,000 to 12,000 bd. ft. (9.4 to 28.3 m^3). The lower half of this range can be met technically by the solar kiln technology demonstrated to date. Larger kiln capacities may also be feasible, multiple units may be employed, or a mix of open-air and solar drying might be used to satisfy more stringent drying requirements.

Another important requirement of small-scale operators is for drying technolgies which can be operated and maintained with modest skill levels. Based on experience to date, solar kilns are simpler to operate, require less supervision than conventional kilns and only slightly more supervision than air drying schemes. Under normal operating conditions, the fans and vents may require adjusting, at most, only a few times a day. Precise monitoring of solar kiln conditions is not necessary and the only sophisticated technology required is electric fans. Since the majority of wood products industries producing products for high-quality applications is probably urban based, the use of electric fans should present neither energy supply nor significant maintenance problems. In certain cases, humidification will be necessary for stress relief. For countries with low labor costs, humidification might be provided by manually-operated misters, although this low technology option has been neglected in the literature. Several researchers consider the low operator skill requirement to be the most significant advantage of solar kilns.(70)

A potentially significant application of solar kilns that has not been investigated is the feasibility of natural-convection firewood drying. As discussed in chapter 2, the performance of wood conserving improved cookstoves is dependent on a continual supply of well dried firewood. Troublesome and dangerous build-up of chimney soot and creosote is also minimized when dry wood is burned. The same problems of extended drying times, large space requirements, and limited reduction of moisture content associated with air drying timber occur for natural firewood drying. Likewise, solar firewood drying has the same advantages as solar timber drying. However, solar firewood dryers will have to be employed primarily in villages, rather than near urban areas and, to be feasible, must not appreciably increase the cost of drying. These requirements dictate that solar dryers would need to operate by natural convection in order to avoid the capital and operating costs and electricity requirements of electric fans.

Natural-convection solar wood drying was demonstrated by Rehman and Chawla in India in 1961 to be feasible on a laboratory scale.(71) This approach was discarded because of the poor quality of the dried wood for commercial purposes. However, wood quality is of little significance when the wood is to be burned. Important factors yet to be investigated are the preferred capacity of natural-convection dryers for family and village scales, and the costs of solar-dried firewood.

SOLAR TIMBER DRYING: STATE-OF-THE-ART REVIEW

Natural-Convection Drying: Forest Research Institute, Dehra Dun, India

Rehman and Chawla tested nine small laboratory-sized solar kiln designs at the Forest Research Station, Dehra Dun, India.(72) These designs included both forced-air and natural-convection models with different types and configurations of external collectors. The final design consisted of a metal box collector installed under a metal drying chamber with a chimney to assist thermal convection. In a 16-day test terminated in June, a 13 to 18°C rise in average drying chamber temperature over ambient temperature and a drying time reduction of 54 percent compared with air drying, and a final MC of 13.6

percent compared with 23 percent MC for air seasoning over the same period was obtained. Quality was poor for some timbers; lack of humidification mechanisms was considered the main cause. Thermal circulation was easily disturbed by changing wind directions. Timber also showed a tendency to absorb moisture from the atmosphere at night if the drying chamber was not closed tightly. Some related research has continued in India, but apparently no reports on natural-convection solar kiln designs in other countries have appeared in the literature.

Sharma and his colleagues have listed several compelling reasons for incorporating forced-air capability in solar kiln designs.(73) With strong air circulation, heat may be directly absorbed over the entire surface of the kiln without localized overheating of timber and the associated problems of quality loss. Because of the relatively low thermal circulation in Rehman and Chawla's designs, the timber had to be shielded from direct sunlight and, thus, external solar collectors had to be used.(74) With forced-air circulation, a more compact stacking pattern of the timber is permissible than with natural convection, which increases the capacity of a given design. The efficiency of drying is also increased, particularly when the timber MC is above the fiber saturation point (about 30 percent MC) and the rate of surface moisture removal is the limiting factor in the drying process.

Wind-Powered Kiln in the United States

Curtis Johnson of the Forest Products Laboratory, Madison, Wisconsin, built a wind-powered solar kiln from readily available materials that utilized a wooden structure with a 400 bd. ft. (0.9 m^3) capacity.(75) The north and south walls were angled toward the center by 22.5° and covered with black roofing felt. There was a 37-ft^2 (3.4-m^2) collector area built into the south wall, constructed of glass windows and flat black galvanized iron sheets. Air circulation was provided by a 14-inch (35.6-cm) diameter centrifugal fan powered by a 54-inch (137-cm) diameter windmill. The windmill power was transmitted through a friction plate and rubber roller drive which both increased the fan speed by a 3 to 1 ratio and allowed slippage to prevent damage to the fan during high wind gusts. The windmill had a cut-in speed of 4 mph (6.4 kmph). Fresh air was drawn through vents in the floor, forced up through the collector, and then passed down through the timber stack. Some recirculation of air occurred during operation.

Internal temperature increases of about 50°F (28°C) above ambient temperature in both summer and winter were obtained. Experiments were performed with mixed loads of cherry and oak. After 52 days, the cherry dried from an average MC of 15.5 percent to 8 percent and the oak from 60 percent to 6.5 percent. No comparison was made for air drying rates. The boards had no evidence of checks (i.e. a lengthwise separation of the wood resulting from drying stresses(76)), but otherwise no details were given on quality. One problem was that there was no means of controlling the fan when air circulation was not desired.

Electric-Powered, Forced-Air Design, U.S.A.

Johnson later developed another expanded model with an electric-powered forced-air design and a 850 bd. ft. (2 m^3) capacity.(77) In this design, the south wall was inclined 40° from the vertical and contained 47 square feet (4.4 m^2) of single-pane storm window as the collector area (see fig. 3.2). Two electric fans were situated over the lumber stack and were thermostatically controlled to run when interior temperatures reached 80°F (26.7°C). The air circulation system was similar to the previous model. Dryer temperatures of 10 to 15°F (6 to 8°C) above ambient temperature with a load of green wood and 20 to 30°F (11 to 17°C) above ambient temperature when the lumber MC was below 20 percent were obtained. The suggested vent control was to run initially with the vents open and gradually close them as the lumber dried. Johnson determined that, on the average, 80 days are required to dry 4/4 (4" x 4") green hardwoods during the summer. The annual output could be increased by using the dryer to finish drying air-dried timber from 12 to 15 percent MC down to a final MC of around 8 percent.

U.S. Forest Products Laboratory Kiln

Edward Peck developed a solar kiln designed at the Forest Products Laboratory, Madison, Wisconsin, that formed the basis for several of the subsequent experiments in the field.(78) This kiln had a capacity of about 425 bd. ft. (1 m^3) and consisted of a wooden rectangular frame covered on three sides and the roof with a double layer of transparent plastic sheet with a north wall which was a solid plywood sheet. Corrugated flat black aluminum sheets were nailed to the insides of the studs and rafters to serve as heat absorbers. Air was circulated by a 5/8 hp, 24-inch (61 cm) fan. Two 8-inch (20 cm) vents were installed in the east and west walls.

Average internal temperatures ranged from 8.5 to 21°F (4.7 to 11.7°C) above ambient temperature with full loads. Drying times of 4" x 4" red oak from over 70 percent to 20

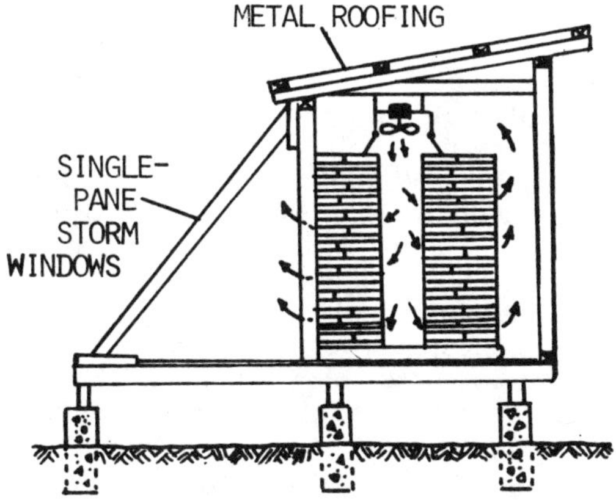

Fig. 3.2. Schematic Diagram of the Johnson Timber Kiln

Source: Paul J. Bois, "Constructing and Operating a Small Solar-Heated Lumber Dryer," Forest Products Utilization Technical Report No. 7, USDA Forest Service, Madison, Wisc., January 1977.

percent MC ranged from 27 to 61 percent of the calculated times for air drying; however, actual drying times ranged from 23 to 105 days depending on the season, weather, and initial MC of the wood. The annual average drying time was estimated to be about 50 percent of the time required for air drying. Timber quality was improved over air-dried timber by having 11 percent less surface checking. Although the dryer was conceived as a predryer, one load was dried to a final 10 percent MC.

Tropical Forest Research Center Kiln, Puerto Rico

Edwin Maldonado(79), and Maldonado and Peck(80) tested a solar kiln built at the Tropical Forest Research Center, Rio Piedras, Puerto Rico. The kiln capacity was 2,000 bd. ft. (4.7 m^3) and was similar in design to the Madison dryer except that the roof had a 16° southward pitch, the floor was cement, heat absorbers were located under the roof only, and four 16 inch (41-cm) fans were used, powered by a 1.5-hp motor. The unit was tested from late 1961 to early 1963.

The average increase in interior temperatures over ambient temperature was about 28°F (15.6°C). The rate of moisture loss during solar drying was twice as high as for air drying. About 1,000 bd. ft. (2.4 m^3) of mahogany were dried from a 50 to 12 percent MC in 23 days beginning late

October, while air-dried lumber reached only 23 percent MC during the same period. No further air drying was possible, but the solar-dried timber MC dropped to 8.5 percent after six more days. End splitting was reduced slightly by solar-drying as compared to air drying.

The dryer was subsequently modified in 1963 to increase the capacity to 3,000 bd. ft. (7.1 m^3) by increasing its length.(81) The number of fans was increased to nine, and three mist sprayers were installed for humidification. A solar water heater was installed with a line running to the mist nozzles located along the base of the north wall. With the fans in operation, the mist was blown up to the roof and then through the lumber stock. A 48-hour mist spray conditioning treatment was found sufficient to remove severe casehardening, that is, stresses associated with shrinkage which may cause splitting or warping when dried lumber is machined.

In a test begun in September, 1963, a full load of mahogany was dried from 47 to 12 percent MC in 40 days; air-dried lumber in the same period reached only 20 percent MC from a lower MC initial value of 41 percent. Solar drying was calculated to attain a 15 percent timber MC level in not more than 39 percent of the time extrapolated for air drying. After conditioning, the quality of the solar-dried timber was not significantly different than that of air-dried timber.

By 1964, the outer 4-mil polyvinyl fluoride sheeting had deteriorated considerably because of excessive flexing and film embrittlement and both layers were replaced by a single layer of window glass. Mixed hardwoods were then tested with similar results as in the previous tests. Mist sprayers were operated daily with the vents closed for the first 20 days until the wettest sample reached a 30 percent MC.

Taiwan Kiln

Tao and Hsiao built a similarly designed kiln in Taichung, Taiwan, with a capacity of 2,500 bd. ft. (5.9 m^3) and two 18-inch (46-cm) fans powered by a 0.25-hp motor.(82) The kiln temperature rose about 9°C above the ambient temperature. In a test begun in December, 1964, 1,800 bd. ft. (4.2 m^3) of schima dried from 44 to 12 percent MC in 48 days. The quality of the timber was acceptable and no serious casehardening problems were noted even though schima is very susceptible to drying defects. This kiln was later modified by increasing the fan size, adding a condenser pipe, and installing a coal-fueled stove for night heating.

Colorado State University Kiln

Troxwell and Mueller built a 1,200 bd. ft. (2.8 m^3) capacity solar kiln at Colorado State University, Fort Collins, without corrugated heat absorbers.(83) All the interior wood surfaces, the north wall, rafters, studs, and fan baffle were painted black to act as heat collectors. The kiln was originally covered with clear polyvinyl fluoride; however, after one year that covering had to be replaced, and corrugated fiberglass-reinforced polyester was installed. This alteration decreased kiln temperatures by about 3°F (1.7°C) because of differences in transmissivity of the materials. Two 0.5-hp, 24-inch (61-cm) fans were used for air circulation. Experiments were performed with 4" by 4" engelman spruce and lodgepole pine. Summer drying times to 19 percent MC were not found to differ between solar and air drying; however, solar drying to 12 percent MC occurred in roughly 65 percent of the time required for air drying. In the winter months, solar drying to 19 percent MC was obtained in about 40 percent of the time required for air drying, while only solar drying was capable of attaining a 12 percent MC level. Drying times varied greatly between seasons. In August, solar drying from 60 to 12 percent MC took six days, while in December-January, drying from 65 to 12 percent MC took 25 days. Splitting was slightly more noticeable in solar-dried timber but was not sufficient to lower the value of the timber appreciably.

Madagascar Kiln

Gueneau built a solar dryer at Tannanarive, Madagascar, based on the Colorado design.(84) The dryer capacity was about 4.5 m^3 (1,900 bd. ft.) with the roof sloped 45° to the west and the timber stack placed on the east side. Two 60-cm fans circulated air from east wall vents along the ceiling and deflected off the west wall through the lumber stack and out north and south wall vents. Experiments were performed with pine and rosewood. The drying times to 20 percent MC were not substantially different between solar and air drying. However, solar drying took only 32 to 66 percent of time required for air drying when drying timber from 20 to 15 percent MC. After 138 days, solar drying could attain final timber MC of 7.5 to 13 percent, while air drying could only attain a 13 to 21 percent MC.

Indian Kiln

Sharma and his colleagues conducted experiments with a 3.5-m^3 (1,500 bd. ft.) capacity solar kiln built at the Forest Research Institute, Dehra Dun, India.(85) The general design is similar to the previous four kilns but included some interesting innovations (see Fig. 3.3). The roof was angled 27° toward the south (0.9 times the latitude) and the kiln was covered with a double layer of clear polyethylene, except for the north wall which was not transparent. A one-hp, 90-cm diameter fan was mounted behind the lumber stacking area by the north wall. A partial false ceiling and north wall were installed. The ceiling is open along the south edge and may be opened or closed behind the false north wall, which houses the fan and forms a plenum chamber. Vents were installed on the lower north and south walls. When the false ceiling is opened along the north, the kiln operates in a recirculating mode, that is, air is forced up the plenum chamber over the false ceiling, down the south wall, and through the timber stack. The lower vents can be opened for partial ventilation. The kiln can also operate in a single-pass mode by closing the false ceiling gap above the plenum, removing the back fan cover from the north wall, and opening baffles located above the false ceiling on the north wall. Air then enters through the baffles, passes over the false ceiling, down the south wall, through the lumber stack, and exits behind the fan. The humidity level in the kiln can be partially controlled by selection of the operating mode. The option of operating in the single-pass mode alleviated problems of condensation encountered in initial stages of drying in an earlier version of the kiln that could operate only in the recirculating mode.

Fig. 3.3. The 3.5-m^3 Capacity Solar Kiln Built at the Forest Research Institute, Dehra Dun, India, in 1971. Photography Courtesy of S.N. Sharma.

The first experiment was performed in the recirculating mode with a full load of toon and mundari timber beginning in May. The results were compared with air drying, unheated forced-air drying, and electrically-heated low-temperature (45°C maximum) drying. Solar kiln temperatures were 10 to 19°C above the ambient temperature. Solar drying was about twice as fast as air drying but twice as long as low-temperature electric drying to reach the same timber MC. Because of the problem with high kiln humidities, unheated forced-air drying was slightly faster above a timber MC of 30 to 35 percent; solar drying was slightly quicker below this range. Final timber moisture contents for open-air drying were only 25 to 35 percent, and for unheated forced-air drying were 20 to 25 percent. Solar drying, however, was able to lower the timber MC to 12 percent in 32 days. An interesting aspect of this experiment is that, out of these 32 days, there were heavy monsoon rains in 12 and normal hours of bright sunshine (11.7 hours) in only three. In over half of the days during the experiment, there were less than 5.8 hours of bright sunshine. The authors concluded that the solar kiln was able to trap enough heat from diffuse and intermittent solar radiation to maintain favorable EMC conditions for drying to low MC levels.(86)

The second experiment, begun in October with hollong timber, was performed with both modes of kiln operation. The single-pass mode was utilized while the lumber MC was above 40 percent. Aluminum reflectors were placed on the east and west sides to increase energy collection while the lumber MC was above 30 percent. These measures served to reduce the humidity in the early drying stages in order to increase the drying rate. After the timber was dried to 40 percent MC, the recirculating mode was used during the daytime. The single-pass mode was operated at night until the timber MC dropped to 25 percent. Kiln temperatures were 15 to 28°C above the ambient temperature. Solar drying rates were higher than unheated, forced-air drying rates for all timber MC levels. The solar drying time to a 35 percent MC level was 51 percent of the forced-air drying time and only 31 percent of the open-air drying time. The final timber MC attained was 12 percent for solar drying (in 28 days) and 25 percent for unheated, forced-air drying. The quality of lumber from these experiments was generally good.

Subsequently, enlarged (7.1. m^3) models consisting of longitudinally joined units of the previous design were tested under commercial operation.(87) These designs have two, 2-hp fans, a spinning disc humidifier, absorbers on the east and west walls to protect against direct sunlight-induced end splitting, and covers made from 5.5-mm glass sheathing above 0.25-mm polyethylene with 12-mm wire netting for protection against hail. The operation and performance of these models were similar to those of the smaller design.(88)

Portable, Demountable Kiln, Philippines

Casin and his colleagues reported experiments with a portable, demountable solar kiln of 480 bd. ft. (1.1 m^3) capacity, developed through the Forest Products Research Institute, Philippines.(89) The roof and east and west walls of this kiln consisted of black corrugated aluminum sheets covered with a 2-mil transparent plastic sheet with a 6 inch (15-cm) air circulation space between the aluminum and plastic sheets. The north and south walls were plywood, and a 0.75-hp 24-inch (61-cm) diameter fan was installed. Tests were performed at several locations with several timber species and board sizes. Average kiln temperatures were 13 to 21°F (7.2 to 11.7°C) above the ambient temperature. Solar drying times were generally 30 to 50 percent of the air drying times to reach the same timber MC. The lowest MC attained through solar drying was 7.8 percent, while air drying reached only 14.5 percent. For most tests, solar drying times were between 37 and 58 days. Slight to moderate checking occurred on solar-dried timber, and the problem could not be eliminated by vent control. Casehardening was a more pronounced problem, depending on the species, and humidification was recommended for stress relief conditioning.

Tanzanian Kiln

The Forest Department Utilization Section at Moshi, Tanzania, reportedly built a solar dryer of 6 m^3 (2,500 bd. ft.) capacity in 1968.(90) This kiln was covered with glass on the roof and polyethylene on the sides. Black galvanized absorbers were placed behind the walls and ceiling. Air circulation was provided by three fans. No performance information is apparently available.

Ghanaian Kilns

Martinka converted a glass-covered greenhouse in Kumasi, Ghana, to a solar kiln with a 1,700 bd. ft. (4 m^3) capacity.(91) Black aluminum sheets were installed for heat absorbers with a single fan for circulation. Drying times of various species to 20 percent MC were compared with air drying and predrying in an unheated, converted oven of 96 bd. ft. (0.23 m^3) capacity equipped with a fan. Solar drying times were 26 to 67 percent of the air drying times, but 1.2 to 2 times greater than the predrying times. The difference of scale and the possible heat contribution of the fan make the latter comparison uncertain.

Ugandan Kilns

Plumptre experimented with three solar kiln designs in Uganda.(92) The first design had a capacity of 1,400 bd. ft. (3.3 m^3) and was covered with two layers of 5-mil weatherable Mylar polyester film. Two 18-inch (46-cm) diameter fans were powered with a 0.75-hp motor. A flat-black false ceiling and the painted floor acted as solar energy absorbers. Curved aluminum reflectors were placed on the east and west sides of the kiln. Maximum kiln temperatures were roughly 20 to 60°F (10 to 30°C) above the ambient temperature. Drying rates were better than air drying, particularly after timber MC fell below 30 percent; however, they were 2.5 to 3 times slower than steam kiln rates. Timber could be solar dried to MC of 12 percent or less in less than a month, while air drying tended to reach only about 20 percent MC. Changes in average daily rainfall had little effect on the rate of drying in the solar kiln. Plumptre explained this result by noting that the heating power of the sun is greater in the clear weather between storms than during the more hazy dry weather.(93) Solar-dried timber quality was as good or better than air- or steam-dried timber. Casehardening was not a problem. The vents were small and closed every night, maintaining high relative humidities within the kiln which probably contributed some conditioning effect. The Mylar covering was expensive and difficult to obtain, but reportedly had a life of about six years.(94)

A second kiln was constructed with a 4,000 bd. ft. (9.4 m^3) capacity.(95) Four 20-inch (51-cm) fans were installed with two 2-hp motors. Reflectors were used again outside the kiln and a solar water heater was connected to a pipe fitted to the kiln wall near the false ceiling aluminum absorber. The drying performance of this kiln was about 10 percent slower than the first model. The solar water heater probably contributed little to the drying process because the capacity of the system was small relative to that of a full timber load.

An enlarged, simplified version of the second model was built with an 8,000 bd. ft. (18.9 m^3) capacity. Black aluminum absorbers were placed on the walls as well as the ceiling. The kiln was covered with 4 and 2 mil polyvinyl fluoride, since weatherable polyester became unobtainable. Operation was simple; besides turning the fans on and off in the morning and evening, the reflectors were adjusted hourly. Performance data was not given.

United Kingdom Oxford Kiln

Recently, Plumptre has been involved in the development of a low-cost, 7.1-m^3 (3,000 bd. ft.) capacity solar kiln in Oxford,

U.K.(96) The design of this kiln departs in several respects from the previous ones. In order to reduce the expense and skills required for construction, the kiln was demountable and utilized the stack of timber to be dried as a structural support. A truss holding an insulated black corrugated metal sheet was placed over the timber stack, and another uninsulated one was angled in front of the stack. Two 24-inch (61-cm) fans with 0.5 hp motors were located under the uninsulated heat absorber. A single-layer tent of UV-inhibited, horticultural-grade polyethylene enclosed the structure. The sides and floor were folded together and buried in a shallow trench dug around the kiln. The doors and vents were also plastic and were fastened by Velcro strips. The single layer of polyethylene was used to reduce costs. In addition, double glazing was felt to have limited benefit because the reduction of entering radiation caused by the second layer often exceeded the gain in nightly heat retention.

The kiln has been tested for several years drying 2-inch (5.1-cm) oak for reconstruction work. During summer tests, the drying time from an initial MC of over 40 percent to a 10 to 12 percent MC was four months and from an initial MC of 60 to 100 percent to a 12 to 14 percent MC was five months. Air drying to 40 percent MC reportedly takes 6 months and kiln drying to further reduce the MC to 12 percent takes another 1.5 months. Solar drying, then, takes approximately 50 to 70 percent of the time required for the normal practice of drying that sized oak in the United Kingdom. The quality of the wood was good, with little or no surface checking and only slight casehardening. Further tests of this kiln were planned in Dominica and Tanzania.

Canadian Kiln

Yang, of Lakehead University, Canada, has built an 800 bd. ft. (1.9 m^3) capacity solar kiln.(97) The south wall was angled 37° from the vertical and the roof was sloped 30° to the south; both were covered with two layers of glass and black metal heat absorbers. A series of pipes were installed to condense and remove moisture from the kiln. Heat storage was provided by a 4-inch (10.2-cm) layer of rock on the bottom of the kiln. Fall and winter kiln temperatures were 40 to 50°F (22 to 30°C) above the ambient temperature. The estimated drying capability was to 12 percent MC in summer and 16 percent in winter. Plans have been drafted for a 4,000 bd. ft. (9.4 m^3) solar kiln equipped with a wood-waste fueled backup burner.

CSIRO Australian Kiln

A radically different approach to solar kiln design involves the use of external solar collectors and opaque drying chambers. Read, Choda, and Cooper reported experiments with this type of kiln operated by the Commonwealth Scientific and Industrial Research Organization in Australia.(98) The maximum capacity of the kiln was 6.5 m^3 (2,750 bd. ft.) and 56 m^2 of solar air heating collectors were inclined 38° to the North. The collectors were glass surfaced with selective-surface coated, copperclad steel vee-corrugated absorbers, insulated with aluminum foil backed rock wool. Heat storage of about 70 percent of the maximum daily heat input, based on the collector area required for maximum kiln capacity (93 m^2), was provided by 15.5 m^3 of crushed basalt located beneath the drying chamber and insulated with polystyrene. The storage system was designed to provide 16 hours of heat recovery per day. The drying chamber walls and roof were constructed of prefabricated insulated panels with aluminum inner surfaces. Two 75-cm diameter fans were located above the timber stack to provide air circulation. A thermostatically controlled system of fan and damper operation circulated air through the solar collectors and the timber stack when the collector temperature was high, circulated air through the rock pile when the air temperature after passing through the timber stack was still high, and blew air through the rockpile into the drying chamber when the solar collector temperature was low. Experiments were performed with 2.8 m^3 (1,200 bd. ft.) of eucalyptus. Timber was dried from a moisture content of 100 to 16 percent in 20 days. Air-dried timber reached a 42 percent MC during the same period. Steam-heated kiln drying to 16 percent MC was estimated to take about half the time as for solar kiln drying. Noticeable collapse (i.e., severe distortion or flattening of single cells or rows of cells in wood during drying) and severe stress affected the solar-dried timber; however, after reconditioning, the finished quality was satisfactory.

External Collector Kiln, U.S.A.

Tschernitz and Simpson have built a 2.4-m^3 (1,000 bd. ft.) external collector solar kiln at the Forest Products Laboratory, Madison, Wis.(99) This kiln was a small-scale prototype of a design proposed to satisfy drying needs of small and medium operations in the Philippines economically.(100) The solar collector had approximately 11 m^2 of absorbing surface area and was built horizontally on the ground, which is a sub-optimal configuration for Wisconsin, but is an inexpensive, effective design for low-latitude countries. The sides and center of the collector were made of wood sills and cement block set into a shallow excavation with the interior surfaces painted with a flat-black coating. The collector frame was filled with sand topped with a thin layer of granulated charcoal to form an inexpensive heat absorbing surface. Fiberglass-reinforced polyester covered the collector 15 cm above the absorption surface. Air was drawn by a 0.5-hp blower from the kiln through the collector in a U-shaped path around a divider and back into the kiln. The plywood drying chamber was insulated with fiberglass and polystyrene and a polyethylene vapor barrier was installed. Two 0.5-hp fans circulated air in the drying chamber. The roof vents and a 0.25-hp, spinning-disk humidifier were controlled by separate humidistats.

An experiment with 2.9-cm red oak lumber was performed in July and August, a period selected to more closely replicate low-latitude conditions. The humidistat was set to open the vents when the kiln relative humidity (RH) exceeded the values of the following schedule: 80 percent RH for timber MC above 50 percent, 50 percent RH from 50 to 30 percent MC, and 30 percent RH for MC below 30 percent. Maximum kiln temperature was 17°C above the maximum ambient temperature for the one day data were presented. Because the timber stock stored considerable quantities of heat, the drying chamber fans were operated for four hours each night after the collector blower was shut off. The timber was dried from 84 to 9 percent MC in 54 days. Air-dried timber MC was only reduced to 20 percent in the same period. The solar drying time was 2.25 times longer that for conventional kiln drying. Air drying was quicker than solar drying when the timber MC was above 40 percent. The high relative humidities maintained within the kiln during the early drying stages were responsible for this effect but also acted to reduce drying stresses, thereby improving product quality. Air-dried lumber had 3.5 times more surface cracks than solar-dried lumber. Stress relief was also successfully effected at the end of the drying period by conditioning at 85 percent relative humidity.

Summary of Solar Kiln Designs and Performance

Design parameters and performance estimates for solar kilns are summarized in table 3.7. Meaningful performance comparisons are difficult because many variables other than design considerations (e.g., timber species, latitude, weather, initial MC, and board thickness) affect kiln temperature, drying time, and the final MC obtained. Many of the entries within the same performance categories are not directly comparable because of inconsistent data recording and presentation practices.

Relative drying times are given, as percentages of matched open-air or conventional kiln drying times, instead of absolute drying times in order to facilitate comparisons between kilns because relative drying times are less influenced by variables other than basic design parameters. The drying rates of most solar kilns relative to air drying are generally much better when the timber MC is below 20 or 30 percent than when the MC is above these values. Differences in the initial MC may affect the relative performance of solar kilns by increasing the period of drying at high timber moisture contents (i.e., at lower drying rates).

All the solar kiln designs have been capable of increasing drying temperatures above the ambient temperature by at least 5°C, and many by 10°C or more up to a maximum reported increase of 35°C. Solar drying times vary between 26 and 88 percent of the time required for air drying, with the typical drying time reduction being about 50 percent. Solar drying times range from 1.2 to 3 times as long as conventional kiln drying processes. Final timber MC obtained through solar drying generally approaches 10 to 12 percent, which is within the range appropriate for high-quality applications. In almost every case, solar-dried timber has a significantly lower MC than air-dried timber. Compared to conventional kiln-dried timber, solar-dried timber quality varies according to these experiments but, even when some degradation was reported, in the majority of cases, loss in quality was not serious.

An interesting observation is the apparent lack of identifiable trends in performance corresponding to changes in design parameters. The one clear exception is the problem of maintaining timber quality when natural-convection dryer designs are used. The lack of other prominent trends may be accounted for by the high relative importance of climatic and timber related factors compared to basic design considerations. Consequently, efforts to improve kiln performance may not be as important as other considerations, such as cost reductions, in expanding the applications of solar timber drying in developing countries.

SOLAR TIMBER DRYING: COSTS

U.S. Forest Products Laboratory (1962)

Peck compared predrying in his solar kiln with air drying to 20 percent MC.(101) Actual costs were not available, but cost estimates were made to provide a basis for comparison. For air drying, charges for capital tied up and invested in the yard and for maintenance were included; for predrying charges for equipment investment, 6 percent interest, 5 per-

Table 3.7. Summary of Selected Solar Kiln Designs and Performance

Design Parameter	India	Wisconsin		Puerto Rico		Colorado	Madagascar
	Kiln Location						
Capacity, m^3	---	0.9	1	4.7	7.1	2.8	4.5
Power	Natural Convection	Wind	├── Electric ──┤				
Collector	External	├── Integral ──┤					
Cover	None	Single Glass	Double Plastic	Double PVF 2 & 4 mil	Single Glass	Single Corrugated Fiberglass-Reinforced Polyester	
Absorber Surface	Metal	Galvanized Flat Iron	Corrugated Aluminum	Corrugated Metal Sheet		None	
Kiln Major Axis	---	├── E - W ──┤					N-S
Structural Slope[a]	---	S Wall Facing S	---	├── Roof Facing S ──┤			Roof Facing W
Average Temperature Above Ambient, C°	13-18	28[b]	5-12	8	15.6[b]	11-17	8-20
Drying Time, % of Air Drying Time	54	---	27-61	50	39	41-67	59-88
Drying Time, % of Steam or Electric Kiln Drying Time	---	---	---	---	---	---	---
Final MC Solar Drying	13.6[d]	6.5-8	10	12[d]	12[d]	12	7.5-13[d]
Final MC Air	23[d]	---	---	23[d]	20[d]	12[f]	13-21[d]
Timber Quality	Poor	Good	Less Surface Checks	Less End Split	Good	Slightly More Splitting	Good
Special Features	Laboratory Model		Predryer		Humidifier		
Source	1	2	3	4, 5, 6	4	7	8

[a]No entry corresponds to vertically and/or horizontally mounted collectors.

[b]These data refer to daily maximum temperature above the ambient temperature.

[c]These values are compared to the total time used in the normal practice of combined air and kiln drying schedules.

[d]These data refer to the moisture content reached for specified drying times.

[e]This value refers to predrying operation only.

[f]This value refers to summer operation. Winter drying values would be higher.

(continued)

Table 3.7. Continued

Design Parameter	Kiln Location: India		Philippines	Ghana	Uganda	U.K.	Canada	Australia	Wisconsin
Capacity, m^3	3.5	7.1	1.1	4	3.3	7.1	1.9	6.5	2.4
Power	Electric								
Collector	Integral							External	
Cover	Double Polyethylene 0.25 & 0.05 mm	Double 5.5 mm Glass Sheathing & 0.25 mm Polyethylene	Single Plastic 2 mil	Single Glass	Double Polyester 5 mil	Single UV-Inhibited Polyethylene	Double Glass	Single Glass	Single Fiberglass-Reinforced Polyester
Absorber Surface	Plywood	V-corrugated Iron	Corrugated Aluminum		Corrugated Iron	Corrugated Metal		V-Corrugated Copper-Clad Steel	Granulated Charcoal
Kiln Major Axis	E - W		N-S	E - W	N-S	E - W		---	---
Structural Slope[a]	Roof Facing S		---	---	---	Roof Facing S	S Wall & Roof Facing S	Collectors Facing N	---
Average Temperature Above Ambient, C°	10-28[b]	---	7-12	---	10-35[b]	15[b]	22-30[b]	22[b]	17[b]
Drying Time, % of Air Drying Time	43-63	40	30-50	26-67	58-75	53-67[c]	---	<50	48
Drying Time, % of Steam or Electric Kiln Drying Time	200-250	Up to 200	---	120-200	250-300		---	200	225
Final MC Solar Drying	12[d]	10-12	7.8-19.3	20[e]	8	10-12	12-16	16[d]	9[d]
Final MC Air Drying	25[d]	---	14.5-28.6	---	20	---	---	42[d]	20[d]
Timber Quality	Slightly More End Cracking, Less Case-hardening	Less Cracking and Splitting, Light to Moderate Case-hardening	Case-hardening	---	Good			Collapse and Stress, Easily Removed	Less Surface Checks
Special Features	Reflectors, Two-Mode Operation	Humidifier, Two-Mode Operation	Demountable	Predryer, Converted Greenhouse	Reflectors	Kit, Demountable	Heat Storage, Condensation Pipes	Heat Storage	Humidifier
Source	9, 10	11	12, 13	14	15, 16	17	18	19	20

(continued)

Table 3.7. Continued

Sources:

1. M.A. Rehman and O.P. Chawla, "Seasoning of Timber Using Solar Energy," Indian Forest Bulletin No. 229, Forest Research Institute, Dehra Dun, India., 1961.
2. Curtis L. Johnson, "Wind-Powered Solar-Heated Lumber Dryer," Southern Lumberman 203 (October 1961).
3. Edward C. Peck, "Drying 4/4 Red Oak by Solar Heat," Forest Products Journal 12 (March 1962).
4. M. Chudnoff, E.D. Maldonado, and E. Goytia, "Solar Drying of Tropical Hardwoods," Forest Service Research Paper ITF-2, Institute of Tropical Forestry, Rio Piedras, Puerto Rico, April 1966.
5. E.D. Maldonado and E.C. Peck, "Drying Lumber by Solar Radiation in Puerto Rico," Forest Products Journal 12 (October 1962).
6. Edwin D. Maldonado, "Solar Radiation Used to Dry Mahogany Lumber in Puerto Rico," Tropical Forest Notes No. 14, Tropical Forest Research Center, Rio Piedras, Puerto Rico, December 1962.
7. H.E. Troxwell and L.A. Mueller, "Solar Lumber Drying in Central Rocky Mountain Region," Forest Products Research J. 18 (January 1968): pp. 19-24.
8. P. Gueneau, "Une Expérience de Séchage Solaire du Bois," Bois et Forets des Tropiques, No. 131 (1970); summarized in Brace Research Institute, "A Survey of Solar Agricultural Dryers," Technical Report T99, Quebec, December 1975.
9. S.N. Sharma, Prem Nath, and B.I. Bali, "A Solar Timber Seasoning Kiln," J. Timber Development Association of India 77 (April 1972).
10. S.N. Sharma, "Solar Timber Drying," Seminar Proceedings: Industrial Applications of Solar Energy, Madras, June 4-6, 1975, National Productivity Council, New Delhi, 1976.
11. S.N. Sharma and P. Nath, "Solar Kiln Field Trials (Commercial Trials on the F.R.I. 7.1 Cu. m. Solar Heating Timber Seasoning Kiln)," Final Research Report, Forestry Research Institute Dehra Dun, India, August 1977.
12. R.F. Casin, "Solar Drying for Lumber," Technical Note No. 76, Forest Products Research Institute, Los Banos, Philippines, 1967.
13. R.F. Casin, E.B. Ordinario, and G.Y. Tamayo, Jr., "Solar Drying of Apitong, Narra, Red Luan and Tanguile," Philippine Limberman 15 (1969): 23-30.
14. E. Martinka, "Predrying of Some Ghanaian Timber," Technical Note No. 11, Ghanaian Forest Research Institute, Kumasi, 1969.
15. R.A. Plumptre, "The Design and Operation of a Small Solar Seasoning Kiln on the Equator in Uganda," Commonwealth Forestry Review 46 (1967): 298-309.
16. R.A. Plumptre, "Solar Timber Kilns: Their Suitability for Developing Countries," Technology for Solar Energy Utilization, Development and Transfer of Technology Series No. 5, UNIDO, New York, 1978, ID/202.
17. R.A. Plumptre, "A Simple Solar Heated Dryer for Timber," Intermediate Technology Development Group Conference on Small-Scale Energy for Developing Countries, January 1979.
18. Frank Hughes, "Prof. Yang's Solar-Heated Kilns," Canadian Forest Industries (March 1978): 62-63.
19. W.R. Read, A Choda, and P.I. Cooper, "A Solar Timber Kiln," Solar Energy 15 (April 1974): 309-16.
20. John K. Tschernitz and William T. Simpson, "Solar-Heated, Forced-Air, Lumber Dryer for Tropical Latitudes," Solar Energy 20 (1979): 563-66.

cent depreciation, maintenance, taxes and insurance were used. Under these assumptions, if the cost of air drying was \$0.10/1,000 bd. ft. per day, then solar predrying was not economical throughout most of the year. If the air-drying cost was \$0.15/1,000 bd. ft. per day, then solar predrying would be economical during most of the year.

Peck also concluded that solar drying was more economical than conventional kiln drying to 10 percent MC (i.e., \$15.66 per 1,000 bd. ft. versus \$30 per 1,000 bd. ft.). This result was based on average drying costs of \$0.27 and \$1.50/1,000 bd. ft. per day, respectively, for solar and conventional drying, with corresponding drying times of 58 and 20 days. The basis of the average daily costs was not given.

India Forest Research Institute Kiln (1976)

Sharma estimated in 1976 the cost of electricity for solar kiln drying 25 mm planking to be Rs 30 per m^3 of lumber and Rs 53 per m^3 of 75 x 100 mm framing lumber for a unit electricity cost of Rs 0.20 per kwh.(102) Drying costs in economically-sized steam kilns were estimated to be not less than Rs 98 per m^3 of planking and Rs 157 per m^3 of framing, approximately three times the solar kiln costs. The construction costs including labor for a 3.5-m^3 capacity solar kiln were estimated to be about Rs 7,700. The construction costs of a steam-heated kiln of equal capacity were estimated to be Rs 25,000. With the assumption that the output of a steam kiln is about double that of a solar kiln, the capital cost per unit volume output of a solar kiln was calculated to be about 67 percent of the cost of a steam kiln. Although replacement of the polyethylene covering on solar kilns may be necessary every 0.5 to 2 years, the polyethylene cover cost was estimated to be well below the normal depreciation costs of an equivalent steam kiln.(103)

Solar kilns were found to require substantially less labor for operation than conventional kilns based on field testing two 7.1 m^3 kilns under commercial conditions.(104) Only one-third of a man-day was estimated to be necessary for normal daily operation of the solar kiln. Steam-heated kilns require continuous monitoring of boiler and kiln drying conditions. Assuming kilns of equal holding capacity and drying times of 18 and 10 days, respectively, for solar and steam-heated drying, the labor requirement for drying a full load in a solar kiln was calculated to be only 30 percent of the requirement for a steam kiln (i.e., 6 man-days versus 20 man-days). The low labor requirements of solar kilns is important for small operators because, by adopting solar kiln technology, drying capabilities can be improved significantly with only marginal increases in labor requirements beyond those currently necessary for open-air drying processes.

CSIRO Australian Kiln

Read and his colleagues compared the costs (1973) of drying by a steam-heated kiln with drying by an external collector solar kiln.(105) Only the capital costs of the steam-heated kiln were based on actual costs; the capital costs for the solar kiln and the fixed and operating costs of both kilns were estimated. Solar drying was not found to be economical, costing \$A3.72 versus \$A3.29 per m^3 of timber for steam-heated kiln drying. They concluded that the overall thermal performance of the solar process must be improved to reduce capital costs. The analysis was based on equivalent performance between the dryers for 37.8 m^3 of timber dried in four days, the then current capability of steam-heated kilns.

United Kingdom Oxford Kiln

Plumptre has compared the costs of drying oak and pine in the Oxford kiln (7.1 m^3 capacity) and in commercial kilns in the U.K.(106) By depreciating the main kiln structure over five years and the polyethylene cover over one year, assuming a 2p per kwh price for electric power, and including labor costs, the cost of drying 2-inch (5-cm) oak was calculated to be £49.56 per m^3 £and £13.30 per m^3 for 1-inch (2.5-cm) pine. These results compare to commercial kiln charges of £63.60 and £11.66 per m^3, respectively. The cost of solar drying was 79 percent that of commercial kiln drying for oak (ignoring the difference in total drying times which favors solar drying), but the costs were higher for pine, which is a softwood.

In another analysis, Plumptre concluded that, when the timber industries first require improved drying facilities, the predominant economic factor affecting the choice of technology often is not the cost per unit of timber dried, but the initial capital cost of the facility.(107) Thus, reducing the initial investment required may be more important to facilitate the adaptation of solar kiln drying than improving performance, unless the improvements are inexpensive to implement.

Capital Costs for Small Kilns

Tschernitz and Simpson, through discussions with Philippine wood industry representatives, estimated in 1976 that kiln costs between \$3,000 and \$5,000 would be acceptable for small and medium operators.(108) Using an estimate of a 1,000 to 3,000 bd. ft. (2.4 to 7.1 m^3) weekly lumber consumption, the maximum acceptable capital cost of a solar kiln with a four-week drying period would be about \$520 per m^3 of holding capacity, and not more than \$350 per m^3 of capacity if the drying time is six weeks.(109) It is important to note that these are very rough cost approximations; costs will probably have to be lower than these to be attractive.

Indian kiln

Sharma and Nath estimated in 1977 the total construction cost of a 7.1-m^3 capacity solar kiln, including humidification and glass sheathing, as Rs 20,000.(110) The cost per m^3 of holding capacity then is about $352.

Oxford kiln for developing country use

Plumptre estimated in 1979 the cost of constructing the Oxford kiln in a developing country as £981, including shipping costs of the kit, local materials costs, and labor.(111) For a maximum holding capacity of 7.1 m^3, this cost is roughly $290 per m^3. The drying time for 2-inch (5-cm) mahogany in the tropics was estimated to be two months.

Philippine design

Tschernitz and Simpson estimated in 1977 the costs of constructing a 4,000 bd. ft (9.4 m^3) capacity external collector solar kiln in the Philippines.(112) The estimates were based on material and labor costs in the Philippines except for the collector panels, humidifier, humidistat, and damper motor which were assumed to be imported, and for which packing and surface shipping charges were included. The total construction cost was estimated to be $5,650 or about $600 per m^3 of holding capacity. Imported components comprise 63 percent of the total estimated capital costs. Substitution or elimination of these parts could substantially reduce the costs of the kiln. For example, if the required cover panels for the collector were reduced by one-half by constructing a single-glazed collector instead of a double-glazed collector, the cost of the dryer would drop to $478 per m^3 of capacity.

An integral roof and wall collector solar kiln design of the same capacity was also proposed.(113) The total construction cost in the Philippines was estimated in 1977 to be $4,605 or $490 per m^3 of holding capacity.

Summary of Cost Comparisons

The majority of economic comparisons between solar and conventional kiln drying have shown solar drying to compare favorably. This result is largely due to the lower capital and operating costs associated with solar drying and diseconomies of scaling conventional processes to small-scale applications.

The only analysis favoring conventional kiln drying was based on the costs of operating a large-scale solar kiln designed to perform with the same drying time and capacity as conventional technology. By designing solar kilns for low-temperature, intermittent drying, the capital and operating costs appear to be a low enough fraction of the costs of conventional kiln drying that, despite increased drying times (commonly by a factor of two or more), the economics of the overall process may favor solar drying technologies.

There is a trade-off between increasing solar kiln performance and keeping capital costs low. Two of the more recent solar kiln designs illustrate divergent approaches to this problem. The Oxford kiln developed by Plumptre minimizes capital costs at the expense of kiln efficiency and drying time performance.(114) The external collector design proposed by Tschernitz and Simpson maximizes kiln efficiency and performance but at higher capital costs.(115) Both of these technologies appear to be expensive when compared to a crude estimate of the drying needs and capital costs acceptable to small and medium operators in the Philippines. In order to improve the economic performance of these kilns, the estimated drying time of the Oxford kiln needs to be improved and the capital costs of the external collector kiln need to be reduced.

Economic analyses of the commercial operation of solar kilns apparently have not appeared in the literature, reflecting the fact that solar kilns are still primarily in the research and development stage. Except for the estimates by Tschernitz and Simpson for the Philippines,(116) the financial possibilities for small operators to improve their drying facilities have not been investigated. The economics of solar kiln operation appear to be attractive. It would be useful to corroborate these cost estimates with analyses of the use of solar kilns in actual business settings, to gain insights into the directions research on solar kiln improvements should follow.

SOLAR TIMBER DRYING: EVALUATION

Cost-Technology Trade-offs

Introduction

The balance between performance and cost of solar kiln design is difficult to resolve and depends heavily on the priorities of the potential kiln users. If reduction in attainable final timber MC is the primary reason for considering solar technology, and the length of the drying period is a secondary consideration, cost reduction may well be more important than performance improvement. As Tschernitz and Simpson showed for

Quezon City,(117) a small increase of only 10°F (5.6°C) in drying temperature over the ambient temperature is sufficient to significantly lower the EMC so that the timber MC can be reduced to 10 percent, a level which is necessary for export and high-quality applications. Although the minimum temperature increment required to lower the EMC will vary with region, large increments are not required. Even the less efficient solar kiln designs are capable of modest temperature increments and, therefore, are capable of satisfying a need for sufficiently dried timber without improvements, given that long drying periods are acceptable.

If, however, reduction in drying time is an important need, then performance and efficiency improvements of solar kiln designs should be emphasized. Increases in drying temperature over that necessary for attaining a final timber MC of around 10 percent are important not for reducing the MC further, which is generally not required, but for decreasing the time required to complete the drying process. Reduced drying times are the primary reason for considering design improvements to increase collector performance, reduce energy losses, or reduce drying rate fluctuations through provision of heat storage.

Reducing energy losses

Wengert evaluated the energy losses of the Colorado State University solar kiln.(118) The convection of solar energy from the translucent fiberglass walls and roof of the kiln was found to be the largest single loss of energy, equal to nearly one-third of the incoming solar energy. Other important sources of energy loss were reflection and transmission of solar energy, ventilation losses, long-wave radiation losses, and conduction losses through the concrete floor. Wengert recommended the use of double layers of covering material to decrease the convection and long-wave radiation losses, and suggested that the inner layer be very transparent so as not to increase solar reflection losses. Translucent outer covering materials were recommended to reduce transmission losses. Wood floors or a combination of concrete over a wooden base were suggested to reduce conduction losses through the floor. Ventilation design improvements and the use of infrared selective coatings and heat-absorption baffles were also recommended. The effects of these improvements on the economics of the kiln use were not estimated. Apparently, the recommendations were not implemented so their effectiveness in improving kiln performance has not been analyzed.

Plumptre argued that double glazing with flexible plastic sheet may have little benefit, the heat retention gain often being less than the reduction of entering radiation.(119) Double glazing may also be inadvisable because good covering materials are expensive and often difficult to obtain in developing countries. Plumptre has cautioned that the economics of kiln improvement must be kept in mind:

> While it is possible, no doubt, to increase the efficiency of solar kilns greatly, the cost of doing so must be considered. The main justification for solar kilns is their low cost and ease of building as compared with conventional kilns. There is no point in losing this major advantage in order to employ some sophisticated method of increasing kiln efficiency.(120)

Improving collector performance

The major actions to increase the amount of energy entering integral roof and wall collector solar kilns have been to angle the roof or wall to intercept the sun's rays more directly,(121) and to supplement the incoming radiation with reflectors.(122) None of these reports give either costs or the actual or estimated effects on kiln performance of these modifications. Reflectors may be useful for extending the upper size limit of feasible integral collector kiln designs since the surface area that can be utilized for collectors increases at a slower rate than the timber capacity with increasing kiln size.

Tschernitz and Simpson presented the results of computer simulations on the performance of different solar kiln designs in the Philippines.(123) The optimal configuration for integral collector kilns was shown to be a combination of roof and east and west wall collectors with the major axis of the kiln oriented north-south. The effect on dryer capability of adding a south wall collector was marginal or negative. The analysis program underestimated the diffuse radiation component which may be quite high in tropical high-humidity areas.(124)

The improvements discussed should be examined for their effects on kiln performance and for their economic benefits. Measurements of the direct effects of the improvements on kiln heat collection and loss could also be made using techniques similar to those of Wengert.(125) This approach would be more rigorous than comparing differences in kiln performance but, also, more expensive to perform. In any case, better field data are needed to evaluate the effectiveness and desirability of kiln improvements.

Heat storage

The provision of heat storage capability has been commonly cited as a useful improvement for solar kiln design.(126) However, few kilns have been built with heat storage systems, except for the CSIRO kiln in Australia which utilized crushed basalt for heat storage.(127) System modeling of solar kiln design at CSIRO showed absorbent-bed storage to be more economical than gravel bed storage, although the total annual costs of the kiln with absorbent-bed storage were still higher than for conventional kilns.(128) This result may be true for high-capacity, high-performance kilns, but it may not be for small-capacity kilns with important constraints on capital. Tschernitz and Simpson concluded that rock or earth storage was the least expensive to use with solar kilns by avoiding the heat-transfer technology necessary for water storage and the expense of salt hydrates.(129)

Since even without heat storage solar kiln drying reduces the drying time and final timber MC below those of air drying, and conventional kiln drying is apparently not competitive for small scales, there is no immediate need for reducing diurnal fluctuations in solar kiln drying rates. The important need, then, may be for providing rainy and cloudy weather drying capability. Experiments at the Forest Research Institute, India, showed solar drying to be uneconomically slow in continuously cloudy weather; however, as long as there was intermittent sunshine, the drying rate was close to normal. According to Sharma:

> Even on cloudy days, the solar kiln successfully makes use of the available intermittent sunlight and diffuse skylight to trap requisite amounts of heat for maintaining temperatures 5°C to 7°C higher and R. H. inside the kiln somewhat lower than the ambient, i.e., conditions suitable for drying of timber to lower moisture contents.(130)

The expense of installing heat storage capacity should be evaluated against the likelihood of occurrence of extended, unbroken periods of cloudy weather.

Humidification

Means of humidification were considered necessary to improve timber quality by several investigators.(131) A few investigators have made special efforts to reduce kiln humidities.(132) Other investigators reported little or no problems with humidity or timber quality.(133) Part of this divergence is probably because of differences in the drying characteristics of various timber species. Operating procedures may also have contributed to this apparent inconsistency. Plumptre indicated that the high quality of the timber dried in the Ugandan kiln may have been due to small vents causing the RH to remain high.(134) By having high kiln humidities at night, estimated to be as high as 90 to 100 percent while the timber MC was high, mild reconditioning of the timber was probably occurring.(135) High kiln humidities also slow the rate of drying, so a balance may have to be sought between timber conditioning and drying times. Where vent control is insufficient to prevent quality loss in the timber, some form of humidification will be required. In order to hold the capital costs of kilns low, simple methods of humidification should be explored, particularly for areas without water supplies.

Cover materials

Another important consideration is the type of covering material chosen for solar kiln construction. The 4-mil polyvinyl fluoride on the Puerto Rico dryer deteriorated after two years and was replaced by glass.(136) The same covering on the Colorado State University kiln had to be replaced in only one year. Clear corrugated fiberglass-reinforced polyester was installed in its place.(137) Sharma and Nath gave six months as the expected life of 0.25-mm polyethylene covering because hail damaged the covering easily.(138) Trials with 5.5-mm glass sheathing successfully resisted hailstones up to 12 mm in diameter. An inner layer of 0.25-mm polyethylene was found to last about 1.5 years.

Tschernitz and Simpson presented 1975 costs of various covering materials in the Philippines.(139) Polyvinyl chloride cost about two times the cost of polyethylene, the most inexpensive and least durable plastic considered. Polyvinyl fluoride and Mylar polyester were about 20 and 24 times the cost of polyethylene, respectively. According to Plumptre, the life of Mylar in Uganda was six years.(140) If the expected life of Mylar in the Philippines is the same, it would be economical to use Mylar only if the expected life of polyethylene is less than three months. Ordinary window glass was 1.5 to 2.8 times more expensive than Mylar. If the kiln life is estimated to be 15 years, then the cheapest glass would be more economical to use only if the expected life of polyethylene is less than five months. This crude analysis ignores differences in kiln performance which could be substantial, but it also ignores the high discount rate which should be applied to transactions involving capital-short entrepreneurs.

Status of Solar Kiln Utilization

All of the solar kilns reported in the literature were initially built for experimental purposes, except the wind-powered kiln Johnson used to supply his workshop lumber.(141) It is difficult to obtain information on the present status and use of many of these kilns. Some of the experimental kilns were used commercially after or during the testing period but only a very few of the designs have found their way into the commercial sector; thus, the majority of the solar kilns appear to have been used for research purposes only. In view of the near uniform success reported for experiments and the apparently attractive economics, it is, at first sight, perplexing that use of small-scale solar kiln technology is not more widespread. This situation may be the result of drawing heavily from published information; but, even if we are unaware of most of the private sector solar kilns, it is apparent that solar designs are presently used vary rarely by the wood-seasoning industry. For example, Tschernitz and Simpson mentioned no existing solar kilns in the Philippines, the target nation of their feasibility study, despite the experimental work done in the 1960s there and the need for low-cost, small-scale kiln technology for the export sector.(142)

Some commercialization has occurred in India. The Forest Research Institute, Dehra Dun, field tested 7.1 m^3 capacity solar kiln models under commercial conditions at different latitudes in India. This effort must have had some success because Sharma and Nath reported in 1977 that four commercial units were in operation.(143)

Plumptre stated that his first and second solar kiln models (3.3 and 9.4 m^3 capacity) were in almost continual commercial use for seven years to dry timber for both the government and private furniture workshops.(144) Two other kilns of the larger design were built by the Ugandan Prison Industries and a private furniture manufacturing company, but neither became functional. One kiln was covered with a light-gauge polyethylene that was quickly ruined and apparently never replaced. The other kiln was never covered because the manager interested in the project left the firm before arranging to obtain covering material.

Plumptre has discussed several reasons why the adoption of solar kiln technology has been slow:(145) (1) demand for properly seasoned timber has been small in many of the countries in which solar experiments have occurred; (2) construction has been too complicated for small operators even though the solar designs are simpler to construct than conventional kilns; (3) there has been a lack of appreciation that solar drying can reduce timber MC sufficiently for high-grade applications; (4) while kits are available for conventional kilns, neither kits nor even plans have been availabile to those interested in solar kilns; and (5) durable glazing materials have been difficult to obtain. Plumptre considered the main factor preventing the expansion of solar kiln use in Uganda to be the unavailability of weatherable polyester after local production of this material ceased.(146)

The Oxford kiln has been in use for several years to dry timber for use in reconstruction work in Magdalen College.(147) One estate near the university built one for their own use.(148) The Oxford kiln design is being commercialized as the Jakrap Solar Kiln Mark I through Jak Forest Fuels, Ltd., U.K. Kits will be marketed containing the motors, fans, wired control panel, sufficient UV-inhibited polyethylene to replace the cover once, and assembly instructions. Purchasers will have to supply the wood, corrugated metal sheet, and electric cable. Prior to 1979, this model had not been tried under tropical conditions. Tests were planned to begin in 1979 in Tanzania and Dominica.

Some research and demonstration of solar supplemented low-temperature kiln drying with large-scale, 24,000 to 50,000 bd. ft. (59 to 118 m^3) capacities, is occurring in the United States.(149) A private corporation has been considering construction of a solar kiln with a 30,000 bd. ft. (71 m^3) capacity in New York.(150) These technologies are far outside the range of the target groups in developing countries and are possibly a long way from being viable economic propositions even in developed countries.

4 Village-Level Rice Bran Processing and Utilization

NEEDS AND OBJECTIVES

The optimum use of rice bran is an important postharvest and rural development problem for rice-growing developing countries. Very little rice bran is used in these countries in any form for human food, although it could represent an important source of quality protein and oil. Most rice bran, by far, is used for animal feed; some is used for non-food and non-feed purposes and much is wasted. It has been estimated that 8 million metric tons of rice bran remain unused in the rural areas of developing countries.(1) The present study was undertaken with three main objectives:(2)

- To review and analyze the state-of-the-art of rice bran processing and utilization, with emphasis on Asian developing countries.
- To develop system concepts for processing rice bran to optimize utilization of the protein and oil for human food, with emphasis on village needs.
- To identify research and development priorities covering all pertinent aspects related to new rice bran processing systems: technological, economic, and cultural.

NUTRITIVE POTENTIAL OF RICE BRAN

Rice is a cereal crop of major importance, especially in Asia, which accounts for 90 percent of the annual world production of about 300 million metric tons of paddy (unhulled rice).(3) The world's seven largest producers are Asian countries: The

People's Republic of China, India, Japan, Pakistan, Indonesia, Burma, and Thailand, in approximate order of decreasing tonnage.(4) Only a small percentage of rice is traded worldwide; nearly all is consumed in the countries that grow it.(5)

About 10 percent of paddy is a mill by-product called bran plus polish.(6) Bran includes the outer bran layers and part of the germ; while polish includes the inner bran layers, part of the germ, and part of the starchy interior of the rice kernel.(7) Bran plus polish contains about 10 to 15 percent protein,(8) and about 15 percent edible oil.(9) Hereafter in this chapter, the term "bran" will be used to signify "bran plus polish."

Based on the above figures, the bran fraction of the 300 million metric tons of annual world paddy production is estimated to contain a potential of 3 to 4.5 million metric tons of protein and about 4.5 million metric tons of oil. A somewhat higher value for the protein, 5 million metric tons, was estimated by Barber.(10) As shown in table 4.1, recent estimates of potential oil vary widely.

The amounts of protein and oil in the world's rice bran are important. Based on 3 to 4.5 million metric tons of protein and 4.5 million metric tons of edible oil annually, 50 grams of protein and 2,000 Cal required per day per person, minimum, and a food energy value for the oil of 9 Cal per gram, world rice bran represents the minimum protein requirements of some 150 to 250 million people, and the minimum caloric requirements of about 50 million people. This is even more significant when it is realized that most rice is of Asian origin, and much Asian rice is grown in areas where there are serious nutritional problems.

However, there are substantial challenges to the realization of the nutritive potential of rice bran. Although protein concentrates (high-protein extracts) can be readily prepared from rice bran, their preparation takes place primarily in research laboratories, rather than operationally in rural areas of developing countries.(11). Although edible oil can be obtained from rice bran in proven processes on a commercial scale, this is often frustrated in developing countries by the rapid development of bran rancidity beginning immediately upon milling.(12) Rancid bran is not a practical source of edible oil, cannot be used as a food ingredient, and is greatly limited as a feedstuff. In short, the nutritive potential of rice bran is seriously downgraded and often lost when rancidity is allowed to develop unchecked.

Some developing countries are beginning to move toward edible oil production from bran produced mainly at large rice mills in urban areas. This requires the solution of the bran rancidity problem because the acid content of oil from rancid bran is so high that refining to obtain edible oil is infeasible. Typically, acidic oils have to be converted to soap, a much

Table 4.1. Published Estimates of Potential Oil in World Rice Bran

Million metric tons of oil	Reference
1.8 (crude oil) = 1.6 (edible oil)	D. Adair, "Rice Bran Stabilization," UNIDO Ad-Hoc Expert Group Meeting, Vienna, December 1976, ID/WG.240/5.
2 (edible oil)	W.H. Tanaka, quoted in "Research and Development of a Small-Scale, Low-Cost Rice Bran Stabilizing Unit," Report of Ad-Hoc Expert Group Meeting, UNIDO, Vienna, December 1976, prepared by The International Centre for Industrial Studies, ID/WG.240/10; and H.G.R. Reddy, "Rice Bran Stabilization," UNIDO Ad-Hoc Expert Group Meeting, Vienna, December 1976, ID/WG.240/3.
5 (edible oil)	S. Barber, "Stabilization of Rice Bran," UNIDO Ad-Hoc Expert Group Meeting, Vienna, December 1976, ID/WG.240/7.

lower value product than edible oil, and the caloric value of the oil is lost for food or feed purposes.

Although conceptually there are simple technologies for stabilization of rice bran against rancidity, these are not widely practiced at present in the developing world. If not wasted, rice bran is used for animal feed, or as a soil amendment or fertilizer of low Nitrogen-Phosphorus-Potassium (NPK) value, or as a fuel along with the hulls,(13) but most of the potential nutritive values are presently lost to the human diet.

PERTINENT TECHNOLOGIES

Rice Milling

Bran quality is greatly dependent on the way the paddy is milled. Some methods yield an intimate mixture of hull particles and bran (see table 4.2). More modern milling methods remove the inedible fibrous hulls first, then mill off the bran in a separate step.

Table 4.2. Major Rice Milling Methods in Developing Countries

Method	Sector	Capacity	Bran Quality
Hand-pounding, by mortar & pestle	Rural/Village	7 kg paddy/hr	Contaminated with hulls
Engleberg hullers	Rural/Village	.25 or .5 metric ton paddy/hr	Contaminated with hulls
Small rubber-roll hullers	Rural/Village	.25 or .5 metric ton paddy/hr	Hull-free
Large, commercial mills	Regional/ Urban	1 metric ton paddy/hr & larger	Hull-free

Source: D.S.C. Spencer, S.C. Dunstan, Ibi I. May-Parker, and Frank S. Rose, "Employment, Efficiency and Incomes in the Rice-Producing Industry of Sierra Leone," African Rural Economy Paper No. 15, Michigan State University, East Lansing, 1976; and Central Food Technological Research Institute, "The Food Industry in India," Background paper for Working Group No. 6, International Forum on Appropriate Industrial Technology, UNIDO, New Delhi, November 1978, ID/WG.282/27.

If hulls contaminate the bran, utilization of the bran for food and oil is complicated by the need to remove as much of the hull particles as possible, probably by sieving. But complete removal of hulls by sieving is apparently impossible. For example, results of an experiment conducted by Duckworth and Dent indicated that the material passing a 40-mesh screen contains 20 percent hull particles.(14) This hull removal obstacle has led many investigators to regard bran utilization from hand-pounding or from huller mills as an essentially intractable problem.(15) They choose, therefore, to focus on the larger, more modern mills that produce bran uncontaminated with hulls.

The percentages of bran produced in admixture with hulls, and bran produced without hull contamination are not accurately known. However, Hawkey has prepared some estimates of contributions to total world rice production by various modes of processing.(16) The percentages in table 4.3 are based on his production estimates.

Table 4.3. Contributions to World Rice by Major Processing Methods

Method	%	
Hand-pounded	12 %	35% of milling yields mixtures of hulls and bran
"Inferior" small mills*	23	
Modern small mills**	14	43% of milling yields bran free of hulls
Modern commercial mills	29	
Parboiled (milling method not specified	22	22% yields bran at least partly stabilized in parboiling
	100%	

*Probably Engleberg hullers.
**Probably Japanese rubber-roll hullers.

Source: R. Hawkey, "The Case for Rural Rice Bran Stabilization," Report to UNIDO, February 1975.

The estimates in table 4.3 suggest that somewhat over one-third of world rice is milled in ways that yield mixtures of bran and hull particles. A 1978 estimate for India indicates a greater value - two-thirds of the total rice in that country was milled by 70,000 Engleberg hullers yielding bran/hull mixtures.(17) Hand-pounding and small huller mills have been prevalent in rural areas of developing countries. Although there have been some indications that these techniques are losing ground to more modern milling methods which separate the bran from the hulls, it seems reasonable to assume that the more traditional methods are still in widespread use at present in rural areas. Better data on the actual situation are needed.

Hand-pounding

A 1970 estimate by the United Nations Industrial Development Organization (UNIDO) indicates that over half of the rice crop in Indonesia and over a third in India is hand-pounded with mortars and pestles made of wood.(18) According to the same source, hand-pounding is also substantial in Nepal and important to a lesser extent in Sri Lanka. Spencer et al. reported that hand-pounding of paddy is the most widespread method of processing rice in Sierra Leone, mainly for home consumption.(19) The typical hand-pounding equipment includes a pestle about six feet in length and mortars of various sizes. Pounding is carried out by up to three people, usually women and children, working simultaneously with the same mortar and pestle. Hulls and bran are winnowed away from the rice periodically, and additional pounding and winnowing occur depending on the desired degree of milling.

Capacity of the typical mortar and pestle hand-pounding operation in Sierra Leone has been measured at about 10 pounds of paddy per hour per person (1.6 persons average) or 16 pounds (7.3 kg) of paddy per hour, after winnowing and rest time are taken into consideration.(20) Collier et al. reported a similar value for Indonesia.(21) Milling recovery (percent of paddy converted to milled rice) can be quite high in hand-pounding, about 68 percent in the Sierra Leone study. Since hulls are generally about 20 percent and bran about 10 percent of paddy, milling recovery usually cannot rise significantly above 70 percent of paddy. However, the percent "brokens" (less than 75 percent whole kernels) is high in hand-pounding, about 47 wt percent from unparboiled and 32 wt percent from parboiled rice, according to Spencer and his colleagues. Brokens are worth much less than whole grains in trade. Parboiling, as will be discussed later, causes the rice kernels to become stronger and more resistant to the impact of the pestle.

Engleberg hullers

As hand-pounding of paddy disappears, small mechanized mills become important. Koga asserts that hand-pounding is diminishing in Thailand and that most farmers now take paddy to one of the 20,000 small mills that will custom-mill the rice for their home use, the miller being paid in bran and brokens.(22) For years, in many developing countries, the main piece of milling equipment in such mills has been the Engleberg type of huller, a single-pass steel cylinder mill that produces a by-product mixture of bran and hulls. Three manufacturers of such mills are Lewis Grant, McKinnon, and Schule. Replacement costs in 1974-75 for the smallest model (.25 metric ton per hour) were about $2,700 with 15 hp electric motor and $3,500 with a 15 hp diesel engine.(23) In Sierra Leone, milling recoveries with Engleberg-type hullers were high in 1974-75, similar to those observed for hand-pounding, but percent brokens was lower - 36 percent for unparboiled and 17 percent for parboiled paddy.(24) In 1974, Collier et al. reported good recoveries of about 66 percent for Engleberg-type hullers in Indonesia.(25)

Despite these generally favorable reports from Sierra Leone and Indonesia, Engleberg milling recoveries must be regarded as significantly lower than commercial mills. For example, a 1978 report of the Central Food Technological Research Institute, Mysore, indicates that milling recoveries from Engleberg mills in India are 4 to 5 percentage points lower than recoveries in commercial mills.(26) This is likely to bring recoveries down to the 60 to 65 percent range. Milling recoveries near 60 percent may be regarded as representing significant losses especially if foreign exchange is earned by export of milled rice. Therefore, when much of a nation's rice is milled by Engleberg hullers, it is understandable that replacement or improvement of these mills is likely to become a matter of government policy.

Thailand, an important exporter of rice, apparently prohibited the sale of Engleberg mills within its borders some ten years ago, to accelerate introduction of more modern mills that provide higher milling recoveries and lower yields of brokens.(27) Other nations have taken less drastic action. In India, much research and development effort has been directed toward the low-capacity "Mini Rice Mill" for rural areas, with good milling recoveries, and a bran by-product that is free of hulls.(28) Several versions are being pursued.

The displacement of the Engleberg by more modern small mills would seem to be a challenging goal. The Engleberg offers lower investment and maintenance costs than any of its mechanized replacements.

Small, modern, rubber-roll hullers

When the Engleberg is replaced, it is usually by a small Japanese rubber-roll huller, plus friction-type whitener that removes bran separately with an abrasive roller, processing .25 or .5 metric ton of paddy per hour. Satake Engineering Company is an important manufacturer of such equipment. Investment costs are nearly twice as high as for Engleberg-type mills, (29) but the bran is not contaminated with hulls, the milling recoveries are higher, and the percent brokens is lower.(30) However, expensive replacement rubber rollers are needed, imposing a maintenance cost obstacle that the Engleberg huller does not have.

The small rural Engleberg and rubber-roll mills of .25 and .5 metric ton per hour capacity would produce at most about 25 and 50 kg of bran per hour, respectively. Larger mechanical mills can be found in more urban areas, but the small single-pass mill together with hand-pounding currently continues to dominate rice processing in the rural areas of the developing world. As hand-pounding gives way to the small mill, these mills should increase in numbers and importance. And, as the Engleberg gradually gives way to the rubber-roll mill of the same small size, an increasing number of small rural mills will produce bran uncontaminated by hulls, which will greatly uncomplicate the problem of bran stabilization and utilization.

Large, modern, commercial mills

Other rice milling technologies found in developing countries, usually at larger scale than .5 metric ton per hour, include rubber-roll multi-pass and the older, disc-cone multi-pass systems.(31) Multi-pass milling involves sequential removal of bran in several steps. Because grains are subjected to less pressure in multi-pass than in less costly single-pass milling, percent brokens is lower (at the 20 to 25 percent level) and milling recoveries are higher (near the 70 percent level).(32) After hull removal in rubber-roll or disc-hullers, the bran is removed in several passes through either a "pearler" or a "whitener." The former utilizes friction between the grains, and the latter applies friction by means of an abrasive roller.(33)

Solvent-extractive milling

Solvent-extractive ("X-M") milling of brown (hulled) rice is a relatively recent innovation, with early patents having been issued in 1966 and 1968 to Truman B. Wayne.(34) These patents, and others issued and pending in many countries, have been exclusively assigned to Food Engineering International, Inc., a wholly-owned subsidiary of Riviana Foods, Inc., of Houston, Texas. Descriptions of the process which has been commercialized at Riviana's Abbeville, Louisiana, plant have been published.(35) Riviana will license in the United States and abroad. At present, the only X-M plants are in the United States.

X-M milling involves "one-break" bran removal in the presence of the solvent (hexane) to extract oil from the entire kernel, bran, and germ during the bran removal process. The by-product defatted (oil-free) bran, called "Protex," has a higher protein content (17-21 percent)(36) than the 10-15 percent typical of conventional full-fat bran plus polish. Riviana asserts that, because almost all of the oil has been removed, Protex will not develop significant rancidity for six months or more, if stored under cool, dry conditions.(37)

Protex is sold as an animal feed, but is also enjoying limited but increasing use in human foods in the United States: in specialty breads, granola-type cereals, as a base for artificial spices, as a protein supplement in meat and sausage products, and in some health food products.

Bran Stabilization

Two classes of enzymes, lipases and peroxidases, are involved in degradative rancidification processes of the oils in rice bran. The lipases catalyze the hydrolysis of the oils, leading to the formation of free fatty acids (FFA) and glycerol; and the peroxidases mediate certain steps in the oxidative deterioration of oils, leading to a host of partially-oxidized compounds with rancid odors and flavors.(38) Stabilization requires destruction or at least substantial partial destruction of lipase and peroxidase activity.

Lipase activity may begin immediately upon removal of the bran in the milling process which releases the lipases and oil from their separate cellular locations, allowing them to mix. If sufficient moisture is present, the lipase catalyzes the reaction with water to form fatty acids (hydrolytic rancidification). The free acids are more susceptible to oxidation than the neutral oil, and the ensuing process (oxidative rancidification) involves the peroxidases.(39)

Control of lipase activity is important if oil is to be extracted for food purposes. The presence of fatty acids makes the oil inedible. Although it can be removed by refining with sodium hydroxide, the presence of even small amounts of fatty acid leads to greatly reduced yields of neutral oil in the refining process. If the FFA content is higher than 10 to 15 percent, processing for edible oil is likely to be considered uneconomic.(40) Under warm, humid conditions, this level can be reached in a matter of two or three days after milling if the bran is not stabilized.(41) Oil that is high in FFA can be used only for non-food industrial purposes, such as soap manufacture, and is not usable, even in part, for edible oil production.(42)

Lipase activity is halted if the bran is dried to a moisture content of about 3 percent, but will begin again if moisture is added.(43) Essentially complete and permanent destruction of lipases is brought about by moist heat treatment for a few minutes at about 100°C, with good agitation. This must be followed by dry heating to lower the moisture content below levels at which microorganisms can grow on the bran. If this is not done, lipases produced by the microorganisms will carry on the formation of FFA.(44)

Peroxidases are more resistant to heat than lipases. Treatment conditions which have lowered lipase activity in rice bran to less than 1 percent have, at the same time, lowered peroxidase activity to only 50 percent.(45)

Stabilization processes can be categorized under either dry-heat or moist-heat treatment. Both may involve a variety of equipment configurations for bringing the bran into contact with dry or moist heat. Table 4.4 describes briefly five categories of stabilization processes that have been proposed or tested in recent years.

Dry heat stabilization processes may utilize direct heating of the bran by hot air and/or hot flue gases; alternatively, indirect heating may be employed. Moist heat processes may utilize low-pressure steam, or application of liquid water fol-

Table 4.4. General Approaches to Stabilization of Rice Bran

	Scale*	Stage of Development
Screw conveyors	Large	Applied
Extrusion cookers	Medium/Small	Tested
Rotating-drum stabilizers	Medium/Small	Tested
Thin stationary beds	Medium/Small	Concept
Fluidized beds	Large/Medium	Concept

*Small = rural/village scale; Medium and Large = regional/ urban scale

Sources: Compiled by authors from a variety of sources.

lowed by heating. Apparently, moist heat is more effective than dry heat in bringing about destruction of lipase and peroxidase activity.(46) But dry heat processing is inherently simpler in concept, and so may gain the advantage over moist heat in many developing country contexts, especially if complete destruction of enzymic activity is not mandatory. This might be the case if only a short period of stabilization were required, to allow collection and transport to a central oil extraction plant not far from the rice mill.

General objectives for a rice bran stabilizer development program were drawn up by an Expert Group convened in December 1976, by the United Nations Industrial Development Organization (UNIDO).(47) These objectives included:

- Capacity. To stabilize 100 kg rice bran per hour. System to be attached to small rural mills that process at least one metric ton of paddy per hour, producing bran uncontaminated with hulls.
- Product quality. Keeping quality to be at least 4 weeks if to be used for oil extraction, and at least 8 weeks if to be used for feed ingredient. FFA not to rise above 10 percent before use as feed or oil source, starting at below 4 percent before stabilization.
- Other objectives. Low-cost, easy maintenance, easy operation with few attendants. Moist or dry heat. Hulls to serve as fuel.

In effect, UNIDO chose to focus on the regional/urban-level, rather than the rural/village-level, by selecting 100 kg bran per hour as the capacity (see table 4.2) One metric ton of paddy (1,000 kg) yields roughly 100 kg of bran. This conversion is used throughout this chapter.

Previously, UNIDO had retained the Applied Scientific Research Corporation of Thailand (ASRCT) to define the most suitable rice bran stabilization technology. ASRCT presented its report at the December 1976 Expert Group Meeting. Although ASRCT felt that moist heat is more effective, it recommended a dry heat system because it would cost less and be simpler to operate and maintain.

Further, ASRCT took the position that the minimum practical capacity should be 75 kg of bran per hour. This was based on an anticipated 6 percent yield of bran plus polish from a mill that processes 1.25 metric tons of paddy per hour, although paddy typically contains 10 percent bran plus polish. This minimum size of mill was chosen by ASRCT because it felt that smaller mills could not produce economic amounts of bran of oil-extracting grade (hull-free).

Screw conveyors

ASRCT proposed a stabilization system in which bran would be heated indirectly by flue gas from a hull furnace, as the bran is moved through a jacketed screw conveyor system requiring 4 to 6 horsepower.(48) The countercurrent flow of flue gas would enter at 300°C and exit at 150°C. The furnace would burn about 8 kg of hulls per hour to process 75 kg of raw rice bran per hour. The bran would enter at 25°C and 12 percent moisture content, and leave at 105-110°C and 4 percent moisture content after being retained in the system for 20 minutes. The discharged bran would be allowed to cool naturally before it would be packed in gunny bags.

The screw conveyor approach has been utilized for rice bran stabilization in Japan by Yokochi, in a large-scale application that comprises three stages: direct steam injection in one conveyor, followed by steam-jacketed drying in two successive conveyors, then forced air cooling in a fourth and final conveyor.(49) Burns and Cassidy hold a 1951 U.S. patent on a rice bran stabilization method including steam treatment at about 100°C, followed by heating to a temperature in excess of 100°C, then followed by cooling. A 90-foot screw conveyor was employed.(50)

The continuous stabilizer developed by the Instituto de Agroquimica y Tecnologia de Alimentos (IATA), Valencia, Spain, includes a screw conveyor as part of the system, preceded by a U-shaped trough mixer with steam injected through small holes in the bottom of the trough.(51) Pilot plant scale was described as 50-200 kg bran per hour, and commercial scale as about 450 kg bran per hour.

Extrusion cookers

Another stabilization approach that has attracted recent attention is the low-cost extrusion cooker, developed originally to cook soybeans and grains by the heat generated when the material is worked mechanically while passing through a compression screw and being extruded through a die. Temperatures well in excess of 100°C can be realized.(52) Table 4.5 provides a comparison of extrusion cookers that may be small enough in capacity to match with the paddy capacities of typical rice mills, rural and urban, in developing countries.

The MFM extrusion cooker appears to be sized about right for matching with the small rice mills typical of rural and village practice. However, the Insta-Pro and the Brady seem to match better with what we have termed regional-urban rice mills in the 2 to 4 metric ton per hour (200 to 400 kg of bran per hour) range (see table 4.2).

Lin and Cater reported a test of the Brady Model 206 on rice bran, conducted in the spring of 1973.(53) Temperatures above 252°F (122°C) were sufficient to stabilize the rice bran for subsequent oil extraction for about four weeks. Although storage tests were not continued beyond this length of time, the FFA had risen from 4 percent to a level of only 5 percent in four weeks, suggesting that stability could have been demonstrated for a much longer period. Lipase activity was, therefore, virtually destroyed. Also, peroxidase activity was reduced to about 50 percent, and moisture content and moisture-absorbing capacity were both substantially reduced. These favorable results were obtained when the stabilized bran was stored in moisture-proof polyethylene bags at room temperature. However, at 95°F and 100 percent relative humidity in the same bags, FFA formation generally increased much more rapidly when stabilizing temperatures were below 252°F, indicating that in tropical areas it may be essential to stabilize at temperatures above 252°F.

A much larger extrusion cooker unit is the Anderson "Expandolex," which has been used to process 100-300 metric tons of bran per day. Water and steam can be added. No change in FFA after 84 days has been reported.(54)

Rotating-drum stabilizers

A 100 kg per hour rotating drum stabilization system for rice bran has been constructed and tested by Hawkey.(55) The project was carried out in the Khmer Republic (Cambodia) for UNIDO in 1973. The objective was to construct and test a low-cost stabilizer for small rural mills producing bran uncontaminated with hull particles. Local materials, labor, and tools were used to assemble a hull-fired furnace and boiler, and three separate rotating drums for steaming, hot-air drying

Table 4.5. Extrusion Cookers for Developing Country Rice Bran

	Capacity, kg/hr	Power, HP	Power Source
Meals for Millions (MFM)(1)	45 (text. veg. protein)	20	Automobile
Triple F Insta-Pro 500(2)	233 (soybeans)	31	Tractor
Brady Crop Cooker 206(3)	436 (soybeans)	53	Tractor

Note: A small rice mill, rural/village-level, .25 metric ton of paddy per hour, will produce about 25 kg of bran per hour, plus associated hull particles. Most of the hull particles will be removable by sifting, but the product may be 20% hulls, 80% bran. Therefore, the stabilizer will probably have to process about 31 kg of bran plus hulls per hour.

Sources: (1) M. Sterner, "Exploration of Potential for Low-Cost Extrusion Cooker--Worldwide"; (2) L.J. Hanson, "Manufacturers' Experiences with LEC's--The Insta-Pro Extruder"; and (3) M. Van Peursem, "Manufacturers' Experiences with LEC's--The Brady Crop Cooker," all in Low-Cost Extrusion Cookers, edited by J.M. Harper and G.R. Jansen (Ft. Collins, Colo.: Colorado State University, June 1976).

and forced-air cooling, respectively, requiring a total of 1.5 horsepower. Steaming was carried out at 100°C for 5 minutes, drying with 50°C air for 40 minutes, and cooling with 25°C air for 15 minutes. If bran were so treated within seven days of production, it was reported that stability could be achieved for three months. The bran was found acceptable as feed by all animals in a traditional compound, and was expected to be similarly acceptable for solvent extraction.

Hawkey estimated plant cost at £100 (approximately $200) without steam, and twice as much with steam generation facilities.(56) These values appear to be low, probably because local materials and labor were used. Operating costs were estimated at £2.5 per ton, mostly labor (£1.0 per ton) and amortization of capital (£1.3 per ton)

Even without the application of steam, a rotating drum stabilizer has been shown to be effective in checking lipase activity.(57) This investigation, conducted at the Central Food Technological Research Institute (CFTRI), Mysore, India, showed that FFA could be held to not more than 10 percent in four months, when the treated bran was stored in a cloth bag in a humidity oven at 37°C. The drum was heated with an open vent in the lid to allow air and some moisture to escape. Then the vent was closed and heating continued until 110-115°C was reached. This level was maintained for about five minutes, then the vent was opened, steam was allowed to escape, and the treated bran was removed and allowed to cool.

Thin stationary beds

The stabilization of bran by steam treatment of thin stationary beds is an approach that appears to be inherently simple. ASRCT investigators steamed bran layers, 3 cm thick, in an autoclave.(58) Five minutes steaming at 0.8 kg per sq cm gage (11.4 psig) and 115°C was sufficient to hold FFA well below 10 percent after 50 days of storage. Ten minutes was needed to achieve stability at 0.3 kg per sq cm (4.3 psig) and 105°C.

A stabilization study was conducted at CFTRI with rice bran steamed in .5 to 1 inch thick layers in a steam cooker, then dried to below 10 percent moisture content.(59) Fifteen minutes steaming was sufficient to hold lipase activity to about 8 percent FFA after 80 days at 37°C.

Adair has suggested that the thin stationary bed, moist heat concept might be carried out on the flat upper surface of a metal chest heated internally by boiling water or exhaust steam.(60) A 2-mm high wall around the upper surface would retain a lightly compacted layer of bran slurry. Technical feasibility of the concept would depend on the heat transfer properties of the system, but if 2 mm proved satisfactory and if the entire labor-intensive processing cycle (slurry spreading, drying, dry bran removal) could be accomplished in six minutes, Adair calculated that about 11 sq meters of surface would be needed to process about 75 kg of bran per hour.

Fluidized beds

The Expert Group convened by UNIDO in December 1976 also considered a dry-heat fluidized system for rice bran stabilization, but the concept did not receive general approval.(61) Bran agglomeration during treatment, with attendant changes in fluidization characteristics, was thought to be a serious drawback. The system concept called for the use of components designed by the International Rice Research Institute (IRRI) for a paddy batch drier: hull furnace, blower for hot air and flue gas mixtures, a venturi for bran introduction, a duct in which the bran and the hot air would travel together for a sufficiently long interval of time to effect stabilization, and a cyclone for air-bran separation. A sketch of a similar concept was provided by Barton at the same meeting.(62) ASRCT also considered fluidized systems in its study of rice bran stabilization, but rejected them on the basis of poor fluidization characteristics, and the need for costly blowers and controllers.

Summary

The stabilization of rice bran to prevent or to slow the development of rancidity requires that certain enzymes be inactivated. Moist or dry heat may be used for this purpose; moist heat appears to be more effective, but dry heat appears to be simpler and cheaper. A variety of stabilization equipment configurations are known, some more suitable for regional/urban-level and some better for rural/village-level applications.

Specifically, screw-conveyors and fluidized beds involve complex, expensive components. Further, their inherently large scale seems to make them less suitable for the village and more suitable for the regional/urban-level applications that have been of previous interest to UNIDO (rice mills of 1 metric ton paddy per hour capacity, or larger). However, the thin stationary bed concept, the rotating-drum stabilizer, and the extrusion cooker might be adaptable to the needs of the village rice mill.

The thin stationary bed concept is inherently simple, has no moving or complex parts, requires no steam generation, and would be labor-intensive in operation. Adair has estimated a capacity of 6.8 kg bran per square meter per hour for a 2-mm thick bed, but we are aware of no tests that have been made. If the estimate is confirmed, a village-level stabilizer (25-50 kg bran plus unremovable hulls per hour) would require about 5 to 10 square meters of heated surface. This might be provided as a flat plate of sheet metal with a 2-mm lip on all four sides. Heat could be supplied by boiling water under the plate. This might be the least expensive technical approach to village-level rice bran stabilization. However, it remains to be seen if performance will be adequate in terms of lipase and peroxidase inactivation, and throughput rate.

Hawkey's rotating-drum stabilization system was constructed for small rural mill use, and tested in Cambodia with hull-free bran. Complete diagrams and experimental details were not given, but the cost estimates based on local materials and labor were attractively low, even though many items of equipment were built: three separate rotating-drums (steaming, hot-air drying, forced-air cooling), and a hull-fired furnace and boiler for steam generation. The quality of the treated bran was reported to be satisfactory.(63)

The extruder cooker, if it can be simplified and scaled properly for village use, is a third technical approach of interest. The MFM unit appears to be of approximately correct scale but was designed for texturization of vegetable protein. It apparently has not been tested on rice bran, nor has the Insta-Pro 500, which is about five times too large for village use. However, the Brady Model 206, at least ten times too large (see table 4.5), has been successfully operated on rice bran and product quality was satisfactory.

Hull Combustion

Hulls have been burned for many years in special furnaces in U.S. Gulf Coast rice mills. However, the adaptation of U.S. hull combustion technology to the small scale required in developing nations has apparently not occurred. Hawkey stated the need for such small furnaces to burn rice hulls in connection with bran stabilization.(64) One organization reporting work in progress is IATA, in Spain, but no details of this work seem to be available yet.

The Expert Group convened in Vienna by UNIDO in December 1976 considered hull furnace specifications and prepared recommendations without endorsing any specific furnace designs.(65)

- Heat output specification: 1,800 kcal/kg hulls
- Furnace design features: Variable control of stoking, continuous discharge of ash, provision for removal of fly ash from flue gas, adjustable primary and secondary air inlets, forced draft.
- Heat exchange features: Exchangers for flue gas and for low-pressure (atmospheric) wet steam generation, with

provision for cleaning, easy replacement of worn parts, removal of condensate.

At this same meeting, Barton presented four rice-hull furnace designs.(66) One design called for two 44-gallon oil drums, joined by a short duct. One drum would be fitted with hopper, feeder, and grate; the other would receive the hot flue gases through the duct and, by means of baffles, serve as an ash trap. The flue gases would then be used to treat rice bran, directly or indirectly. Another design would essentially accomplish the same functions in a furnace lined with firebrick, without the cost-cutting expedient of oil drum construction.

A third hull furnace design presented by Barton involved a downflow combustion chamber with louvers on all four sides. Air would pass by the furnace in cross-flow, and the mixture of air and flue gas would then be passed through a baffled ash trap. Barton pointed out that proper burning of rice hulls will lead to a clear flue gas with no noxious fumes. Smoke can be minimized by starting with a hot fire, and regularly adding fuel in small amounts.

The fourth design given by Barton was for a boiler fired by rice hulls, to provide steam for moist-heat treatment of rice bran. Water would be preheated by passing it through a jacket on the flue, then the hot water would be converted to steam by passing it through a jacket built into the upper wall of the furnace. Steam would then be conducted to a pivoted stabilizer pan that could be rotated so the lid would be at the top for emptying and filling.

The stabilization system proposed by ASRCT would burn 8 kg of hulls per hour to process 75 kg of raw rice bran per hour by dry heat treatment, with flue gas as the indirect source of heat.(67) Few design details were given for the furnace in the ASRCT presentation. The furnace was simply represented as a firebrick-lined box, top-fed at one end, equipped with a sloped grate, showing 300°C flue gas exiting from the top of the firebox at the other end.

Packaging and Storage of Stabilized Bran

The benefits of stabilization can be lost if the packaging and storage of the product are inadequate. If moisture penetrates the package, molds may grow and lipase activity may be regenerated. Improper packaging and storage can permit insect infestation and rodent attack.

A 1977 study conducted by ASRCT demonstrated the value of polyethylene bagging for keeping moisture content below the point at which molds will grow. Stabilized and dried bran could be kept for at least six months in polyethylene. In cheesecloth bags, stabilization could be maintained for only 20 days in a warm, humid environment. Costs and local availability of polyethylene bags were not discussed.(68)

The UNIDO Expert Group considered the problems of proper packaging, storage, and transportation of stabilized rice bran, and offered these conclusions:(69)

- There is an urgent need to study the physical properties of rice bran related to packaging, storage, and transportation.
- There is an urgent need for studies of microbiological changes in stabilized bran stored under normal commercial conditions in the humid tropics.
- Alternatives to the reusable gunny bag need to be studied, including simple bulk storage and transport.
- Good housekeeping in storehouses and control of infestation by rodents and insects are essential.

Specific materials for bagging and specific ideas for simple bulk storage and transport were not given.

Oil and Wax Extraction

Solvent extraction is most generally employed in the LDCs to obtain oil from rice bran. The bran is treated with hexane, and the oil-rich hexane is removed by filtration. The solvent is then stripped from the oil, and the oil is refined by caustic treatment and bleaching in the same manner as in the refining of vegetable oils from seeds.(70) After removal of residual solvent by heating, the deoiled bran is generally used as a feed component. A useful wax can be recovered from the oil by chilling, followed by centrifuging. The recovered wax is about 1.5 percent of the oil.(71) The X-M process of milling in the presence of hexane results in a bland, stable edible oil similar to oils obtained through conventional solvent extraction.(72)

Bran of low FFA (below 10 percent) will provide yields of crude oil as high as 18.5 percent, about 80 percent of which may be recovered as finished oil - edible as a salad or cooking oil or as a material for oleomargarine manufacture.(73) The wax may be used as an ingredient of furniture, floor, or automobile wax products.

Hydraulic pressing at 1,000 to 4,000 psi offers lower yields of crude oil, an average of 10 to 13 wt percent in comparison with 17 to 18 wt percent via hexane extraction. Pressing requires that prior softening of the fiber be carried out by steaming.(74)

Yokochi has described the use of batch solvent extractors in developing country applications. Such systems are well-suited for bran, but generally require bran that is relatively uncontaminated with hull particles. On the assumption that sieving will not remove all hull particles from Engleberg milling, it is noteworthy that batch extraction equipment specified by Yokochi includes special filtration and cleaning facilities for removal of impurities, as well as preliminary conventional sieving and air separation steps. His typical batch system with three extractors will process 0.5 tons of rice bran per hour, or the combined bran output of 10 to 20 typical small rural mills of .25 to .5 ton of paddy per hour capacity.(75) Continuous solvent extraction, according to Yokochi, is not necessarily unsuitable for rice bran, which tends to be powdery in structure. He indicates, however, that proper pretreatment is mandatory, involving steam heating, drying, and pelletization. Only a few plants have adopted this continuous approach.(76)

Comparison of batch with continuous solvent extraction reveals that initial costs of batch plants are less than two-thirds those of continuous plants of the same capacity.(77) Also, fewer moving parts lead to lower maintenance costs for batch plants. Continuous plants must be located so that machinery can be easily transported to repair facilities having skilled labor. Batch plants are more labor-intensive than continuous plants, as might be expected; but the continuous plants need fewer technically skilled employees because the machinery is automatically operated. Batch plants must have some skilled labor; valves must be opened and closed manually ten times for each extraction. Failure to manipulate valves properly may result in accidents, including explosions. By contrast, continuous plants are relatively accident and explosion-free, according to Yokochi.(78)

Protein Concentration and Fiber Removal

Rice bran contains proteins that can be extracted and precipitated in the form of a gel somewhat akin to soy protein concentrate. In the wet alkaline process, extraction of the protein is typically carried out with aqueous alkali, then the protein is precipitated by acid or heat. The protein is then separated by centrifugation and dried, or used in the gel form. Using a wet alkaline process, Saunders and co-workers at the Western Regional Research Laboratory, U.S. Department of Agriculture, obtained concentrates containing 23 to 31 percent protein in yields of 14 to 20 percent from full-fat bran, and higher protein contents from defatted bran.(79) When starch was removed in a variation of the process, a concentrate containing 33 to 38 percent protein was received in 11 to 12 percent yield. The protein concentrates typically contained 1 percent fiber, or less, so the wet alkaline process is effective for fiber removal as well as for protein concentration. It was also found that the oil content of protein concentrate from full-fat bran was in the range of 33 to 48 percent, and 49 to 55 percent when starch was removed.

Stabilization severely limits the wet alkaline extractability of protein. Betschart and his colleagues found that about half as much protein could be extracted from moist-heat treated Spanish bran (95-100°C), compared to unheated samples.(80) The losses were mostly in the water-soluble (albumin) and salt-soluble (globulin) protein fractions which apparently are especially susceptible to denaturation and aggregation. Ball milling was effective in enhancing protein extractability, but only in the case of samples that were not treated with moist heat.

A second general approach is wet fractionation (water extraction). Bran is ground in water with substantial separation of the cellulosic fibers from the protein, starch, oil, and other nutritive components which are transferred into the liquid phase in various colloidal forms. Fibers are then removed by centrifugation, and the liquid phase is treated to obtain a flour that is higher in protein and lower in fiber than the original bran. Also obtainable are an oil fraction and an aqueous concentrate of vitamins and minerals. Such processes may be applied to raw or stabilized full-fat brans or to defatted brans.

Barber and his co-workers have described a wet fractionation process resulting in a low (3 percent) fiber fraction, and a high (17.8 percent) fiber fraction.(81) The process does not result in a significant concentration of protein; the low-fiber fraction is only about one-sixth higher in protein than the original bran. The process, therefore, appears to be mainly a means for lowering the fiber content to levels appropriate for human consumption. Typical yields, according to Barber et al. are about 57 percent of the 3 percent fiber fraction, and 32 percent of the 17.8 percent fiber fraction.(82) The remainder, 11 percent of the original bran, is an aqueous nutritive concentrate of vitamins and minerals.

Mihara has also reported a wet fractionation process similar to Barber's in some respects, but involving chemical treatment of the low-fiber product to achieve a higher percentage of protein, near 80 percent, in the concentrate.(83) Starch, oil, and a vitamin syrup concentrate were also obtained. In effect, the Mihara process combines wet fractionation, wet alkaline extraction, oil extraction, and vitamin concentration.

Simple sieving and dry grinding plus sieving represent a third general set of related approaches to the problem of protein concentration and fiber removal, termed "dry fractionation." However, the literature is ambiguous about its efficacy. Houston and Mohammad reported that sieving was effective in the preparation of a bran fraction suitable as a food ingredient.(84) They found that 29.3 percent of a defatted bran sample passed a 100-mesh sieve and contained 11.9 percent protein and only 5.2 percent fiber. The original bran contained 10.6 percent protein and 11.4 percent fiber.

The same authors later combined dry grinding with sieving to obtain improved results, starting with defatted bran. The low-fiber fraction passing a 140-mesh screen, about one-fourth the weight of the original bran, contained 15 to 16 percent protein and was regarded by Houston as an acceptable food ingredient.(85)

Barber and de Barber also studied dry fractionation but characterized it as without merit.(86) Based on a consideration of their own and other's data, their conclusions appeared to be:

- dry sieving of defatted bran provides protein enrichment and fiber removal values that are unattractively low,
- dry sieving of previously ground defatted bran produces similarly unfavorable results, and
- dry sieving of full-fat bran (apparently unstabilized) is impractical because particles stick together during sieving, impeding the segregation of fractions of very fine particle size. Full-fat bran contains its natural oil content, that is, it has not been deoiled (defatted).

Published literature on dry grinding and sieving of stabilized full-fat bran has not been found.

Parboiling of Paddy

This process of parboiling paddy is of ancient origin and is applied to about 25 percent on the world paddy production. It is of special importance in the Indian subcontinent, where over 60 percent of India's paddy is parboiled, but is generally not employed by the Chinese, Koreans, or Japanese.(87)

The essentials of indigenous and modern parboiling technology need to be understood, mainly because adequate parboiling can at least partially deactivate lipases responsible for the development of FFA in the bran. The typical indigenous parboiling process in the developing world is applied to the unhulled rice before it is milled. It includes three steps: steeping in hot water (60-70°C is best),(88) steaming with nonpressurized steam,(89) and drying in the sun.(90) The product is easier to hull than unparboiled paddy. Molds and spores are destroyed by the heat and the polished grains are harder and tougher than unparboiled polished rice.(91)

Bran from parboiled paddy does not usually develop FFA rapidly because the steaming suspends lipase action for many days.(92) However, parboiling tends to destroy certain natural antioxidants so the oils are subject to a somewhat accelerated oxidative degradation, normally a slower process than FFA development.(93) Therefore, it is not clear that parboiling would arrest the total set of degradative reactions sufficiently to make practical the collection and shipment of bran to central oil extraction plants if such centers were far distant.

Modern parboiling processes are based on the three steps of the classic Asian approach, with two main improvements. First, steaming is done in closed vessels with adequate mixing to ensure that heating is uniform. Second, because of the unreliability of the weather, sun drying is replaced by drying with forced hot air, or on a heated surface, or by other artificial means.(94) Incorporation of such improvements into indigenous parboiling technology would seem to be imperative even at the village level to improve steam/paddy contact and the thoroughness of drying. This has already been done in developing countries to some extent, but apparently only in large processing plants. A disadvantage of parboiling becomes apparent if protein concentration is to be considered in an overall processing scheme. Parboiling alters proteins in such a way that they are less extractible.(95)

USES FOR BRAN AND BRAN PRODUCTS

A variety of uses and products which can be derived from rice bran include:

- Food: as a food ingredient; as a source of food products - edible oil, protein concentrate, bran flour, vitamin concentrate
- Feed: as a feed component
- Other uses: source of wax; fertilizer or soil amendment; seed bedding; or a source of industrial fatty acids

Bran as a Food Ingredient

Although bran plus polish represents about one-eighth of the brown (hulled) rice, it contains one-fourth of the protein. The quality of the protein in bran plus polish is somewhat better than that of the protein of the rice kernel, due to a higher content of lysine,(96) the first limiting amino acid. (The human organism requires all of the "essential" amino acids, but in differing amounts. These requirements are not usually matched exactly by the contents of essential amino acids in a given food protein. The first essential amino acid to be exhausted in the digestive process is termed the "first limiting amino acid." Typically, cereal grains such as rice are short of lysine, one of the essential amino acids.)

The PER (protein efficiency ratio) of X-M defatted bran has been given as 1.7 - 1.9, based on a value of 2.5 for casein.(97) Protein Efficiency Ratio is the weight gain in grams per gram of protein fed to a growing test animal, with adequate caloric intake to drive the protein biosynthesis process, and with the protein being fed at an adequate level for a sufficiently long period to assess the value of the protein. The PER of X-M defatted bran implies a nutritive value for the bran protein that is significant, but somewhat less than that of the comparison protein, milk casein, regarded as a protein of superior quality.

Edible oil is the other bran component of greatest nutritive importance. It also is polyunsaturated, with about 45 percent oleic and 30 percent linoleic content. Linoleic acid is an essential fatty acid in human nutrition. In addition, rice bran contains substantial amounts of vitamins B1, B2, B6 and niacin, and the mineral phosphorus. About 90 percent of the phosphorus is present as phytin, which makes the phosphorus nutritionally available.(98)

Despite the fact that the foregoing nutritive values have been well known for a long time, rice bran continues to be little used as a food ingredient mainly because of high fiber content, cultural unacceptability, and potential rancidity problems. Even if the bran is extracted with hexane, the small amount of residual oil will cause flavor problems if the bran is not stabilized.

Some ways to introduce bran into the diet directly as a food ingredient have been studied and reported. Rice bran was substituted for up to 25 percent of the wheat flour in bread as early as 1941.(99) Muffins have been baked with equal amounts of rice bran and wheat flour; and pancakes, cookies, cakes, and pies have all been made with significant substitution of rice bran for wheat flour.(100) Other efforts to use bran as a food ingredient have undoubtedly been made in Asia, but we have no information on them at present.

Oil

Food uses for refined, edible rice bran oil include: cooking and salad oil, deep-fat frying oil, oleomargarine, shortening, mayonnaise and salad dressing, and other emulsified products. The refined oil is light-colored and bland. It is very stable because of a low content of the unstable linolenic acid, and a relatively high content of alpha-tocopherol (vitamin E), a natural antioxidant.(101) It performs well in frying tests.(102) In most other respects, rice bran oil is comparable to other high-quality edible oils. Because of rancidity problems, most rice bran oil is not refinable and is converted to soap.(103) Crude or semi-refined rice bran oil can be used as an anti-rust or anti-corrosive oil.(104) Rice bran oil can also be sulfonated and used in the finishing of textiles and leather.(105)

Wax

The wax that can be recovered from the oil by chilling is similar to carnauba wax. Such hard vegetable waxes are used in polishes, candles and leather dressings.(106)

Protein Concentrate

An off-white, bland protein concentrate paste may be pressed to form a cheese-like material, or it may be dried to a stable film.(107) Rice protein concentrates have been used with rice flour to form high-protein extruded pasta products; and, with rice oil and emulsifiers, it is possible to form imitation milks similar to the "Vitasoy" beverage introduced in Hong Kong.(108) The imitation milk made from rice bran protein concentrate closely approximates the composition of cow's milk.

Rice Bran Flour

Specific studies on uses for rice bran flours have not been found. Presumably, rice bran flours would be at least as acceptable as rice bran to use as a wheat flour substitute, and probably more acceptable because of the lower fiber content and improved color compared to rice bran. Fiber is a beneficial dietary factor, but it is irritating to the intestinal tract if present at the high percentages that are typical of rice brans.

Vitamin Concentrate

Rice bran was formerly an important source of B-vitamin concentrates before synthetic vitamins became available, and some such use does still continue.(109) Extraction with ethyl alcohol or isopropyl alcohol will include these vitamins.(110) However, when hexane is used, the defatted bran, usually used for animal feed, is almost as rich in the B-vitamins as full-fat bran.(111) An integrated process has been developed at the Central Drug Research Institute, Lucknow, India, for the recovery of liquid B-vitamin concentrates and other nutritive substances.(112)

There is virtually no vitamin A, C, or D in bran. Vitamin E is present in the full-fat bran, but is extracted with the oil and is not in the defatted bran. If the paddy is parboiled, B-vitamin concentrations in the bran will be lowered. For example, about two-thirds of the thiamine, about half of the riboflavin, and about one-eighth to one-half of the niacin are lost in parboiling.(113)

Bran as a Feed Ingredient

Mixtures of rice bran and hulls are usually fed to cattle, water buffalo, bullocks, sheep, and other ruminants, but can also be fed in smaller amounts to swine, horses, and other monogastric animals.(114) Hull-free bran has been used successfully in fish farming; and if not rancid, can replace up to half of the total ration for poultry.(115) Houston has warned against the use of rancid bran on the basis of nutritive losses and possible toxicity.(116) There are other limits to the use of rice bran in rations. Up to 30 percent of hull-free, full-fat rice bran in dairy or beef cattle feed, and up to 20 percent in swine feed is permissible, but larger percentages lead to excessively soft and fatty meat.(117)

Bran as a Fertilizer or Soil Amendment

NPK values of full-fat and defatted rice bran samples have been reported as 1.56 - 2.6 - 0.8 and 2.6 - 5.0 - 1.8, respectively.(118) The defatted bran is clearly superior to the full-fat bran. Also, the oil content of the full-fat bran is of no value as a fertilizer, and is even regarded as harmful to the fine roots of plants.(119)

Miscellaneous

Defatted rice bran can be used as a seed bed for sweet potatoes and nursery garden seedlings.(120) Industrial fatty acids can be produced by hydrolysis of rice bran oil. About 40-50 percent of the fatty acids is oleic acid, and about 20-40 percent is linoleic acid.(121)

Fatty acid mixtures from rice bran oil can be used in alkyd resins for paint products. Plasticizers for synthetic polymers can be made from esters of unsaturated fatty acids and from epoxidized rice bran oil. Fatty acids are also used to make synthetic sufactants and emulsifiers. Various metal soaps of fatty acids can also be produced for the rubber and other industries.(122)

USES FOR RICE HULLS

Fuel

About one-fifth of the paddy is accounted for by the hulls. In a good boiler, each kilogram of hull will produce 2.5 to 3 kilograms of steam. Parboiling requires about 70 to 150 kilograms of steam per ton of rice.(123) Therefore, it appears that the hulls are more than sufficient as a fuel to raise the necessary steam for parboiling. A moist heat stabilization process similar to the steaming step of parboiling, but applied to the bran only, should require even lesser amounts of heat.

The heating value of hulls is lower than that of wood (6000 versus about 8,400 Btu/lb.).(124) Hulls represent a nearly sulfur-free source of thermal energy. If all power requirements of a rice mill were met by burning hulls, only 80 percent of the hulls would be utilized if parboiling were not involved, releasing the remainder for other uses.(125) If parboiling were to precede milling, the total power needed might require all the hulls, but with little or no additional fuel requirements.

Cost comparisons made by the Food Corporation of India in 1974 indicate that hulls had a small economic advantage over coal, even though hulls have one-fourth the heating value of coal and are generally burned at a somewhat lower boiler efficiency. The advantage fell to rice hull fuel because it cost only about one-seventh as much as coal per unit of weight.(126)

Feed Ingredient

Hulls can be added in small amounts to ruminant feeds in the form of whole or ground hulls, or as a mixture with bran and polish in millfeed. The latter is generally sold on a guaranteed basis of 5 percent (min.) fats, 6 percent (min.) protein, 16 percent (max.) ash, and 30 percent (max.) fiber.(127) The ash would be mostly silica. Rice millfeed has been rising rapidly as an animal feed component in the United States in recent years.(128)

Ammoniated hulls are produced by treating rice hulls with anhydrous ammonia under heat and pressure. This is a commercial product that is apparently more valuable as a feedstuff than untreated hulls.(129)

Soil Amendment

Although rice hulls have very low NPK values and cannot provide significant fertilizer values, they are effective in conditioning clayey soils, and in providing some organic matter that will biodegrade in the soil, although the degradation takes place very slowly. Good mulches can also be made from rice hulls, hull ash, and charred hulls, and hulls can be used as diluting material in commercial fertilizers.(130)

Building Material Component

Rice hulls have a low thermal conductivity, comparable to rock wool,(131) and a lightweight insulation made of cement and hulls has been patented in Italy.(132) The Tropical Products Institute (London) has developed a cement-hull block that may be suitable for wall construction if the walls are non-load-bearing.(133) White ash from the burning of hulls at high temperatures is a raw material for fire-brick manufacture.(134) Hull ash has been mixed with soil to produce mud bricks with insulating properties.(135)

A variety of architectural boards can be made from rice hulls and certain resins.(136) A water-resistant composite board, for example, can be made from rice hulls and phenol-formaldehyde resins using procedures and equipment that are typical in the particle board industry.(137) The boards are potentially usable for furniture, wall paneling, and floor underlayment.

Miscellaneous

Hulls can be carbonized to give lightweight activated carbon for decolorizing and other purposes. The process of carbonization also yields tarry products and a low-heating-value fuel gas.(138) The polysaccharide, xylan, can be extracted from rice hulls and converted to the sugar, xylose.(139) Furfural is another chemical that can be obtained from rice hulls, by treatment with sulfuric acid.(140) Hulls have been shown to be successful as bedding materials for animals, and as poultry litter and nesting materials.(141)

Hulls are unusual plant materials in that the silica content is high, about 20 percent and, because of this, ground hulls can sometimes be used as abrasives.(142) Further, the silica is readily extractable in the form of sodium silicate for industrial uses.(143) This is done by treating the "gray ash" from hull burning at moderate temperatures (90-95 percent silica) with sodium hydroxide or sodium carbonate solution. The Paddy Processing Research Center, India, has proposed that this process be developed as a cottage or small-scale industry.(144) Alternatively, silica can be extracted as a powder and used as a water purification material, similar to Diatomite.(145)

ECONOMIC AND GOVERNMENT POLICY ISSUES

The importance of economics and government policy on the choice of technique needs to be stressed. Some of the issues were brought out at a recent UNIDO Forum on Appropriate Industrial Technology, held in New Delhi, India, in November 1978. One of us (Morgan) attended the sessions of the working group on food storage and processing. Rice milling and rice bran utilization were discussed at those sessions. Participants in the working group included individuals active in this field in Thailand, India, Burma, and other countries. The following is a summary of information and opinion pertinent to the rice bran problem that emerged from discussions of the working group.

Impact of Rice By-Product Price

The price of rice by-products has an important impact on the demand for and uses of bran and hulls. The by-product usage pattern could be greatly affected if prices continue to move upward. For example, if the demand for rice hulls as raw material for particle-board manufacture increases, this would tend to withdraw the hulls from more traditional agricultural uses.

On the other hand, it is possible that some governments will adopt or continue policies that encourage farmers to use rice by-products as domestic cattle feed. Exports of rice bran oil and/or defatted bran (the residue from oil extraction) might be discouraged or prohibited, or prices could be controlled. Keddie has emphasized the importance of low cost if the bran and its oil and protein values are to be accessible to villagers.(146)

If government price and export restrictions were not imposed, considerable export earnings might be developed, but at the expense of less oil for domestic food purposes and less meal for domestic feed. For example, India exports defatted bran to Europe for feedstuff purposes, so defatted bran available to domestic feed milling is reduced but foreign exchange is earned. Further, in 1978 the Indian government embarked on a major program to stimulate production of edible-grade rice bran oil.(147) Clearly, the importance of government policy cannot be overemphasized, and policies differ significantly among the rice-producing nations.

Impact of Selection of Scale of Milling

The scale of rice milling and rice bran processing is extremely important. Timmer has carried out an analysis of rice milling alternatives for Indonesia which points up some of the pros and cons of small-scale vs. large-scale mills.(148)

At the UNIDO Forum, some participants indicated that the single rubber-roll mill developed in Japan is rapidly displacing the Engleberg huller and hand-pounding in some countries, in spite of government efforts to shore up the labor-intensive hand-pounding industry. The feeling was that the rubber-roll mills are making strong inroads because they provide much better bran-hull separation than the Engleberg; furthermore, the drudgery of hand-pounding is eliminated. A rubber-roll mill was said to be available in Thailand in 1978 for $2,500.

The eventual disappearance of hand-pounding in favor of rubber-roll mills raises the issue of who controls the rice by-products and enjoys the benefits when the scale of milling is increased. If the farmer is no longer in control, is the miller? What are the relevant political and social relationships in the village? Are the rubber-roll mills sufficiently small in scale so that the villagers will derive considerable benefits? Is the Brady crop cooker, under present study as a rice bran stabilization device, too large to be appropriate for village use?

The Appropriate Technology Development Association of India (ATDA) has indicated that rice milling in India is becoming organized increasingly as a medium-scale industry located in urban areas.(149) According to ATDA, the effect of this trend on the village is negative because, first, the paddy must be sold to urban mills at a low price due to monopolistic tendencies; second, rice is milled in urban areas and supplied to rural people at a higher price; and third, employment potential is reduced in villages and a rural economic activity is lost. The result is the draining of capital resources as well as personnel from the rural areas. ATDA has as an objective of one of its studies the identification of suitable technology to reestablish rice milling in rural areas on a cottage basis to generate economic activity, create employment, and foster capital formation.

Burmese Experience in Rice Bran Stabilization

The Burmese have a number of rice bran solvent-extraction mills in operation. Most produce edible oil, but some produce oil for soap manufacture. Rancidity problems occur and there is continuing need for 24-hour transport. Therefore, Burma is building bran stabilization units for rice mills. Although it was generally felt at the UNIDO Forum that there is no commercially successful rice bran stabilizer available at this time, the Burmese are confident that their approach will be satisfactory.

Issues Requiring Further Study

The following is a condensation of rice-related issues needing attention, according to the UNIDO Forum working group on food storage and processing.(150) Four general objectives were identified:

- Maintaining or even lowering the cost of the product, by increasing processing efficiency.
- Upgrading nutritional quality to correct protein-caloric deficiencies.
- Generating employment.
- Increasing consumption.

Concerning the choice of rice milling technology, the working group expressed concern about the displacement of labor caused by mechanization and raised the issue of constructive compensation. However, it was recognized that mechanical milling has increased the quality and quantity of milled rice, in comparison with hand-pounding, and has opened up possibilities for utilization of the bran. Concerning rice bran stabilization, the working group felt there is a need to continue research and development on low-cost, small-capacity units that the medium and small-scale rice mills can afford.

Concerning international cooperation, the working group strongly recommended encouragement of interinstitutional cooperation on the model of a network for exchange of experiences, R&D activities, design and engineering activities, information dissemination, and promotion of technical cooperation among developing countries at both the regional and subregional levels, utilizing an interdisciplinary approach. Cooperative R&D projects were recommended for development of new technology and further refinement of existing appropriate techniques in activities such as "food grain storage, grain milling, and recycling of wastes and by-products of the food processing industry." The United Nations system and the developed countries could provide assistance to developing countries, it was suggested, in the supply of information on choice of technologies for specific food processing industries, and in providing specialized equipment and know-how.(151)

SYSTEM CONCEPTS FOR PROCESSING RICE BRAN

Problem Definition

Even though it contains attractive percentages of edible oil and quality protein, rice bran is not well utilized for food purposes. Encouragement of food uses for rice bran is an especially difficult problem in rural areas of developing countries where there is typically no provision for stabilization against rancidity, and where bran is often contaminated with hull particles. However, several successful responses might be made to the rural development challenge of optimization of rice bran utilization. But a number of technical, economic, and cultural problems must be addressed:

1. If freshly milled village bran is not processed or used immediately, it must be quickly stabilized against rancidity by wet or dry heat treatment at the village level. Rancidity develops so rapidly that it is impractical to collect and transport the bran to large processing centers in urban areas. Here, the key technical need is a simple, effective, light-capital stabilizer that will process 25-50 kg of bran per hour, plus any contaminating hull particles that cannot be removed by simple sieving, if bran/hull mixtures are involved.

2. Bran/hull mixtures or hull-free bran that is stabilized may go to regional centers for recovery of edible oil by solvent extraction, an efficient process but probably too complex to be adaptable to village use. However, it is not known if either batch or continuous solvent extraction can be carried out on bran contaminated with significant amounts of hull particles. Tests may indicate that solvent extraction is not adaptable to bran/hull mixtures without substantial modifications.

3. Stabilized hull-free bran or stabilized bran/hull mixtures to be collected and transported to regional centers must be adequately packaged and stored while awaiting collection, to avoid mold growth and attack by insects and rodents. Traditional packaging materials, e.g., the gunny bag, are unsatisfactory; improved materials and methods are needed.

4. In addition to edible oil production, the regional processing centers might upgrade the protein content of the bran from feed to food uses by one or more processes that increase the protein/fiber ratio. Protein concentrates and high-protein, low-fiber bran flours might be produced. The choice of process for improvement of the protein/fiber ratio would depend on a number of technical, economic, and cultural considerations, and each processing technology would need to be carried forward from present experimental and pilot stages of development.

5. If the objective were to retain as much as possible of the edible oil for village consumption and to develop village industry, the bran might be hydraulically pressed there. Oil yields would be about two-thirds as high as in solvent extraction, but pressing appears to be readily adaptable to village requirements. No oil would be lost. Oil appearing in the residual meal would probably be recovered as feed for village animals. Here, the key technical need is a simple, effective, low-cost hydraulic press developing 1,500 to 4,500 psi, with the associated equipment for prior softening of the bran by steaming, and adequate to handle bran/hull mixtures as well as hull-free bran.

6. Also, the protein value of the bran might be allowed to go to village animal feed, after deoiling by pressing; or an attempt might be made to recover both protein and oil for food uses. The same processes that might be employed at regional centers for improvement of protein/fiber ratio might be adapted to village uses in simplified forms.

7. The edible oil would certainly be acceptable to village consumers, although it may not be in demand presently in certain countries which are not currently oil-short (Sri Lanka, Malaysia, Indonesia, Philippines). This could change considerably as population levels continue to increase. The residual meal would also seem to be completely acceptable for feeding village animals. But it is not clear that the protein concentrates or the high-protein, low-fiber bran flours would be accepted as food ingredients by the villagers, especially in home food preparation. Perhaps such high-protein products would be in demand for local food product manufacture if the entrepreneur could use them to reduce costs or improve functionality. The nature of the high-protein product will also be crucial. If it resembles bran too closely it will probably suffer from association with animal feed in the thinking of the consumers. On the other hand, if it is too dissimilar from any known food ingredient, its acceptance might suffer due to lack of familiarity.

8. If parboiling of paddy prevails in a given area, there may be a substantial impact on the choice of bran processing options because parboiling is known to limit yields of protein in alkaline extraction and perhaps other protein concentration methods as well. It is necessary to learn how parboiling affects all processes for improvement of protein/fiber ratio. This will be a significant factor in the Indian subcontinent and in Nigeria, where parboiling is prevalent.

9. In all of the foregoing, attention should be given to analyses of costs and benefits, impacts, local infrastructure requirements, etc., that must be carried out before successful systems can be fully conceptualized, specified, and implemented. It is recognized that each country will present unusual circumstances and that total systems will probably

have to be specified on the basis of special situations in specific regions.

Concepts for Village-Level Processing Independent of Regional Centers

Systems might be developed for the processing of rice bran entirely at the village level, without any bran processing at the regional center level, to produce edible oil and high-protein products of low-fiber content The appearance of the high-protein products would be different from bran, which is widely associated with animal feed and, therefore, is not well accepted.

In this context, there are two classes of bran products that will be considered: bran flours and bran protein concentrates. The flours would probably have protein contents similar to or somewhat higher than those of wheat flours, and fiber contents that would be acceptable for routine use. The concentrates would be virtually fiber-free, and might be used as produced in gel form or as dried powders. The flours and the concentrates both appear significantly different from raw bran. The flours would resemble wheat flour, although they would not be white. The concentrates in gel form would resemble soy tofu, a widely used protein food in Southeast Asia.

Two known processing approaches that yield bran flours are wet grinding and dry grinding. The former, also called "wet fractionation," has been studied by Barber and his colleagues.(152) The latter, also called "dry fractionation," has been studied by Houston and Mohammad(153) as well as by Barber. Two other approaches yield protein concentrates, alkaline extraction(154) and wet grinding-coagulation.(155) Each of the four processes has advantages, disadvantages, and limitations, but may be adaptable to the demands of a village enterprise. Characteristics of each are summarized in table 4.6, and applicability considerations are given in table 4.7.

If village processes for improvement of protein/fiber ratio could be integrated with partial deoiling via hydraulic pressing, raw bran might be utilized without prior stabilization, if the steaming pretreatment before pressing could be relied upon to provide sufficient stabilization. In this instance, dry grinding would probably not be selected for improvement of protein/fiber ratio because unstabilized bran cannot be ground and sieved effectively through small-mesh screens. So, if dry grinding were to be used for improvement of protein/fiber ratio, stabilization would have to precede the dry grinding. If alkaline extraction, wet grinding-coagulation, or wet grinding were chosen, it would seem that the steaming pretreatment needed to soften the bran before pressing should also serve as an adequate stabilization treatment.

The village processing systems that should probably receive first consideration for edible oil and high-protein food production are diagrammed in fig. 4.1. The wet grinding and dry grinding approaches, because they would reduce the hulls to the consistency of a flour without means for protein removal, would be considered applicable only to hull-free bran. The alkaline extraction and the wet grinding-coagulation approaches, because they involve specific steps to extract protein, are presumed to be applicable to bran/hull mixtures. This would have to be demonstrated, however, as no data have been found on bran/hull mixtures for any of the four processing approaches for improvement of protein/fiber ratio.

The integrated systems of figure 4.1 possess the advantages of retention of food and feed values in the village, a shift toward food rather than feed uses, and the generation of new employment opportunities in the village. Although these systems have been conceived of based on published information, it should be recognized that applications studies are needed to answer the technical questions that have been raised, and to demonstrate the technical and economic feasibility of the concepts.

Concepts for Linked Village-Level and Regional-Center Processing

If it were necessary to increase the national production of edible oil by linking the village to the general market system through solvent extraction in regional centers, it would seem clear that stabilization at the village-level would be mandatory, to allow time for collection and transport of bran to regional centers where efficient deoiling could be carried out via solvent extraction. This method of deoiling is apparently not practical for village application because of inherent complexity and need for safety precautions to avoid explosions. It is also clear that bran/hull mixtures would be unacceptable unless village-level sieving could be shown to remove substantial percentages of hulls, and solvent extraction processes could be shown to be adaptable to the presence of the remaining hull particles.

In addition to edible oil production, further processing to increase the protein/fiber ratio might also be considered at the regional level. Alkaline extraction would appear to be an unavailable option because the bran would have been stabilized, and wet-grinding coagulation might not be acceptable for the same reason. But bran flours could be produced by wet or dry grinding of the stabilized defatted bran, if no hulls were present. Bran/hull mixtures are not likely to be acceptable unless wet-grinding coagulation is proved feasible for stabilized bran/hull mixtures.

Table 4.6. Characteristics of Processes for Upgrading Bran by Increase of Protein/Fiber Ratio

Product Type	Protein Concentrate		Bran Flour	
Process	Alkaline Extraction	Wet-Grinding Coagulation	Wet-Grinding	Dry-Grinding
Product Composition, %, dry basis				
protein	23-31	31	17	12
fiber	1	nil	3	5
fat	33-48	60	17	*
Protein Yield % of dry bran	6	9	10	3

*Bran had been defatted prior to dry grinding.

Table 4.7 Applicability of Processes to Village-Level Improvement of Bran Protein/Fiber Ratio

	Full-Fat Bran + Hulls		Full-Fat Bran (Hull-Free)	
	Raw	Stabilized	Raw	Stabilized
Alkaline extraction	Maybe[a]	Not applicable[b]	Applicable	Not applicable[b]
Wet grinding-Coagulation	Maybe[a]	Maybe[c]	Applicable	Maybe[c]
Wet Grinding	Not applicable[d]	Not applicable[d]	Applicable	Applicable[e]
Dry Grinding	Not applicable[d]	Not applicable[d]	Not applicable[f]	Maybe[g]

[a]Depends on satisfactory demonstration that 20 percent hull particles will not interfere with protein extraction.

[b]Stabilization greatly diminishes alkaline extractability of protein.

[c]Depends on satisfactory demonstration that stabilization does not significantly limit protein extractability in this process.

[d]Hull particles would be expected to be ground to flour-sized particles with the bran, and become extremely difficult to separate from the bran particles.

[e]Probably more economical to apply the stabilization to product solids, rather than the bran.

[f]Raw bran cannot be ground and sieved due to agglomeration of particles.

[g]Depends on satisfactory demonstration that stabilized bran can be ground and sieved without agglomeration of particles.

Note:
"Applicable" = all process steps have been demonstrated in laboratory tests.

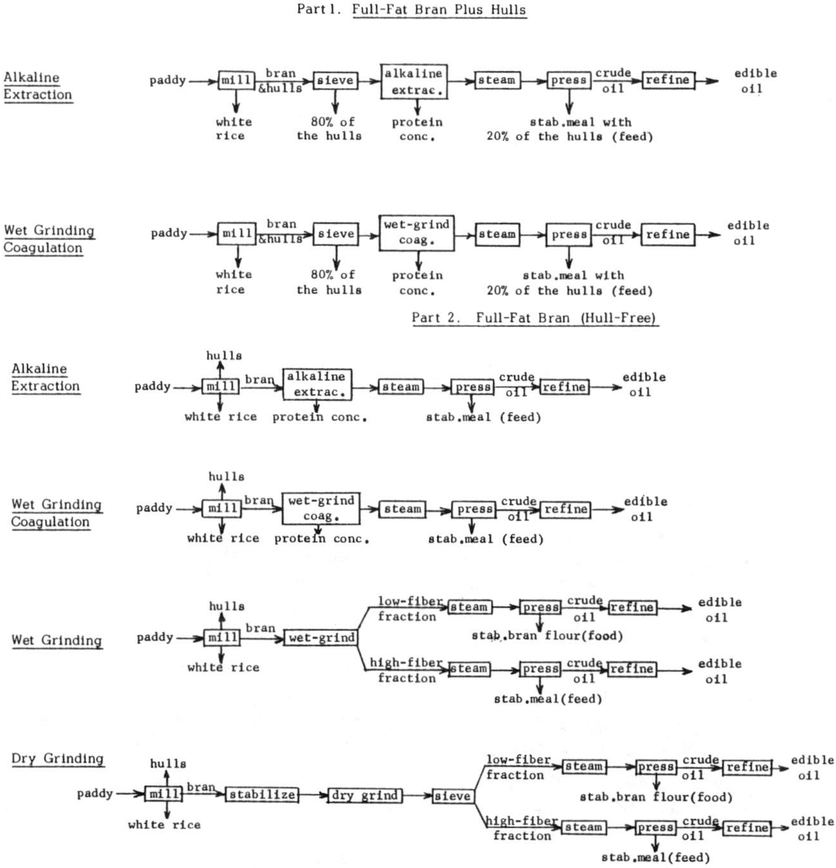

Fig. 4.1. Concepts for village-level systems with improvement of protein/fiber ratio, integrated with edible oil production.

Impact of Parboiling on Bran Processing Options

Parboiling, applied to 25 percent of world paddy, and especially important in the Indian subcontinent, will influence the choice of bran processing system at any level, village or regional. Parboiling would eliminate protein concentration by alkaline extraction from consideration and, perhaps, wet-grinding coagulation as well; parboiling is known to limit protein extractability severely. Wet or dry grinding could be employed to yield bran flours, but only on hull-free bran. Therefore, if only bran/hull mixtures were available, no processing to increase the protein/fiber ratio could be carried out in the village. Only steaming and pressing to yield edible oil would be a viable village processing option, with the meal and all of its protein content going to feed village animals.

If the village were not oil-short, as presently occurs in certain areas where there are ample supplies of coconut and other edible oils, then it would seem that parboiling would effectively foreclose any initiative to shift bran toward food uses and away from animal feed, if only bran/hull mixtures were available. This curious conjunction of rice parboiling, ample coconut and other edible oils, and only bran/hull mixtures may be rare, limited perhaps to Sri Lanka.

OUTLINE OF A RESEARCH AND DEVELOPMENT PROGRAM

Technological Factors

Separation of bran from hulls

Duckworth and Dent studied mixtures of rice bran, hull particles, and broken rice obtained from "inferior mills," and removed most of the hull fragments with a machine-agitated hemicylindrical 40-mesh screen.(156) The material that passed through the screen was estimated to contain about 20 percent hulls, from measurements of fiber content. The material that did not pass the screen was described as a mixture of hulls and broken rice. Apparently, little if any bran failed to pass the screen, but some hull fragments also passed.

In another series of tests, Duckworth and Dent studied hand-sifting with a nest of sieves (30,40,and 60-mesh).(157) Most of the hull particles were removed by the 30 and 40-mesh screens, and the residue on the 30-mesh screen was described as "almost entirely" hull particles and broken rice.

If typical bran/hull mixtures are 61 percent hulls,(158) and if the material passing a screening system is 20 percent hulls as reported by Duckworth and Dent, then by material balance it can be shown that 84 percent hull removal can be accomplished by simple screening procedures. This calculated value should be verified by experiment, and hopefully improved upon in further tests with typical bran/hull mixtures from Engleberg mills, hand-pounding, and other significant sources of bran/hull mixtures such as the wooden "chakki." Also, parboiling may have an effect on hull particle shape, so representative bran/hull mixtures from parboiled rice should probably be studied.

Stabilization

The development of at least four simple, light-capital stabilizer prototypes for village application should be explored:

- A moist-heat (steam) rotating-drum unit, with externally-generated steam, or with steam in situ by addition of water to the bran.
- A dry-heat rotating-drum unit, satisfactory for short-term stabilization.
- A thin, stationary bed stabilizer following the suggestion of Adair.(159)
- An extruder-cooker, substantially smaller than the Brady Model 206 or the Insta-Pro Model 500.

A technically-feasible village rotating-drum unit might take advantage of the almost universal availability of 55-gallon drums and animal or human labor to provide power for agitating the bran during the steaming process. For example, a bicycle might be used to operate a pulley. It would appear that agitation is absolutely essential to avoid channeling of the steam.

The drum might be stationary with its axis horizontal, equipped with a cover, mixing paddle, and crank. Steam could be injected at several points simultaneously in the back end of the drum, and released through several small ports in the cover at the other end. These ports might be covered by burlap or similar material that would allow steam to leave, prevent the bran from blowing out, and provide a slight back-pressure on the drum. A simple bearing for low rpm would be provided where the paddle enters the cover. Simple means for protection against over-pressuring of the drum would have to be provided, such as a spring-loaded port. The use of steam poses a potential safety hazard that must be considered, however.

Further, a satisfactory method for drying the stabilized bran by external heat must be devised that is not susceptible to weather variations, and that does not consume much fuel. This might be done by means of a steam-heated drying surface, perhaps of indigenous material such as locally-made brick. Alternatively, a 55-gallon drum configuration might be employed, with steam heat applied from an external jacket.

A satisfactory rotating-drum unit using steam might also serve as the steaming device needed prior to hydraulic pressing for oil production. The two functions, stabilization and steaming, might even be performed by the same piece of equipment in one operation.

A satisfactory extrusion cooker approach might be based on experience with available commercial low-cost extrusion cookers. However, the Model 206 Brady Crop Cooker, which processes 436 kg soybeans per hour,(160) would require considerable scale-down to match the bran output of typical village mills processing .25 to .5 metric ton of paddy per hour. Since bran is about 10 percent of the paddy, only about 25 to 50 kg bran per hour would have to be processed at typical village mills. Even if unremovable hull particles were mixed with the bran, the Brady would still be about 10 times too large, assuming that its bran capacity is similar to its soybean capacity, and that the Duckworth and Dent value of 20 percent unremovable hulls(161) is valid. The Insta-Pro Model 500, which has about half the capacity of the Brady Model 206,(162) is about five times too large for village rice bran use.

The extrusion cooker developed by Meals for Millions (MFM) for textured vegetable protein applications may be of more appropriate scale and capital cost for adaptation to village rice bran than either the Brady or the Insta-Pro models. The MFM extrusion screw is only about 6.3 cm in diameter,(163) considerably smaller than the 9.4 cm of the Insta-Pro,(164) and the 12.7 cm of the Brady.(165) The MFM unit texturizes protein at about 45 kg per hour.(166) If its rice bran capacity is similar to its protein capacity, then little modification might be required for service at a typical village rice mill. In 1976, Sterner estimated that the MFM unit might be built in a developing country for $4,000 to $5,000; a second generation design was expected to cost even less.(167)

As part of this study of stabilization techniques, it is important to discover the relevant experience in the world's largest rice producer, the People's Republic of China. The Chinese may have experienced some success in this area that could prove useful to others. However, we came across no readily available information about rice bran processing or stabilization in China during the course of our study.

Alkaline extraction

Simplified means for preparing protein concentrates appropriate for village application should be studied. Bran/hull mixtures as well as hull-free bran would be tested. Sodium hydroxide solution would be added to raise the pH to about 9, using a mixer, to extract the protein from the bran. The juice containing the protein would be separated from the solid residue by hydraulic pressing at low pressure. The pH of the juice would then be lowered to about 4 by addition of hydrochloric acid, and the protein concentrate product collected by centrifugation.

Protein recovery should be about 40 percent of the bran protein. Also included in the protein concentrate product would be about half of the bran oil, and some of the bran starch. The solid residue might be neutralized, stabilized, dried, and fed to village animals; or it might be neutralized and softened by steaming and then hydraulically pressed to recover up to two-thirds of the remaining oil. The final residue would go to animal feed.

The feasibility of the foregoing scheme will need to be demonstrated in a village context using appropriate equipment. Further, bran/hull mixtures do not appear to have been studied, so it is important to establish that the presence of hulls does not significantly interfere with protein extraction.

Wet-grinding coagulation

The Nakataki Water (NW) process described by Mihara(168) includes certain key features that might be adaptable to a village wet-grinding coagulation process. Raw and stabilized full-fat bran and bran/hull mixtures would be ground in water, liberating protein and oil in colloidal form. The solid residue would be separated from the colloidal solution by centrifugation, and the protein concentrate product precipitated from the solution by addition of a coagulant, then removed by filtration. The product would probably resemble the oil-rich protein concentrate produced by alkaline extraction, and the solid residue might be treated as the residue from that process. Mihara's yield data for the NW process suggest that protein recovery may be higher than in alkaline extraction. Less oil may remain in the solid residue, so it may be uneconomical to press it to recover edible oil. If it is true that the protein concentrate from wet-grinding coagulation is more oil-rich than similar concentrates from alkaline extraction, this may be an undesirable feature.

Wet grinding

In the case of wet grinding, the adaptive studies would be based on the Spanish IATA process reported by Barber and his co-workers.(169) Their approach is totally mechanical, with no chemical substances involved in the processing. In a version that should be acceptable in a village application, bran would be soaked in water and selectively ground in water suspension in a disk, rotating, or colloid mill. The high-fiber, low-protein fraction would be separated from the low-fiber, high-protein fraction by sieving. Each fraction would then be dewatered and dried. The most appropriate mill and sieving technique will need to be determined for village use.

Dry grinding

Another totally mechanical approach is dry grinding. In comparison with wet grinding, lower protein recoveries and products of higher fiber content are expected, but energy is not required to dewater and dry the product fractions. Adaptive studies might follow the work of Houston and Mohammad,(170) using screen sizes from 20 to 200 mesh, and analyzing the fractions for protein, ash, fat and fiber, but extending their work on defatted brans to full-fat stabilized (hull-free) brans. The sievability of such brans through fine screens has not been established. If aggregation is shown to be a problem with these full-fat stabilized brans, as it is known to be for full-fat unstabilized brans, then dry grinding would have to be limited to defatted brans in regional center processing only, making the anticipated low-cost advantages of this approach unavailable to the village.

Hydraulic pressing

Following the procedures described by Yokochi,(171) the following oil-containing solids should be steamed, then pressed at 1,500 to 4,500 psi: hull-free bran; bran/hull mixtures; solid residues from alkaline extraction of hull-free bran, and bran/hull mixtures; solid residues from wet-grinding coagulation of hull-free bran, and bran/hull mixtures; the low-fiber and the high-fiber fractions from wet grinding of hull-free bran; and the low-fiber and the high-fiber fractions from dry grinding of hull-free bran. In detail, the steps include steaming, drying, prepressing, pressing in a ring or cage hydraulic press, oil cake crushing, and oil refining by caustic treatment. Further, tests for lipase and peroxidase activity and rate of free fatty acid (FFA) formation should be conducted on the solids after pressing, to establish the ability of the steaming step to bring about an adequate level of stabilization against rancidity. This phenomenon has apparently never been studied.

Solvent extraction

It is necessary to explore the ability of existing batch and continuous solvent extraction processes to handle brans contaminated with hull particles that cannot be removed by sieving. It may be that 20 percent hull content will be confirmed as a practical limit on hull removal by sieving, as the Duckworth and Dent work suggests.(172) If costly modifications are needed to cope with bran/hull mixtures, or if hull particles cannot be tolerated in any significant amount, this result will have an important impact on planning.

Parboiling

Studies are needed to establish the limits parboiling imposes on the four processes for increasing protein/fiber ratio, summarized in table 4.6. It is known that parboiling adversely affects yields from alkaline extraction, but no information is available on the other three approaches. Studies should also be made to determine lipase and peroxidase activities and rate of FFA formation in brans from parboiled paddy. Further, if parboiling results in increases in silica content of the rice or the bran, the health effects of high silica levels will need to be investigated.

Packaging and storage of stabilized bran

Studies are needed to test polyethylene and other nontraditional bagging materials for stabilized and dried bran, to test new ideas for bulk storage and transport, and to determine the costs of these new approaches. Treatment of traditional packaging materials (gunny bags) might be considered also. Moisture penetration, which allows molding and regeneration of lipase activity, must be prevented, as must insect infestation and rodent attack. Stabilization must be maintained for sufficient time to allow for a regular collection cycle, plus a safety factor. Use of synthetic polymers to achieve such objectives may be necessary, not only for bagging, but also for bulk storage containers. Testing in an environmental chamber at various combinations of humidity and temperature will probably be necessary.

Hull combustion

Designs for efficient, low-cost hull combustor-boilers are needed, primarily to serve the energy needs of the stabilization system and the steaming stage prior to hydraulic pressing. The use of 55-gallon drums should be explored, as Barton has proposed for the burning of rice hulls.(173)

Socio-Economic Factors

Product acceptance studies

The problems of gaining acceptance for edible bran and bran products, other than edible bran oil, are unquestionably challenging. Among others, the following apparent limitations must be addressed:

- Bran is widely associated with animal feed in the thinking of many potential consumers.
- Bran is visually discernible and imparts aesthetically undesirable color to many food products.
- Bran has a high fiber content (10-20 percent) that is readily detectable by the palate, and irritating to the digestive tract if consumed routinely in large amounts.
- The introduction of bran or bran products into many processed foods to enhance protein content is often unacceptable because the potential consumer cannot afford such foods; those in most need of protein are usually the least able to pay the price.

Although attempts should continue to find acceptance for edible bran, more success might be realized with bran flours and protein concentrates. The bran flours may be sufficiently different from brans in appearance and fiber content to escape significantly the first three limitations listed above. Also, bran flours might be accepted into indigenous foods requiring flour, so that the need for involvement with expensive processed foods might be avoided.

However, it is not clear how bran flours might be accepted into the villager's diet. Studies are needed to explore the extent to which bran flours might be utilized in indigenous foods that are widely consumed. This should be done for a number of major cultures in the rice-growing LDCs. Research kitchen facilities would be needed, and sensory evaluation panels of villagers would need to be organized and trained.

Bran protein concentrates should also escape many of the limitations of bran because concentrates do not look like brans, and have little or no fiber content. Concentrates may become well accepted wherever soy tofu is widely used, but his postulate needs to be tested. Also, imitation milks similar to Vitasoy but made with rice bran protein concentrate instead of soy protein should certainly be tested for acceptance in several cultures.

Product acceptance studies should be directed to small-scale food product manufacturers, such as bakers, as well as to the ultimate consumer. If bran flours or bran protein concentrates can improve product functionality or lower cost of operation, the village entrepreneur might adopt these bran products more quickly than a village consumer might include unfamiliar ingredients in home-prepared foods.

Economic, infrastructure, and impact studies

Techno-economic studies are needed for each of the technical development steps that have been outlined in this program. Analyses of benefits and costs, foreign exchange impacts, government incentive/disincentive options, financing options, the appropriateness of cooperatives, the question of who will do the investing, and others need to be considered as well as the technical problems. Studies are needed to determine costs of stabilized bran storage, collection, and transport systems if these are to be required. Major political-economic questions of local vs. national interest need to be studied, e.g., whether edible oil production for export should be maximized to earn foreign exchange, or whether local optimization of food value utilization is the preferred goal, or some combination of the two.

<u>Identification of potential cooperating research institutions</u>

A cooperative research and development program might be conducted by a group of institutions, each making its special contribution. Modern food engineering skills and facilities exist in most developed countries and in some developing countries; expertise in local food problems is generally available locally. It is believed that one useful approach to the rice bran problem might be expected to come from a joint effort mounted by a group of research institutions, some from developing and others from developed countries, pooling individual abilities and shared interests. It is important to identify potential contributors to such a program as early as possible, and establish the nature of the contribution that each might make.

There are many centers of expertise in this field, particularly in Asia, where rice-related activities are presently being pursued or might be pursued. Among others, there are the following: Food Corporation of India, Central Food Technology Research Institute (Mysore, India), Indian Institute of Technology (Kharagpur), Paddy Processing Research Center (India), Appropriate Technology Development Association (India), Applied Scientific Research Corporation (Thailand), Kasetsart University (Thailand), and the International Rice Research Institute (the Philippines). There are undoubtedly active centers in Indonesia, Burma, and other countries as well. The strengthening of these institutes and the recognition of their contributions in these areas is important.

5 Materials and Products Based on Agricultural Wastes and Natural Fibers

NEEDS AND TECHNOLOGICAL INITIATIVES

Historically, it has been pointed out that progress toward increasing the use of agricultural wastes and natural fibers in developing countries has been slow, even in countries with insufficient timber and nonrenewable resources for domestic needs. Many such countries, while not self-sufficient in wood, mineral, and other resources, have substantial quantities of agricultural residues and nonwood fibrous materials which, more often than not, they fail to utilize.(1) If suitable products can be identified which use these materials and, as a result, new industries can be established, former dependence on decreasing world timber reserves and nonrenewable resources can be reduced and opportunities for employment can be created.

In various UN seminars and conferences over the past decade,(2) it has been noted that developing countries cannot continue to depend indefinitely on many of the raw materials that are currently used, and that agricultural wastes and natural fibers which are plentiful in most developing countries can substitute for these currently-used raw materials. Estimates of the amounts of some of these wastes in 1972 are presented in table 5.1.

The primary nonfeed, nonfood, nonfuel products from these wastes are materials for building and construction which are greatly needed in developing countries. Most developing countries continue to rely heavily on imports of building and construction materials, which account for 14 to 40 percent of their construction costs.(3) The scale on which these materials are imported and their effect on foreign exchange may be gauged from information covering a number of developing

Table 5.1. Estimated Availability of Specific Nonwood Plant Fibrous Raw Materials, 1972.

Raw Materials	Worldwide Quantity[a] (bone dry metric tons, thousands)	
Sugar cane bagasse	55,000	
Wheat straw	550,000	
Rice straw[b]	180,000	
Oat straw	50,000	
Barley straw	40,000	
Rye straw	60,000	
Flaxseed straw	2,000	
Misc. seed straws	3,000	
Subtotal, straws		940,000
Jute fibers	4,425	
Kenaf and roselle fibers	1,674	
Sisal fibers	648	
Abaca fibers	92	
Henequen fibers	164	
Subtotal, fibers		7,003
Reeds	30,000	
Bamboo	30,000	
Papyrus	5,000	
Esparto and Sabai Grasses	700	
Cotton staple fiber and second-cut cotton linters	14,500	
Subtotal, grasses, seeds, etc.		80,200
Estimated Total		1,027,203

[a]Values are for potential availability with present collection methods.

[b]Does not include the 60 million tons of rice hulls generated annually.

Source: Adapted from Joseph E. Atchison, "Agricultural Residues and Other Nonwood Plant Fibers," in <u>Materials: Renewable and Nonrenewable Resources</u>, AAAS, Washington, D.C., 1976, p. 178. Copyright 1976, the American Association for the Advancement of Science, reprinted by permission. More recent (1978) estimates available from Mr. Atchison differ somewhat from these 1972 estimates.

countries over various periods between 1955 and 1965. These data show that, in most countries in Africa, Asia, and Latin America, the value of "imported building materials ranged from 5 to 8% of the total value of imports, yet expenditure on building and construction materials is from 3 to 5% of the gross domestic product (GDP)" in developing countries.(4) For these data, it is clear that building and construction materials use up a disproportionate share of foreign exchange when compared to other industries.

The production within developing countries of building materials and materials used in construction from indigenous agricultural residues and fibrous materials is also necessitated by demands in those countries for housing and low-cost shelter, caused by both population growth and urbanization. Building programs which utilize such materials will not only cut down on the import content of structures, but, in most instances, their absolute cost as well.(5)

While the technologies reviewed or suggested for R&D in this study are basically those for small-scale, local, or rural use, the appropriate technology in the building materials industry may not necessarily be the smallest. Part of the reason for this is because some wastes, e.g., sugar cane bagasse and rice hulls from large mills, are deposited centrally in large quantities; to make use of these wastes a correspondingly large facility is required. Otherwise most of the waste will go unused.

> It has never been claimed that small scale is automatically better than large. The intention is that these rural industries will develop into larger enterprises as the economic and social conditions of the area change - in particular that they show growth in step with agricultural development. The numbers of people employed in them will vary from place to place according to a number of factors including the supply of raw materials and closeness to large industrial centers, but particularly to the ability of local agriculture to support them.(6)

This whole question of scale in the building materials industry is relative. To quote the building materials industry monograph published by UNIDO in 1969, "Few building materials are manufactured on a large scale; many are produced efficiently on a relatively small scale. The industry is usually characterized by a large number of small producers."(7) The most important aspect, probably even more important than the foreign exchange or balance of payments aspect, is that the use of these wastes and natural fibers in the building materials industry has an important multiplier effect on both employment and income and, thereby, plays a dynamic role in a country's development. In addition, building materials from residues can provide another market for those wastes which may not be useful for food, feed, or energy purposes.

Along with encouraging new industry at the village level, the building materials industry can also provide employment during periods when less labor is required for agricultural production. Construction output may fluctuate considerably, especially at the local level. The supply of building materials must be adjusted to meet the situation. In agricultural areas, the trend is to build, or at least to purchase, materials during the slack season after harvesting so that, after several years or seasons, the villager will have acquired enough materials to start building. Normally, demand fluctuations are accommodated in the building materials industry by adjusting the level of output, by stock changes, and by importing or exporting to meet requirements. Adjustments of output are often difficult to make in industrialized countries owing to the use of capital-intensive production processes which require high output for economic operation; stocking adds cost, and importing because of shortages or exporting surpluses require established channels and markets, none of which are generally found in most rural areas of most developing countries, and often not even in their metropolitan centers.(8) However, rural, agriculturally-based industries which use crop residues for building materials manufacture may not suffer from these constraints.

Most of the world's developing countries lie in the tropics and subtropics. In these areas, local organic materials are the main sources of products used for housing, for other forms of shelter, and for the components of other industries, such as transportation and storage. Traditional use of organic materials is based on their availability, renewability, and low cost. A living tradition of methods and skills has grown up enabling families to satisfy their basic shelter and other needs with simple tools and devices, with little cost for labor. The built products of these activities are usually remarkably sensitive to the demands of climate, and to the cultures and needs of the users.

The problems associated with continued use of renewable resources are caused primarily by population growth, rural-to-urban population shifts in growth situations, and, technically, the inability of organic materials to respond to performances required under changed conditions of use. In cities and urbanizing areas, organic materials are no longer plentiful and cheap; and urban areas impose more stringent demands on both material performance and the methods and skills traditionally utilized. In the shelter industry, for example, public concern for health and safety results in new building codes; the urban dweller operates under socioeconomic conditions that discourage the use of self-help and mutual aid in building; the skills required to build with traditional materials are lost or diffused through urban migration; user-families have little time, energy, or interest to invest in the necessary cyclical routine of maintenance, repair, or replacement. Thus, in urbanizing areas, housing quality suffers with the continued implementation of rural techniques, forcing those responsible for construction, cost, and maintenance of housing and other shelter to search for new materials and methods based on new use requirements - for example, concern for the spread of fire in high-density areas has resulted in the proscription of grasses for roofs in most cities, and the substitution of manufactured materials such as metals and cement-based products.

In attempting to meet the different performance needs of urban areas, the substitution of manufactured for organic materials has created its own severe set of problems. For one, to use the example of the shelter industry again, building costs have risen in direct relation to the higher cost of manufactured materials, both to the user and in terms of foreign exchange. Since cost considerations are of dominant concern, the reduction of material content in a building more often than not overrides other important planning and design criteria, such as adequate space standards, functional requirements, structural stability, and other qualities of building performance. (Again the roofing example: wide overhangs are essential for comfort in the tropics, but cannot be afforded in low-cost housing built with manufactured materials.) And, while the properties of manufactured materials may meet certain performance requirements, such as fire-safety and durability, better than those of organic materials, they suffer by comparison in other important respects. For example, the low thermal mass of plant materials makes them excellent for tropical use. Building methods based on the use of traditional materials are less costly than those that involve a higher technology, and overall building quality is usually more suitable and responsive to the use patterns and needs of the local users.

This chapter contains recommendations for specific research and development initiatives which are intended to encourage continued use of local renewable resources for building materials and other products, responding to varying needs over the rural to urban continuum.

At the rural end of this continuum, emphasis is placed on techniques that improve certain properties of organic materials - principally, their resistance to fire and their durability - without significant alteration of their physical form. The initiatives recommended in this area are directed at improved preservation techniques and at the breeding of plants so as to improve their physical properties. Emphasis is placed on the retention of physical form for several reasons. Traditional building forms are, to a large extent, the result of familiar materials used in familiar ways. Vernacular building has learned to capitalize on the superior physical properties of materials in response to cultural and climatic requirements, and to minimize their less desirable characteristics. In rural areas, cost constraints will dictate for years to come that building and other products be based primarily on natural or modified natural materials - evidence of this can be seen in India, Bangladesh, and other Eastern countries, where governments have established plantations of timber and bamboo, either to provide a cost-stabilized source of materials, or to provide sources for emergency housing.

Emphasis at the urban end of the continuum is placed on initiatives directed at developing new materials for specific urban uses, in response to the more severe requirements and the needs and opportunities for greater mechanization and industrialization of the production processes. New materials and unfamiliar processes will be required. The specific research and development initiatives recommended are the development of composite materials which utilize carbonized plant materials and other organic materials as lightweight, low-cost fillers coupled with the development of nonpetroleum-based binders from wood and agricultural waste products. A more detailed set of initiatives is included in table 6.5.

STATE-OF-THE-ART REVIEW

Background and Limitations to the Review

The review started with a perusal of current appropriate technology publications for articles relating to the use of agricultural wastes and natural fibers to make products of a nonfood/feed or nonenergetic nature. What few articles were available were widely scattered and not indexed under the heading of this chapter. Further investigation has shown that this field is a relatively new one for concentrated, coordinated research even though modern research efforts have been occurring since World War II into materials from agricultural wastes. The current preferred use for these wastes is for the production of energy, feed, and food stocks. Less than 20 percent of the technologies listed in the 1978 United Nations Food and Agricultural Organizations publication, Agricultural Residues: Compendium of Technologies, deal with nonfuel, nonfood/feed processes. Most of the technologies listed are in the conceptual, laboratory, or pilot stages, and those in the industrial stages are primarily in the developed countries.(9)

It is also apparent that only a relatively low proportion of agricultural residues and fibrous materials have been tested and utilized on a commercial scale. Successful large-scale commercial production from agricultural wastes and natural fibers has been limited to bagasse, flax, and linseed residues, and to European reeds.(10)

The primary reason for lack of information on this subject appears to be the result of the paradigms of industrialized countries which have left most of their agricultural roots behind. Most developing countries have económies based predominantly on agriculture. And with agri-industrialization increasing during the past decade, many countries find themselves with growing surpluses of agricultural wastes available in large quantities and presenting serious problems of disposal.(11)

Some of these residues are at present used as cheap fuels and only recently has research on and development of uses for certain agricultural wastes progressed to the point where the establishment of industries using them for the production of building materials has been seriously considered. Primary limitations to the use of these wastes and natural fibers for products has come from requirements for storage, which are greater than for other traditional raw materials, such as sawdust, and their need for preservation and extra or special fire-retardation treatments.

Another limitation, which received considerable attention at the United Nations' Meeting of Experts on the Use of Agricultural, Industrial and Consumer Wastes in Low-Cost Construction is that imposed by public opinion: the term "wastes" has a pejorative connotation to most people, and could evoke consumer resistance to panels or other building materials known to be manufactured from these residues. The report of that meeting stressed that a product's quality was the only valid criterion for its selection for given end uses, and that those who are concerned with specifications and purchasing should not be influenced by the raw material used in producing these products. This point was stressed both in the introduction and in most of the chapters dealing with marketing and promotion of such products, as well as elsewhere in the report, and earlier in UNIDO's panel on the Production of Panels from Agricultural Residues in Vienna in late 1970.(12)

Selected Material Areas

The balance of this section will review the state-of-the-art in research, development, processing, and industrialization for certain specific waste and nonwaste fibrous materials. These materials have been selected on the basis of data availability, prevalence in developing world areas, and potential for viable research and development initiatives. More detailed information can be found in the thesis-report by Kick.(13)

Rice hulls

Rice, one of the world's primary cereal crops with over 300 million metric tons produced annually in the world is also a prime generator of wastes.(14) Over 60 million metric tons of rice hulls, frequently mixed with rice bran, are left after milling; and over 180 million metric tons of straw are left in the fields after harvesting. Utilization of the bran for food and feed uses was considered in chapter 4.

Research activities aimed at identifying potential uses for rice hulls have been going on for over 60 years. Most of these efforts have attempted to take advantage of some properties of the hulls themselves, such as their fuel value, their abrasive character, or the absorbent quality of the fuel ash as a filter. Recently, studies examining the technical and economic feasibility of various uses which involve more than one of the properties of rice hulls have been completed.(15) These uses may be classified as follows:

- Measures which make use of both the energetic and silica contents of the hulls;
- Measures which make use of both the textural and structural characteristics of the hulls; and
- Measures using both of the above concepts.

"The most promising approaches to the solution of the problem of rice hull disposal fall into (these) two basic categories."(16)

Research into the first concept has been directed primarily toward the manufacture of cement or pozzolana (a siliceous material which, when finely ground and mixed with lime and water forms a compound possessing cementitious properties) by the controlled combustion of the hulls, using the heat to produce such materials from the ash of previously burned hulls. In the United States, VITA Volunteer Phil Cady has shown through his research that rice hulls can be used to make hydraulic cement,(17) and Professor P. Kumar Mehta of the University of California, Berkeley, has developed a furnace to control the burning of the husks which produces a "highly reactive black ash which, when mixed with lime [without heat](18) becomes a rich, black cement that . . . is . . . structurally as good as portland cement."(19) Additionally, the lime content of the rice-hull-ash cement is less than a quarter of the lime content of normal portland cement, making it suitable for acidic environments.(20) The capital requirements for a basic furnace, according to Professor Mehta, "are very modest" and can be "applicable to small-scale, rural-based plants."(21)

A pilot-scale commercial plant was opened near Sacramento, and a full-scale plant was scheduled to open in April 1975 near Stuttgart, Arkansas. To date no reports on the success or failure of this plant have been obtained. The process has been patented in this country jointly by Professor Mehta and the University of California, and patents have been applied for in the major rice producing countries. The major drawback to the Mehta process is its use of fuels other than rice hulls so that, unless some heat recovery system is included, the excess heat used in burning the hulls is lost. To our knowledge, no system has been developed to produce a high quality cement when rice hulls are used as fuel because impurities result from presently available, uneven, and incomplete combustion methods.(22) The Arkansas plant was scheduled to supply 50,000 lb/day of 200 psi steam, while burning 7.5 tons/hr of rice hulls. The steam was to be supplied to a food-processing concern, freeing the latter from having to use other fuel sources.(23)

Research conducted by the Asian Institute of Technology (AIT) in Bangkok has shown that village-burned rice hull ash was a poor pozzolanic material, but that rice hull ash produced in a controlled burn - 450°C for four hours - conformed to ASTM specification C618-72 for pozzolanas.(24) While AIT's report cited Professor Mehta's paper, no mention was made of mixing the ash they obtained from their laboratory with lime, but rather with portland cement. Their experiment was apparently aimed at determining whether such an ash could be used as an additive to ordinary portland cement, thereby reducing the need for importing or transporting that material. Results indicated that the laboratory-burned ash can replace up to 20 percent of portland cement without any significant adverse effect on concrete properties.

In India, at the Central Building Research Institute (CBRI) at Roorkee, and at the Council of Scientific and Industrial Research's (CSIR) Regional Research Laboratory at Jorhat, research has shown that mortar cements, which need not be structurally as strong as most concrete used for high quality applications, but which are better binders, can be made from rice hull ash.(25)

The CBRI has developed two processes. One process requires the mixing of rice husks and clay with water, and then forming the mixture into balls which, after sun-drying, are fired in an oven-shaped brick kiln. After firing, the balls are ground into a fine powder, which can then be mixed in suitable proportions with lime and sand for mortar. The second process calls for mixing the husks with lime-sludge by-products from sugar processing, fertilizer production, tanning, or paper manufacture. The mixture is then fired and ground as the first, but does not need the addition of lime to produce mortar.(26) Both processes use rice hulls for fuel, and neither requires sophisticated plant, machinery, or techniques.(27)

The Jorhat process consists of burning the husk and then mixing the ash with binders to produce a mortar cement which has a seven-day compressive strength of 26-35 kg/cm^2. The laboratory-scale process showed projected costs (1976) of 100,000 rupees (approximately US $12,500) for a plant producing five to eight metric tons per day.(28)

While the economic feasibility of using rice hulls to manufacture cement may be somewhat affected by economies of scale, cement production on a small scale using these or similar processes may prove to be economical in many regions because transportation of cement frequently accounts for over 50 percent of the retail price.(29) The scale of modern cement production processes in industrialized countries requires plant capacities of 5 to 10 million barrels a year, containing approximately a quarter ton of concrete, while the "annual production of a plant based on the amount of rice hulls typically available in a particular geographical area may be limited to 2 million to 3 million barrels."(30) With the aforementioned transportation difficulties, increasing fuel costs and, normally, lower labor costs, developing countries' production factors may enable them to produce cement at lower cost, or at the same cost but on a smaller scale, than present practices dictate.

The Jorhat labs have found that the CBRI processes can be expanded to include the manufacture of building bricks from the mortar. The first CBRI process has been shown, at laboratory-scale, capable of producing 5,000 bricks/day for an investment of (1976) 130,000 rupees (US $16,300), and the second process can produce the same quantities for about 8 percent less.(31)

The Ghanaian Building and Road Research Institute at Kumasi has produced bricks from lime and rice hull ash without autoclaving. "The elimination of autoclaving lowers capital investment, and allows for savings in fuel," the result of which, in areas of Ghana where brick clays and firewood or other fuel is scarce, are bricks "which are cheaper than clay bricks."(32)

Another attractive approach to rice hull utilization and/or disposal "appears to be a manufacturing complex that converts part of the hull into water-glass (sodium silicate) which is then used in the bonding of the remainder of the hulls into panel boards."(33) While this process may have a similar limitation to those of cement manufacture from rice hulls (that is, a significant portion of the hulls may go unburned or incompletely-burned), it is felt that the technologies are feasible, since furnaces which completely burn the husks have been designed, and unconventional processing techniques used in other manufacturing processes, such as fluidized bed reactors, may be successfully applied to rice combustion.(34) Although this approach appears to be a relatively large-scale, capital-intensive one, it illustrates one of the many possible uses for rice "wastes"; in addition, no external fuel source is required.

Rice hulls mixed with cement can be used to manufacture hollow blocks that are comparable in strength and serviceability to traditional commercial hollow concrete blocks. These blocks can be formed by pressure molding in light-capital equipment like the CINVA-RAM. Research into the production of these blocks, as well as similar blocks made from sawdust, wood chips, and coconut trunk particles, is being undertaken at the Forest Products Research Industries Development Commission (FORPRIDECOM) of the Philippines at Laguna.(35)

At the Institute of Standards and Industrial Research of Iran, research has been undertaken to develop a light-weight cement aggregate from rice husk ash.(36) Conventional concrete aggregates cannot suffice for this unless air, hydrogen, or carbon dioxide is introduced (entrained) into the cement paste, and special, low-density aggregates are often expensive.(37) The Iranian research showed that an existing composite-roof company, or rice factory, with a "low" capital investment of (1975) U.S. $3,000, could produce 2,000 light-weight concrete roof blocks per day from 10,000 tons of hulls per year. This investment is primarily for a press.

The United Kingdom Ministry of Overseas Development's Tropical Products Institute (TPI) has developed a method of producing lightweight concrete aggregate using unburned husks which have been thoroughly soaked in water.(38) This method results in a product which has a somewhat low strength/weight ratio because the amount of cement in the mix is high - four times the air-dry rice hull weight - in order to obtain a workable mix;(39) and the high moisture content of this concrete requires special consideration.(40) TPI chose rice husks (we use the terms "hull" and "husk" interchangeably) as their first materials for investigation "because their disposal is difficult, they are frequently to be found in large quantities in one place and they contain only very small amounts of water-soluble cement 'poisons' - chemicals [such as sugar or tannin] which slow down or retard the reaction of the cement fermentitious materials."(41) It is the intention of the Tropical Products Institute to continue this type of research into various other low-bulk density vegetable wastes. Rice hulls can be ground into a flour and used as a filler in polyester mouldings for plastic furniture. While rice hull flour as such a filler appears only marginally satisfactory, it does compare favorably with cellulosic fillers currently in use.(42)

Rice hulls, as is the case with many other agricultural residues and natural fibers, can be made into panelboards, particle boards, fiber boards, pressed boards, and sandwich panels. Tests conducted under the standard methods of ASTM specification D2017-63 have shown that such boards perform significantly better than standard wood particle boards in resisting rot, both with and without treatment.(43) However, according to Cook, further testing is needed into fire and sound retardance, into minimizing volume change due to water absorption in those processes which use water, and into water resistance of the finished boards.(44) The manufacture of rice husk board, which can be mechanized to various degrees, may, according to some feasibility studies, prove to be a valuable industry in both developed and developing countries.(45) It is clear from both Cook's research and from the United Nation's meetings that to improve the versatility and marketability of such boards, development trials are necessary to fully define the performance characteristics of binders (cement, resins, sodium silicate) and of the production processes, as well as the different types of boards.

It appears that the opportunities for combining several technologies in one manufacturing complex merits further research.

> One [complex] would consist of a sodium silicate (water-glass) manufacturing facility producing sodium silicate from rice hulls as one of the raw materials. The output of this facility would be used in [another facility] where the remainder of the rice hulls would be bonded into architectural board using sodium silicate as binder.
>
> Operational economy would accrue from the reduction of raw material input to a few basic materials, chiefly rice hulls and sodium carbonate, as well as from the savings in packaging, shipping, and sales cost of sodium silicate which would be manufactured in the captive facility. In addition, the plant would be flexible since it would produce two independently salable products.(46)

Although such industries may be somewhat large, capital-intensive, and complex, they may have their place as rural industries evolve in some areas.

Rice husk cellulose, in addition to providing insulation materials, has been found useful in reducing the costs of athletic equipment: "Another delegate [to the Second International Olympic Course for Overseas Sports Administrators] from the Dominican Republic found that with canvas and rice husks, his team could make judo mats, saving £675 (US $1280) on each mat."(47)

Southeast Asia Technology, Ltd. (SEATEC) in Bangkok investigated the use of coconut husks and burned rice hulls as primary and secondary materials respectively for water systems in various areas of Southeast Asia. By using such free or cheap local filter materials, it was felt that expensive, not always available, imported or manufactured filter materials or sand could be done away with or reduced significantly. Trials run in four rural villages showed that, under normal circumstances, no coagulant was needed (as was the case with sand or imported filter materials) and, in abnormal instances, only small amounts of a coagulant were needed. Water from these systems met World Health Organization (WHO) standards and lowered operational and maintenance costs and down-time of the water supplies. The report pointed out that research is still needed for individual family-size jar filters, and why and how these materials perform so that a proper system to positively achieve desired results can be designed. The effluent quality obtained by using these filters is comparable to that obtained from the best slow sand filter prior to disinfection, and it was found that the burned rice husks performed similarly to activated charcoal.(48) Disinfection by chlorination was not used during the tests so that the taste of the water would not be affected.

Coconut wastes

Coconut as a crop has contributed at one time or another every bit of its fruit and fiber to man's welfare, and presently contributes to the livelihood of over 20 million families.(49) The coconut, in the languages of many of the areas in which it grows, is known as the "tree of life." However, in industrialized societies, the only products which have earned income are the meat of the fruit (copra) and the oil from this meat.

The primary "waste" from the coconut are the fibers of the husk (coir) which surrounds the kernel holding the meat. Husk fibers are about 15 to 35 centimeters (6 to 14 inches) long, and consist mainly of lignin and cellulose, along with approximately 10 percent pectins, tannins, and water soluble and insoluble substances.(50) The fibers are embedded at the top of the nut in a "soft core-like ground tissue" called pith, which contains the same ingredients as the fibers (in different proportions) along with hemicelluloses.(51) Both the husk fibers and the pith are chemically active, the pith more than the fibers.

Research at CBRI has shown that particle boards made of chipped coconut husks without the pith removed need little or no resin additive. The most resin shown to be needed was 0.5 percent as compared to 8 to 10 percent in ordinary wood boards. The pith contains reactive ingredients which undergo chemical change during the board pressing process and impart sufficient bond strength.(52) CBRI has developed a machine which chips the husks without separation of pith and which can be fabricated with locally available parts by village blacksmith, machine, or welding shops. The machine can produce 600 kg/day with a five hp power supply, and can be fabricated for an estimated (1973) 5,000 rupees (US $650).(53)

Further research has shown that the cost of particle boards made from coconut husks could fall between 55 and 60 percent of ordinary wood particle boards. Additionally, coconut husk particle boards are moderately resistant to termites and fungi.(54) Fire resistant boards have been developed at the CBRI from coconut pith and banana stems. Based on the CBRI laboratory-scale system, a production-scale plant could be constructed for an estimated (1976) cost of 33,000 rupees (US $4,200) for a capacity of 50 sq. meters (538 sq. ft.) of board per day.(55)

Corrugated roofing sheets can be made from coir waste in a process developed at CBRI. The process consists of soaking the fiber in mineralized water for two hours, followed by draining and then mixing with dry cement. This mix is then made into a mat of the required thickness and slid onto a corrugated mold where it is held under pressure for four to eight hours. After removal from the mold, the sheet is cured and dried. The process needs neither heavy machinery nor high capital - the corrugated molds are the only special equipment. After a special coating is applied, the sheets perform comparably to conventional sheets, are 50 percent cheaper than asbestos cement sheets, and provide good thermal insulation.(56)

Coir pith is traditionally extracted by retting the husks for periods of months, and then shaking the pith out of the husks. This pith has good insulation qualities, and can be used as the filler in sandwich panels(57) or as an additive to cement.(58) When adding the pith to cement, a lightweight concrete is formed which, however, exhibits relatively high drying shrinkage (0.57 percent) when mixed and laid, resulting in small cracks. A plant which could produce 20 sq. meters (215 sq. ft.) per day has been estimated (1976) to cost 20,000 rupees (US $2,500).(59)

Retted coconut pith has also been developed into an expansion joint filler at CBRI for use between concrete slabs in roofs, roads, runways, etc.(60) The technology consists of drying the pith, mixing it with cashew nutshell liquid resin, forming the mix into a mat, pressing it, and finishing by cutting, etc. The material is resistant to termite attack and fungal decay.

Coconut coir pith can also provide about half of the ingredients in gaskets for automobile and other engines. The Indian National Chemical Labs at Poona have developed a technology to produce rubberized coir-pith sheets for an estimated (1976) cost of 133,000 rupees (US $15,250) for a capacity of 5.32 tons per annum.(61) The process involves mixing the pith with nitrile and other rubbers, followed by vulcanizing and pressing. The product to residue ratio is 5.7:3.(62)

An interesting product from the water in the coconut may develop because the fluid taken directly from the nut has been successfully used in emergency situations as a substitute for parenteral solution. The composition of the water has been found to be close to isotonic glucose serum.(63)

As with the wastes from rice hulls, coconut wastes can also be used in combination with cement and soil or sand to produce lightweight building blocks.(64)

At present, no commercial-scale processes have been developed which use the coconut's shell because an efficient chipper has not been developed and because of a lack of inexpensive adhesives.(65)

The remainder of the coconut tree, once it is past its bearing stage, is a nuisance in many instances, and frequently villagers simply put it out of their way because there is no immediate traditional need for it. The Ministry of Forests, Fiji, at its research station at Nasinu, near the capital of Suva, has been investigating the economics of producing sawn timber from these coconut logs. In August 1978, they started the construction of an all-coconut house (except for the roofing, footing, window glass, plumbing, and wiring) in order to gauge the wear and strain characteristics of the hard coconut wood.(66) A similar house has been built in the Philippines.(67)

Sugar wastes

One of the largest noncereal crops in the world is sugar. Considerable effort has been expended in developing commercial production of sugar cane bagasse panels. However, less expensive adhesives for these boards need to be developed in order to bring cost levels down.(68) Processes for making paper from bagasse have also been developed. However, many sugar mills use most of the bagasse as fuel for the sugar-making process.(69)

Sugar-mill press mud, a lime sludge, is a waste by-product from both carbonation and sulphitation milling processes. Press mud amounts to about 6 to 8 percent of the sugar cane, by weight. The primary use for the mud from sulphitation factories is as a fertilizer; but from the carbonation process, the mud, primarily calcium carbonate, is thrown out, frequently at high cost.(70) An old process, developed by researchers at the Shri Ram Institute for Industrial Research, in Delhi, India, in 1952, produced a sound and heat insulating mineral wool from the carbonation mud for .5¢ a pound, but with high capital costs (1952 US $45,000). More up-to-date details or reports of this process have not been found.

Tribhuvan University in Kathmandu, Nepal, has developed a process to obtain a low-cost cement substitute or building lime from the carbonation mud by mixing the dry mud with a sawdust fuel and burning it in a controlled furnace at less than 900°C.(71) Another process has been reported which produces lime from this mud for 47.48 rupees (US $6) per metric ton. This process requires high pressure (6,000 psi) compression of the mud into briquettes which are then fed in mixed-feed kilns and fired at 950°C - 1000°C. The briquettes are then ground in a ball mill.(72)

Other wastes and products

Animal wastes. The Australian Leather Research Group of the Commonwealth Scientific and Industrial Research Organization (CSIRO) has developed acoustical tiles which are made from leather residues. Their pilot stage process consists of grinding the leather, mixing it with a PVA resin, pressing the mix, and then letting it dry. Using this process, 3.6 kg (8 lb) of leather waste can make a tile approximately 53 cm x 41 cm (21" x 16").(73) Cost estimates are unavailable at present. Press boards can also be made from the manure of cattle;(74) no details have been obtained.

Resins and adhesive extenders. Resins or other adhesives are seen as the central problem in the wider use of cellulosic agricultural waste composites, be they boards, blocks, panels, or furniture, according to the Materials Science Panel of the Science Committee of NATO which had proposed a meeting for late summer of 1979 to address this subject.(75) The significant foreign exchange costs of resins and other adhesives is an important consideration for many countries.

The Indian Plywood Industries Research Institute (IPIRI) in Bangalore has been working over the years to develop alternative resins/adhesives or extenders for currently used resins/adhesives. To date, they have produced adhesive extenders from coffee seed meal, sal meal, groundnut cake, soya meal, and tamarind seed.(76) The tamarind seed itself has binding qualities, and when ground to a powder can replace 20-60 percent, by weight, of synthetic or tannin adhesives.

IPIRI has developed a pure adhesive from cashew nutshell liquid.(77) The adhesive contains 75 to 100 percent condensed cashew nutshell liquid, depending upon the product (plywood, particle board) and its use. Some products require the addition of phenol and formalin hardeners. Also, the CBRI has developed a process which uses cashew nutshell liquid to produce a sealant for joints and cracks in buildings, and a water- and weather-proofing compound. This process is estimated (1976) to cost 25,000 rupees (US $3,200) for a 100 kg/day capacity output.

CBRI has also investigated the use of cement as a binder in producing wood-wool type fiber boards, particle boards, blocks, etc. As mentioned earlier, boards can be made with cement from rice husks and coconut pith as a binder; additionally, groundnut husk and wood chips have also been found not to exhibit inhibitory action on strength development in the cement. Panels of adequate strength are obtained when mixed in a ratio of 1:2 by weight and pressed to a density of 1,100-1,200 kg/m^3 (61 lb/cu ft)(78)

Cassava, plantain, and cotton wastes. The Technology Consultancy Center of the University of Science and Technology in Kumasi, Ghana, in 1972 developed a paper glue made from cassava starch and alkali from plantain peels. The pilot plant they developed produces 40 gallons a day, employs 25 rural dwellers, and supplies the greatest portion of Ghana's glue. This project required about $5,000 for plant and machinery; and, relying upon local supplies of raw material, it facilitated the rapid and easy horizontal transfer of technology as well as a relatively high profitability: the annual turnover per worker was equal to the total investment in the plant and machinery. Further, cassava starch can also be mixed with certain plant alkalis to make plain laundry starch.(79)

The director of the Technology Consultancy Center, in a tour of the Indian subcontinent in late 1977, observed a process in Kathmandu, Nepal, and Lucknow, India, for making paper by hand from cotton wastes and tree bark. The material, primarily cotton wastes, was broken up and then soaked in a digester and run over a roller to make a slurry. The slurry was removed by sieve and placed in a stack between cloths, then pressed, dried, and cut. The paper produced was thick, but inconsistently so, and used for file covers and documents. Kaolin could be added to obtain a glossy surface.(80)

Nonwaste plant fiber materials

Literature documenting work in the area of nonwaste plant fiber materials appears to be meager, but growing. There is a good body of information available on the botanical, physical, and mechanical properties of timber and, to a certain extent, plant materials in tropical and subtropical areas; but little scientific effort seems to have been directed at the identification and analysis of species which have potential for improving these materials for specific end uses. The vernacular uses of specific plants offer some valuable clues, but there is no organization to the literature or sources; thus information is hard to extract. In the area of natural durability (resistance to rot, fungal attack, and insects), in West Africa, for example, techniques have been recorded for interlaying thatch roofs with parts of the Cassia Nigracamf plant, and mats used for building elements are known to be impregnated with an extract of certain riverside trees - both techniques providing some degree of termite protection.(81) In the West Indies, the Genip tree (Sapindaceae; the Soapberry family) is known locally to have insect-repellent properties.

Preservative treatments for timber have been developed extensively, but these have evolved primarily in response to the large-scale market demands of the industrialized countries. Little work has been done on developing processes suitable for rural village-scale use in developing countries. Specific preservation processes are commercially available to treat against termites and against fire, but little information is available on work toward a single process or simple series of processes appropriate for village use that would treat against both.

There is considerable technology and information on methods of preserving wood and other products on farms in the United States which might be applicable in developing countries. Much preservation can be accomplished in a passive way, i.e., by building properly and not letting moisture build up. Where treatment is necessary, small-scale methods such as hot and cold tanks or diffusion processes have been used.(82) These processes aren't nearly as good as pressure treatment but they can double the life of the material being treated. The Timber Development Association of India has issued a paper on "Preservative Treatment of Green Bamboos by the Diffusion Process."

In Japan, techniques are known for the mechanical modification of the cross-section of bamboo, primarily for aesthetic reasons, by growing the culms through a framework that acts as a die so that, in effect, the plant extrudes itself into a square or rectangular shape.(83) Considering the great difficulty of fastening round bamboo at the joints of frames and trusses, it is surprising that work hasn't been directed at the modification of shapes to improve and simplify building methods with bamboos and heavy canes.

The UN publication, The Use of Bamboo and Reeds in Building Construction, documents traditional use of these materials over many parts of the world, records experience in preservative materials and methods, and recommends further research, principally in preservation for various end uses, in better and easier fastening methods, and in improved design for building based on bamboo, canes, and grasses.(84) Recently, the Central Building Research Institute in Roorkee, India, has published technical notes on renderings and treatments for making thatch fire-retardant.

In the area of plant breeding, it is well established that savannah and prairie plants develop fire resistance as the result of slash and burn farming techniques and prairie fires. Research is in progress to develop fire-resistant species of plants for the California chapparal.

Work on the carbonization of plant materials for aggregates and fillers is of fairly recent origin. To date, it has been motivated by a need to find substitutes for insulating mineral aggregates in concrete for road and building construction - the potential for making lightweight composites for other uses seems to have been largely ignored.

Prior to the growth of the petrochemical industry and the extensive development of synthetic resins for binders, considerable work had been done on natural binders produced from agricultural waste products. A number of patents were issued for processes based on the extraction of lignin and furfural. With these now in the public domain, and with cost and availability constraining the use of petroleum-based resins, there would appear to be good reason to redirect this early effort to the material needs of developing areas, where there are often ready supplies of timber and agricultural wastes.

Panels and Board from Agricultural Wastes

Research to date has indicated that it is, in principle, possible to manufacture agglomerated boards or panels from almost all agricultural residues and fibrous raw materials. Unfortunately, the properties of boards from many of these resources are unfavorable, and satisfactory properties are frequently obtained only through the use of large amounts of binder. However, it is not clear from the literature whether the residues tested for board making had been subjected to any preprocessing technique to eliminate deleterious constituents, if any, of those residues. Nor is it clear from the literature how many varieties of a given species were tested.

Table 5.2 contains a listing of the more viable raw materials for board production. Flax, hemp shives, bagasse, and cereal straws have been used for the commercial production of 579,000 metric tons of particleboard annually in parts of Europe, although production there has diminished because of an insufficient raw material supply.(85) The other product technologies may also be designed for similar scales of production, although the smallest production scale in recent (1970) use, according to the FAO, is 1,500 metric tons annually.(86) Details of many of the processes in the table were not available to this study.

In addition to the board making processes mentioned in the crop-specific sections earlier in this chapter, one process which is not residue-specific was noted. The Indian CSIR Regional Research Laboratory in Jorhat has developed a simple process for the production of particleboards from rice husks,

Table 5.2. Agricultural Residues Tested for Board Making

Raw Material	Insulation Board	Hardboard	Particleboard	Fiberboard	Other
Abaca (Manila)			E*		
Areca nut husk	E	E			
Bagasse	E, I	E, I	E, I	I	E - Plasterboards
Bamboo	E	E, I	E	E, I	E - Plywood E - Concrete Panel Reinforcement
Banana		E		E	
Cashew nut shell		E	E		
Cassava Stalk	E	E	E		
Cereal Straw	E, I	E, I		I	
Coffee Husk			E		
Coir-dust	E	E, I			
-fiber & husk	E, I	E, I	E		E - Concrete Panel Reinforcement
Cotton-husk			E		
- stalk			E, I		
Esparto			E		
Flax and Linseed Residues	E	E	I	E	
Groundnut (peanut) shell			E, I		
Hemp	I	I	I	E, I	
Jute		E, I	E, I	I	E - Asphalted Roofing
Kenaf		E	E		
Maize (corn) cob		E	E		E - Plywood
Mustard stalk		E			
Palm fronds, fruit stems (raffle)	E		E, I	E	
Papyrus**	E, I	E, I			
Rice-husks	P		E	E	E - Cement Panel Aggregate
- straw			P	E, P, I	
Sisal		E			E - Concrete Panel Reinforcement

*"E" indicates that the material has been the subject of laboratory experimental work only. "P" indicates pilot plant stage. "I" indicates that the material has been, or is being, used industrially.
**Production now ceased.

Source: Adapted from C. G. Kick III, "Appropriate Technology for Building Materials from Agro-Wastes and Natural Fibers," Report No. THA/CDT 80/2, Center for Development Technology, Washington University, St. Louis, Mo., February 1980.

groundnut husks, coconut pith, bagasse, seeds, and grass, as well as sawmill residues, without the use of synthetic binders. The boards have acceptable tensile strengths and moduli of rupture. Although a larger (50-100 ton per day) plant is thought to be more viable commercially, a 10 ton per day plant would require an investment of (1978) $200,000.(87)

Summary and Conclusions

Listed in table 5.3 are the residue-product technologies reviewed elsewhere. As the table shows, more than one type of residue can be used to produce a given product, much the same as table 5.2 presented the many different residues that have been shown to be viable raw materials for board production. While this may indicate other residue-product interrelationships, it cannot be considered as more than an indication until R&D confirms each given residue-product. However, the variety of materials which can be produced with appropriate, light-capital technologies from the small number of residues covered in the state-of-the-art review augurs well for the ability of agro-wastes/fibers to meet the variety of building materials needed in many developing countries. This is especially apparent when one considers that the R&D efforts to date reflect a wide range of thoroughness, from the quite detailed study of rice hulls' pozzolanic properties, to a much less thorough study of the waterproofing properties of banana stems, sheaves, and fronds.

NON-TECHNICAL FACTORS

Community Participation

The primary goal of international technical assistance should be to support the efforts of the assisted state to improve the quality of life of its citizens. Since people live in communities, proposed improvements must consider not only the effect of technical changes on the individual, but also the responsibility that the individual holds to other individuals and to the community at large. The community, in turn, has the right and responsibility to participate in its future by choosing ways in which it will develop, and methods for implementing development. Development implies change, whether by gradual transition, reformation, or radical transformation. Since change usually disrupts social patterns to some degree, the presumed benefits of technical improvements must be weighed against the uncertainties of these disruptions. In rural areas especially, if the community is not involved in planning and

Table 5.3. Summary of Residue Product Technologies

Residue	Product
Banana stems, sheaths, etc.	Corrugated roofing panels Paint material Paper
Cashew nut shell	Adhesive Joint and crack sealant
Cassava	Glue Starch
Coconut husks	Active carbon Building blocks Corrugated roofing panels Expansion joint filler Filler for plastics Gasket material Lightweight concrete aggregate Panel filler Particle boards
Cotton wastes	Paper
Groundnut (peanut) husks	Panels, blocks, etc.
Leather wastes	Acoustic panels Boards
Rice husks (unburned)	Boards Building blocks Filler for plastics Insulation and padding Lightweight concrete aggregate
Rice husks (burned)	Hydraulic cements or pozzolanas Lightweight concrete aggregate Rubber reinforcement Sodium silicate (water-glass) Water filtration medium
Rice straw	Fiberboards Paper
Sisal fibers	Corrugated roofing panels Paper
Sugarcane bagasse	Boards Corrugated roofing panels Paper
Sugarcane mill press mud	Lime Masonry cement Mineral wool
Wheat straw	Paper
Coffee seed meal Groundnut cake Sal meal Soya meal Tamarind seed	Adhesives and/or adhesive extenders

Source: Adapted from C.G. Kick III, "Appropriate Technology for Building Materials from Agro-Wastes and Natural Fibers," Report No. THA/CDT 80/2, Center for Development Technology, Washington University, St. Louis, Mo., February 1980.

implementation of projects, individuals will have little stake and interest in their outcome. It follows that, while the feasibility of a proposed new or improved technology must of course be clear from the outset to policy implementers, the need and proposed applications of the technology must be based on the shared desires and agreements of the community. Any development policy that fails to consider these factors is likely to fail to produce long-term and lasting improvements in life quality, and runs the risk of becoming one more "demonstration" project.

In general, and especially in rural areas, villages are not well organized to implement development projects, and education and training of local people to work toward their identified goals will be required. Individuals will need to be helped to participate, and this will be facilitated if project goals are clearly recognizable, and if the project itself is of a scope and scale of undertaking that encourages the organization of small groups of participants. Since development projects are costly in time, effort, and money, the small-scale project stands the best chance of avoiding waste, and can be abandoned more easily should failure be anticipated along the line. While one given technology will not necessarily be suitable over the range of climatic, cultural, and social conditions likely to be encountered within a country, the project should serve more than local interest - that is, it should form part of an overall government policy, as uniform as possible over the country, so that national development goals can be more easily perceived, and so that intercommunity jealousies can be avoided.

The production of materials from agricultural wastes and natural fibers involves social, economic, political, and environmental factors which affect the development and transfer of such technologies to the areas where they will be of most use. The discussion which follows will attempt not to reiterate those factors which are somewhat universal, but to focus upon impacts which are particularly relevant to the area of renewable resource utilization.

Economic Factors

The successful production of glue in Ghana from cassava and plantains, and the pilot plant's indicated success in using the Mehta process to produce cement from rice husks, along with other examples cited in the previous section, indicate that the conversion of agricultural wastes into useful materials can be profitable and generate employment. An agricultural economist at the University of Hawaii has developed an economic framework for analyzing by-products utilization.(88) While Huang's framework deals with the use of agricultural residues to produce food/feed or fuel (fertilizer) products, and is evolved from considering only two farm wastes (pineapple and dairy wastes), it contains flexible parameters which can be extended to materials production from multiple farm types. The model is stated to have value for use in "evaluating utilization, in searching for optimal use, and in planning future use of the by-products under different economic settings."(89)

According to the report of UNIDO's Expert Working Group on the Production of Panels from Agricultural Wastes: ". . . the need was stressed for careful economic surveys and feasibility studies prior to introducing processing plants."(90) While the importance of such analysis and surveys cannot be overestimated, their value is directly proportional to the output capacity of a producer. Small, rural producers may have a ready understanding of their market and costs and can be quite flexible in their response to them without the expense of surveys. The United Nations Department of Economic and Social Affairs endorses the small-scale production of residue materials as a first step in industrialization so that mistakes in product manufacture can be reduced.(91)

Environmental Factors

In traditional construction, the climate determines the materials which are readily available for use in building and has helped shape local methods of building. These have, in turn, governed the form and character of traditional structures.(92) The complete classification of climatic regions was worked out over 25 years ago by Koppen and Geiger.(93) While classifications such as this are capable of increased detail and accuracy through the use of earth observation satellites, available information can indicate areas for concentration of research in specific types of products. By ascertaining what types of agricultural products and natural fibers are or can be grown in which areas, and what types of building and other materials are used in which areas, the types of specialized/concentrated research which could be done to start or expand the production of agricultural residue materials can be indicated.

Political Factors

While there appears to be a need for all building materials industrial output and production levels to be increased to a large extent,(94) to do this, "a more comprehensive developmental and industrial strategy designed to meet basic socio-economic needs and to achieve adequate growth of technological capability in developing countries is needed."(95) Yet, in many countries there is opposition to industrialization which uses local materials and is small in scale. This opposition frequently comes in the form of "existing policies . . . [which have] hitherto favoured the growth of large-scale and medium industries in the organized urban sector,"(96) and may be alleviated by the development by each developing country of an overall technology plan which embraces the evaluation and improvement of traditional technologies, the application of modern science and technology through national research and development programs, along with the acquisition, use, and adaptation of foreign technology.(97)

A typical policy obstruction arises from building codes which, for reasons of fire protection, decay prevention, and/or structural soundness have frequently been written in product- or technique-specific language rather than performance-specific terms.(98) Such language inhibits the introduction and use of new and non-standard products, including materials from agricultural residues. Work toward the modification of inadequate and restrictive building codes has been recommended by at least two United Nations conferences.(99)

Another source of political opposition has developed from international trade in industrial products. An example of this is provided by one multiuse agricultural crop, the coconut. As has been shown in the state-of-the-art review, coconut plants and their fruit can be made into many products of use to developing countries, but these have been inhibited by demand from industrialized nations. "Coconut products were, separately, taken away 'upstairs' as raw materials; and arguably less suitable and certainly more expensive synthetic finished products were sold back. . . . "(100)

Social Factors

As mentioned earlier, a major factor governing the introduction of technologies which convert fibers and residues into usable products is likely to be product acceptance. The pejorative connotations associated with the term "wastes" could impede/inhibit product acceptance and, consequently, the introduction of the technologies. Various answers have been proposed, which range from training of sales and technical personnel and the publishing of product technical and performance specifications to the substitution of the word "residues" for "wastes." These solutions are of little value to the small-scale, low-volume producer, but the principle behind them should be communicated to prospective producers.

The purchase of building materials in most rural areas is usually limited to those manufactured items which fill a need that natural local materials - which are free or can be obtained through barter - cannot fill, e.g., corrugated roofing, cement, and connectors (nails, hinges, etc.). In introducing technologies such as those we have reviewed, the initial products will probably need to be ones which capture existing markets.

As has been pointed out by M. K. Garg, the social problems caused by resistance to disruption of traditional life-styles, including traditional construction techniques and materials, must be noted not only by industrial designers, architects, engineers, and materials scientists, but also by the regional agriculturalists.(101) A national or even a regional agricultural policy must take into account not only the basic need to increase total food supplies, but also the need to improve employment prospects in rural areas especially during off-seasons and the need to earn more foreign exchange through agricultural export or reduction of imports. These other aspects of agricultural policy relate directly to the need for useful products from agricultural wastes. Additionally, it is essential to encourage the establishment of small-scale industries to support the agricultural sector and to meet the basic needs of the local population.

RESEARCH AND DEVELOPMENT INITIATIVES

Modification of Materials for Improved Performance

In rural areas, as noted above, cost constraints and traditions of building dictate that materials should be used in familiar forms and by familiar methods. Although vernacular builders have learned to select materials carefully for subtle variations in plant materials and, often, to work with them in sophisticated ways so as to optimize their performance, the working life of organic materials remains generally very short, and frequent repair or replacement continues to be necessary. In some areas, bamboos and grasses and leaf materials have become so scarce that, once free for the taking, they are now market commodities. Thus, to improve the physical, mechanical, and building characteristics of traditional materials - the reeds, canes, grasses and bamboos - would be a major contribution to improving the conditions and economics of rural people throughout most of the developing world. Extending the life of traditional materials would be likely to have important long-term effects on rural communities and their development and growth. There is little incentive to build amenities in housing whose longevity can be predicted to be only two or three years. If housing life could be extended to 10 or 20 years, householders would be much more likely to consider solutions to their housing needs more carefully and systematically, and adopt such health, time, and labor-saving facilities as integral water, sanitary, and cooking facilities; authorities responsible for rural village planning would be able to set more rational land-use standards for housing and community development as functions of village and town growth needs.

Traditional materials can be modified by chemical methods (preservative treatments), or by the selection and breeding of species to obtain superior physical and chemical characteristics for particular use.

Preservation

The primary objective in the area of preservation should be to develop materials and methods that treat against fire, insect, and fungal attack, that are suitable for use in rural village conditions.

There are a number of well-known methods for the treatment of timber against termites, borers, and fungal attack, including pressure impregnation with various chemicals, the Boucherie method and its adaptations, and soaking, dipping, brushing, and spraying for surface treatments. Pressure treatments are by far the most effective, are used the world over, and have been adapted successfully for the treatment of bamboo and canes. These are not directly transferable as treatments for other commonly used plant materials such as leaves and grasses used for roof and wall construction. Vernacular builders have developed local techniques such as coatings of limewash or tar, but these are not particularly effective against insect and fungal damage through penetration of the splits and defects that occur as natural materials age and weather.(102)

To date, even in the industrialized countries, separate treatments are specified, promoted, and marketed for preservation against insect attack and rot on one hand and against fire on the other. This adds considerably to builders' costs and logistical problems. Stake tests have shown that a specific treatment for fire retardancy, which utilizes the organic resin amino phosphate, also provides a high degree of protection against termite attack and decay.(103) It would be a giant step forward if these investigations were extended geographically to developing world areas and included a search for other organic compounds available regionally that would provide a single, comprehensive treatment for wood and other organic building materials, with processes designed to be compatible with village scale, skills, and needs.

Plant breeding

There may well be opportunities for the breeding of plants to optimize specific material characteristics, just as cereal grains and other food-bearing plants have been improved. The process occurs in nature, such as with plants that develop survival methods against forest and prairie fires or in savannah areas where slash and burn farming is practiced. The concept is being applied in research on fire retardant plants for the California chaparral, where wildfires annually cause heavy property loss.(104) Since it is well known that certain trees and plants are more naturally resistant and durable than others, research should include a systematic search for plants with natural immunity to insect attack, resistance to fire through slow burning, inability to support flame, or charring characteristics that consitute self-insulation in fires. If exotic to developing areas, their ability to grow under local seasonal wet/dry cycles, their propagation, and their ability to compete with native species would require study. If monoculture is implied for economic or logistic reasons, problems of disease susceptibility should be weighted against the benefits derived from plant selection.

Optimization of desirable characteristics would necessarily be a long-term undertaking, and it would be essential to carefully describe the performance required of the end materials for specific uses. Since it is unrealistic to expect to render cellulosic materials "fireproof" by any method, it would be important to understand the behavior and characteristics of fires under household conditions in the cultures concerned. Fire resistance in plants is a complicated subject, depending not only on the source, type, and character of the fire itself, but the moisture and chemical content of the plant, its mass and structure; a joint effort involving both building experts and botanists would be necessary to formulate the needs, acceptable performance standards, and codes that would apply in the utilization of such modified materials.

New Composite Materials

Carbonized plant materials for aggregates and fillers

The steady depletion of world mineral resources, together with associated transportation and handling costs of heavy materials, suggests that increased attention should be placed on the use of renewable resources for the materials and components of construction and other manufactured products. In some areas of the industrialized world, mineral aggregates have become so scarce that roads are now designed to eliminate the deep layers of gravel traditionally used for roadbeds; shortages of traditional insulating materials such as vermiculite, perlite, expanded mica, shale, and clay, as well as the need to improve over their performance in the face of the need for energy conservation have motivated research into substitutes from synthetic resins and, more recently, from natural sources. Work at the University of Toronto has shown that carbonized cereal grains could be produced for fill-type insulation much more cheaply than, for example, polystyrene beads now widely used as a mineral substitute.(105) Carbonizing refers to the controlled combustion of an organic material by which flammable compounds are burned off, leaving an inflammable mass, primarily carbon.

The renewability and production economics of carbonized plant materials make them especially promising candidates for composite materials in developing countries, many of which either lack suitable mineral resources or cannot afford the investment required for exploitation. Many such countries have underutilized secondary species of timbers, or sources of timber and agricultural wastes, some of which might be developed into carbonized fillers and aggregates. These countries are often natural habitats for fast-growing species of plants, such as grasses, canes, and bamboos. The thermal characteristics, low density, and fire-resistance of carbonized fillers make them especially attractive as components of building elements in urbanizing areas where the industrialization of the building materials industry is an important development goal.

In work done so far, cereal grains, because of their high starch content, have been the focus of research. Kernels of wheat, corn, or rice are expanded by applying heat and pressure, releasing it to "puff" the kernel, and then converting it to carbon by slow burning. The resulting product has a fine cellular structure containing trapped air, making it an excellent insulator. Carbonization renders the material biologically inert and highly resistant to water and most chemicals. Since volatile gases have been driven off in the carbonization process, the material will not support flame, turning to ash under an external flame source at about 800°F; when combined with a suitable matrix such as gypsum or portland cement, the carbon cannot be oxidized and the resulting composite is stable to temperatures of about 3,600°F.

Research in North America has centered on the use of edible cereal grains, especially wheat, to produce the aggregate. For food-deficient world areas, it would be necessary to identify indigenous nonedible plants or wood or agricultural wastes suitable for expansion and carbonization.(106) Failing such sources, it is likely that exotic plant species could be introduced and cultivated specifically for the purpose. A comprehensive study in one or more developing countries would, ideally, be combined with an investigation of compatible binding agents, especially binders from renewable resources.

Obstacles to the use of plant materials as fillers include the loss of the plant for other purposes (e.g., food/feed) and the loss of planted acreage for alternative uses. Also, the research efforts to date indicate that the development of simple, low-cost processes which can be handled by local people may be a difficult task.

Natural fibers as fillers

The basic concept of adding thin fibers to a material in order to improve its properties has been known for some time. Chopped straw and other fibrous matter have been added to clay walls for thousands of years. In recent times, linoleum, asbestos, cement, reinforced concrete, reinforced rubber, and fiber-glass resin systems have assumed considerable importance. Although composites have been used as structural materials for many years, coordinated research and development in this field is quite recent and has occurred mostly during the past 15 to 20 years for defense and space applications in industrialized countries.

There have been a number of field projects, carried out in connection with organizations such as the Peace Corps and VITA, in which attempts have been made to add indigenous fibers such as grass thatch, bamboo, abaca, and the like to soil, clay, concrete, and other common building materials. The success of these attempts often hinges on the quality of the reinforcement, the stability and compatibility of the constituents, and the method of fabrication.(107)

By a judicious choice of fabricating technique, by a proper choice of components, and by a careful study of how to protect small diameter fibers from deleterous reactions, it might be possible to produce improved building materials using composite materials techniques without resorting to carbonization. Objectives to be sought include: (1) maintenance of low cost through maximum utilization of indigenous resources; and (2) improvement of impact strength, compressive strength, shear strength, and tensile strength so that the material can have potential for use as load-bearing structural elements.(108)

Natural binders for composite materials

If it can be established that new composite materials are needed for specific end uses in urban conditions, and that carbonized aggregates and fillers as well as other fillers can be used to create viable products in given areas, it follows that investigations will be necessary to identify compatible binders from local nonpetroleum resources. In one recently-completed USAID-funded research and development project, visits to a series of developing countries revealed the need to drastically reduce the cost content of binding resins used for local plywoods, hardboards, and other composite panels used in industry. For example, at the National Housing Corporation prefabrication facilities in the Philippines, in 1974, urea formaldehyde binders accounted for 65 to 70 percent of the cost of particleboard produced for their houses; resin costs have further increased with raw material and transport costs in the last several years, and substitute binders are considered necessary by NHC if their housing obligations are to be met.(109)

Nonpetroleum binders may be found in both organic and inorganic resources, and will be country or region-specific. Possible raw materials include various hydraulic-setting cements (perhaps from industrial wastes such as fly ash), limestone and gypsum, sulfur, asphalt and coal-tar derivatives, rubber and the latexes of other plants, animal proteins, clays, vegetable oils and starches. Once viable candidates have been identified, preferably in parallel with preliminary research on carbonized fillers and aggregates, a comprehensive technical program can be directed at developing composites for specifically identified industrial needs.

Because many developing areas have large, underutilized stocks of wood and agricultural wastes containing lignin and furfural (two natural binders) in very high quantities, these appear to be particularly attractive candidates for binders from renewable resources.(110) Lignin is found in all trees and the higher plants; it binds the cellulosic fibers together to form woody structure. Lignin content varies roughly from 20 to 40 percent of plant material. The furfural content of common agricultural wastes is also high, and in some wastes the combined content of the two is noteworthy: corncobs, 30.4 percent lignin and 22 percent furfural; bagasse, 20.3 and 17 percent; peanut shells, 28 and 12 percent; rice hulls, 40 and 12 percent.(111) Considerable data are available on lignin use in particleboards and other molded products, including a number of U.S. patents now in the public domain.(112) Among these appear to be some relatively uncomplicated methods for extracting and treating lignin, such as might be suitable for consideration as low-capital investments by developing areas. Since the resin cost component of pressed and molded products is so high (30 percent and up),(113) dramatic improvements could be made in local availability and costs to users and in foreign exhange if resin source and processing methods were made local and simple. For tropical countries, it would be necessary to provide integral protection against fungal and insect attack in most products. To this end, research should include the potential for utilizing natural resources known locally for such characteristics, comparing them with additives already used industrially for these purposes elsewhere.

ORGANIZATIONS: INVOLVEMENT AND CAPABILITIES

Table 5.4 through 5.6 contain a list by country of organizations which have been or are involved in research and development on building materials and other nonfood, nonfuel products from agricultural and other wastes. The list was compiled from several sources, including surveys by the Food and Agricultural Organization (FAO) and the Department of Economic and Social Affairs (ESA) of the United Nations. The principal products are used in building and construction. A wide variety of agricultural wastes are used as raw materials.

In addition to the organizations and activities listed in these tables, there are many other organizations which have been involved in or have the capability for involvement in R&D on building materials and other products from agricultural wastes. For example, the Asian Institute of Technology, Bangkok; the Technology Consultancy Center of the University of Science and Technology, Kumasi, Ghana; and the Institute for Small Industries at the University of the Philippines are but a few of the universities in developing countries in this category. In the United States, the Center for Development Technology (CDT), Washington University, collaborated in the early 1970s with the Universidad Industrial de Santander, Colombia, the Universidad Nacional Autonoma de Mexico, and with Monterrey Tech, Mexico on research in the development of composite roofing materials for tropical areas.(114) Other CDT work has included collaboration with the University of Science and Technology, Kumasi, Ghana on timber farming systems,(115) and with the Monsanto Research Corporation on tropical roofing systems.(116) Other U.S. organizations with interest in this area include Lehigh University, which sponsored a conference on New Horizons in Construction Materials,(117) and the Southwest Research Institute.

Another significant resource in this area are the appropriate technology organizations. In the United States, Volunteers in Technical Assitance (VITA) answers many requests for assistance from developing countries and produces handbooks on low-cost construction. In 1976, a VITA expert panel was convened to consider the topic of wood waste utilization. A visit to the VITA offices by one of us (Kick) indicated that the headings "agricultural wastes" and "waste utilization" account for a relatively small number of requests for information. However, many requests are received in the housing and construction area. Other organizations active in this area include Intermediate Technology Development Group (ITDG) in London, England; GRET in France; the Tropical Products Institute in England; and Cor-Tech, Ltd. in Vancouver, Canada.

Table 5.4. List of U.S. Organizations Dealing with Research on Manufacturing Building and Other Materials from Agricultural Residues and Natural Fibers

Name and Address	Raw Materials(s)	Product(s)
USDA-ARS Eastern Regional Research Center 600 East Mermaid Lane Philadelphia, PA 19118 (A,B,C)*	Hides	Acoustic tile
USDA-ARS Richard B. Russell Agricultural Research Center P. O. Box 5677 Athens, GA 30604 (B,C)	Fruit and vegetable wastes	Plywood adhesives
USDA-ARS Northern Regional Research Center 1815 N. University Street Peoria, IL 61604 (Attn: C. W. Hesseltine) (A,B)	Straw Corncobs Manure	Paper Industrial products Hardboard
USDA-ARS Naval Stores Laboratory P. O. Box 1 Olustee, FL 33880 (B)	Resins, resin acids, etc.	Naval stores
Mr. Poo Chow Dept. of Forestry College of Agriculture University of Illinois Urbana, IL 61820 (A,B)	Cornstalks, cobs, husks; peanut hulls; sunflower seed hulls; kenaf stalks; cotton stalks	Particleboard, fiberboard, hardboard
Mr. M. R. Overcash N.C. State University Dept. of Biological and Agricultural Engineering P. O. Box 5906 Raleigh, NC 27607 (A,B)	Manure	Insulation board
Dept. of Agronomic Crop Science Oregon State University Corvallis, OR 97331 (B,C)	Grass and cereal crop residues	Particleboard
Department of Forest Products Forestry Research Laboratory Oregon State University Corvallis, OR 97331 (B)	Rye grass straw	Corrugating paper
American Agricultural Industries, Inc. 9505 West Devon Avenue Rosemont, IL 60018 (Attn: Thomas J. Army, Ph.D., Vice President) (B)	Agricultural wastes generally	Fibers
Mr. Eldon C. Beagle The Beagle Corporation P. O. Box 874 West Sacramento, CA 95691 (A,B)	Rice husks ash/char	Insulation, filter media
URS Corporation 1811 Trousdale Drive Burlingame, CA 94010 (B)	Rice husk ash	Water filter
URS Research Company Environmental Systems Division 155 Bovet Road San Mateo, CA 94402 (B)	Rice husks	Cement, board, sodium silicate
The National Center for Appropriate Technology (NCAT) P. O. Box 3838 Butte, MT 59701 (G)	Cellulose, general	Insulation

*See notes at end of Table 5.6 for sources of information cited.

Table 5.5. List of Developing Country Organizations Dealing with Research on Manufacturing Building and Other Materials from Agricultural Wastes and Natural Fibers (by Country)

Name and Address	Raw Materials(s)	Product(s)
BANGLADESH		
Bangladesh Sugar Mills Corp. Shilpa Bhaban Motijheel Commercial Area Dacca, 2 (B)*	Bagasse	Paper
COLOMBIA		
Universidad del Valle Chemical Engineering Dept. Apartado Aero 2188 Cali (B)	Rice husk Bagasse	Activated charcoal
CUBA		
ICIDA Via Blanca y Carretera Central Apartado 4026 La Habana (B,C)	Bagasse	Board, industrial textiles, paper pulp, furfural, carbon
FIJI		
Ministry of Forests Principal Utilization Officer Government Buildings Suva (B,D)	Coconut Wastes	Fiber mats, activated charcoal, building materials generally
GHANA		
University of Science & Technology Kumasi (C)	Cassava, plantain wastes Banana wastes	Glue Waterproofing
Building and Road Research Unit Kumasi (C)	Rice husk ash	Bricks
HAITI		
Institut de Dévelopment Agricole et Industrial P. O. Box 1313 Port-au-Prince (B)	Whey, buttermilk, cotton linters, and seed cake	Baseballs, mattress stuffing
INDIA		
Council of Scientific and Industrial Research (CSIR) Rafi Marq New Delhi 110001 (A,B)	General	General
CSIR Regional Research Lab. Jorhat 785006 (A,B)	Rice husk Corn cobs	Bricks, cementitious material, board Furfural
CSIR Regional Research Lab. Canal Road Jammu Tawi (A,B,C)	Rice straw, pine needles, menthe grass, bagasse, general	Fiberboard, particleboard, furfural
National Chemical Lab Poona 411008	Coconut pith Sugar cane press mud	Gaskets Waxes
Indian Plywood Industries Research Institute Bangalore 560022 (A,B)	Various wastes Cane wastes	Extender for synthetic resins, resin Treated roofing mats

(continued)

Table 5.5 (continued)

Name and Address	Raw Materials(s)	Product(s)
INDIA (continued)		
Central Building Research Institute (CBRI) Roorkee, U.P. (A)	Rice husks	Mortar
	Jute sticks	Particleboard
	Coconut pith	Particleboard, expansion joint filler
	Coconut coir	Roofing sheets, boards, insulation concrete
Jadavpur University Dept. of Food Tech. and Biochemical Engineering Calcutta 700032 (B)	Rice husk	Silica, compressed board
INDONESIA		
Asian and Pacific Coconut Community P. O. Box 343 Jakarta (A,B)	Coir dust	Particleboard, insulation board, charcoal, activated carbon
Institute of Technology Bandung Dept. of Chemical Technology Jalan Ganesya 10 Bandung (A,B)	Rice hulls	Furfural, waterglass
	Bagasse	Paper pulp
IRAN		
Engineering Laboratory Institute of Standards and Industrial Research (E)	Rice hulls	Blocks
MALAYSIA		
Department of Forestry Forest Research Institute Kepong, Selangor (A,B)	Empty fruit bunches from oil palm	Paper pulp
MAURITIUS		
Sugar Industry Research Institute Reduit (B,C)	Bagasse	Particleboard
MOROCCO		
Ministry of Agriculture and Agrarian Reform Rabat (B,C)	Rice bran	Glue
	Maize germ	Paint
	Fruit and vegetable wastes	Board, glue
NEPAL		
Tribhuvan University Kathmandu (F)	Sugar mill wastes	Cement
PAKISTAN		
Pakistan CSIR Labs Off: University Road Karachi 39 (B,C)	Rice husk, jute cuttings, rice straw, wheat straw, bagasse, rubber wastes	Various boards, paper
Forest Products Research Div. Pakistan Forest Institute Peshawar (B)	Rice and wheat straw, bagasse	Chipboard, paper
PHILIPPINES		
Philippine Council for Agriculture and Resource Research (PCARR) Los Baños Laguna (A,B)	Cassava stems	Particleboard

(continued)

Table 5.5 (continued)

Name and Address	Raw Materials(s)	Product(s)
PHILIPPINES (continued)		
NACIDA 2 (P.C.A.) Los Banos, Laguna (B)	Coconut husks	Insulation board
FORPRIDECOM College, Laguna 3720 (A,B,C)	Rice straw, bagasse, banana fibers, abaca fibers/stalks, coconut husks	Fiber and particleboard, tannins, adhesives, blocks, tile, activated charcoal, paper pulp
National Grains Authority 1424 Queson Blvd. Extn. Quezon City (A)	Rice husks	Carbonized blocks
THAILAND		
Ministry of Industry Scientific Department Rama VI Road Bangkok (B)	Rice straw	Paper pulp
TRINIDAD AND TOBAGO		
Caribbean Industrial Research Institute University P.O. St. Augustine (B)	General	Building materials
URUGUAY		
Chemical Faculty Cátedras de Proyectode Fabrica y de Química Industrial y Control Analitíca General Flores 2124 Montevideo (B,C)	Rice husk, bagasse, linseed and sunflower seed husks	Building material

*See notes at end of Table 5.6 for sources of information cited.

Table 5.6. List of Non-U.S. Developed Country Organizations Dealing with Research on Manufacturing Building and Other Materials from Agricultural Wastes and Natural Fibers (by Country)

Name and Address	Raw Materials(s)	Product(s)
AUSTRALIA		
CSIRO Division of Protein Chemistry Leather Research Group 343 Royal Parade Parkville, VIC. 3052 (A,B,C)*	Tanned leather	Acoustic tile
CSIRO Division of Forest Research P. O. Box 4008 Canberra, ACT 2600 (B)	Branches, tops & leaves, chips	Boards, pulps for paper
CSIRO Agro-Industrial Research Unit P.O. Box 1666 Canberra, ACT 2601 (A)	Straw	Pulp
Southern Sugar Experiment Station Division of Mill Technology P.O. Box 651 Bundaberg, QNSLND 4670 (b)	Bagasse	Packaging
Sugar Research Institute P.O. Box 21 MacKay, Queensland 4740 (B)	Bagasse	Paper
CANADA		
Cor-Tech Research Ltd. 426 Vanguard Rd. Richmond, B.C. V6X 2P5 (A,B,C)	Rice Husk Cereal straw	Boards: composite boards, hardboards, fiberboards
DENMARK		
Bioteknisk Institute Tilknyttet Akademiet for de Tekniska Videnskaber Holbergsvej 10 6000 Kolding (A,B,C)	Cereal straw	Particleboard, paper board
FINLAND		
Rintekno Oy Consulting Engineers Kotkapolku 2 02620 Espoo 62 (B)	Bagasse, rice husks, corn cobs, almond sheels	Furfural, active carbon
The Finnish Pulp and Paper Research Institute P.O. Box 136 00101 Helsinki 10	Straw	Pulp
JAPAN		
University of Tokyo Faculty of Agriculture Dept. of Forest Products 1-1-1 Yayoi, Bunkyo-ku Tokyo (B)	Bagasse	Fiberboard, particleboard

(continued)

Table 5.6 (continued)

Name and Address	Raw Materials(s)	Product(s)
JAPAN (continued)		
International Projects, Ltd. 3rd Floor, Komiyama Bldg. 2-32 Roppongi 3-Chome Minato-ku, Tokyo (B,C)	Jute stalks	Particleboard
SOUTH AFRICA		
Huletts Sugar Limited P.O. Mount Edgecombe 4300 Natal (B)	Bagasse, sugarcane filter mud	Board, paper, furfural, wax
Hulsakane Ltd. Private Bag 3 Glenashley 4022 (A)	Bagasse	Particleboard
UNITED KINGDOM		
Tropical Products Institute 127 Clerkenwell Road London ECIR 5DB England (A,B)	Bagasse, groundnut husks, rice husks, arrowroot "bitty"	Particleboard, lightweight blocks, packaging

Sources for Tables 5.4, 5.5, 5.6:

A - Agricultural Residues: Compendium of Technologies, FAO, Rome, 1978.

B - Agricultural Residues: World Directory of Institutions, Revision 1, FAO, Rome, 1978.

C - Use of Agricultural and Industrial Wastes in Low-Cost Construction, UN (ESA), New York, 1976.

D - "The All Coconut House" FIJI 1 (September-October 1978): 2.

E - "Agricultural Waste as a Civil Engineering Material," Small Industry Journal 7 (March 1975): 3-33.

F - J.W. Powell, Appropriate Technology in India, University of Science & Technology, Kumasi, Ghana, 1978.

G - Response from NCAT to study survey.

6 Initiatives, Organization, and Mechanisms

INTRODUCTION

A primary objective of this study was to define international development initiatives for the United States in the area of appropriate, light-capital technology for renewable resource utilization. By initiatives we mean efforts that the U.S. government might support to assist and collaborate with developing country organizations and individuals in utilizing appropriate technology for renewable resources to benefit poor people in those countries. Study objectives also included identification of the potential roles of U.S. universities, research institutions, private industry, voluntary organizations and appropriate technology groups in contributing to those initiatives, as well as identification of mechanisms for coupling U.S. resources to developing country institutions in order to implement promising initiatives.

This chapter begins with summaries of the state-of-the-art reviews and analyses contained in chapters 1 through 5. The initiatives derive in part from these reviews and in part from the results of the VITA-CDT panels on the first three topical areas summarized in Appendix A. The remainder of the chapter is concerned with organizations and mechanisms for implementing the initiatives and with possible next steps to take as follow-on to this study.

SUMMARY OF STUDIES IN FIVE TOPICAL AREAS

Wind Energy Utilization

Wind energy systems have considerable potential for adoption at the village level because the performance characteristics of windmills are well suited to many of the basic energy needs in developing countries. In particular, small-scale, simply constructed and operated, light-capital, low power-output wind machines are appropriate for many low-lift water pumping applications. One of the principal needs for power in developing countries is for pumping agricultural (irrigation and stock watering) and domestic water supplies. In many areas, windmills may be able to meet these needs and may also offer the potential for cost, time, and labor savings in regions where other water lifting methods are currently used; however, many uncertainties still remain. Windmills may also be used to supply mechanical power for applications such as grain grinding or milling, cane crushing, oilseed pressing, air circulation for crop drying, refrigeration, sawing of wood, and water aeration.

Windmill designs which offer the potential for isolated, small-scale generation of electrical power are typically more complex technically than the designs suitable for water pumping and rotational power production. Wind-powered electric generators may be suitable for use in developing countries to provide domestic or public lighting, to power refrigeration systems, to power communications equipment, to supply small-scale commercial or industrial power needs, and to drive pumps to supply domestic, stock watering, or irrigation needs. However, because of the high initial costs and relatively low electrical power outputs of windmills for electricity generation, these designs will have more limited applications than mechanical power wind machines in rural areas of developing countries but may still be an important technology for special applications such as providing electricity for communication needs.

There are a number of candidate windmill designs for application in developing countries; for example: Savonius, sail wing (horizontal and vertical axis), multivane, medium-speed, high-speed propeller type, and advanced vertical axis. Savonius and vertical-axis sail designs provide low power outputs and require relatively large amounts of materials; however, they operate independently of the wind direction and are well suited to village-level construction capabilities. Multivane and horizontal-axis sail windmills are probably the most extensively tested and widely available designs for water pumping and other mechanical power applications. Both of these designs are suitable for village-level construction. Although limited field testing, design work, and hardware development have been carried out on medium-speed windmills, this basic design may have considerable potential for both water pumping and small-scale electricity production. High-speed, propeller-type wind machines are well tested, primarily in industrialized countries, and are best suited for electricity generation, but may be difficult to fabricate at the village level because of the need for high precision. Advanced vertical-axis designs may be suitable for electricity generation applications, but these designs are typically complex; viable designs for developing country applications are yet to be demonstrated.

Important factors which will affect the implementation of windmills in developing countries include: materials availability, construction skill requirements, needs for precision fabrication, initial capital costs, ease of local operation and maintenance, ability to match power and demand profiles, and wind resource availability. In addition, a windmill installation must be compatible with the local social and cultural setting in order to be permanently accepted into the village life-style. Some examples of basic windmill designs operating successfully in developing countries are presented in chapter 1. The water lifting windmill designs developed at Las Gaviotas in Colombia and by the Arusha project in Tanzania seem especially noteworthy.

Responses to a wind energy survey conducted at VITA indicate that substantial work to develop designs and to implement them remains before windmills achieve widespread use in developing countries. According to the survey, much of the necessary technical knowledge and experience is frequently unavailable to individuals or institutions attempting to install windmills in developing countries. Existing windmill designs are often too expensive and are not properly adapted to the local conditions typically found in rural villages. However, design modifications and testing efforts similar to those being carried out by the Intermediate Technology Development Group and the Arusha Appropriate Technology Project, along with more emphasis on windmill information dissemination, should speed the introduction of wind energy technologies into developing countries.

Improved Cookstoves

Wood fuel serves as the main energy source for more than one-third of the world's population that lives primarily in rural areas of the developing world. Rapid deforestation is occurring in many developing regions, partially as a direct result of firewood use for cooking, which may account for as much as 90 percent of the total energy budget in some rural areas. Traditional cooking methods over open fires utilize firewood at only 5 to 10 percent efficiency, which is about half that obtainable by improved cookstove designs. In many areas, rapid growth of village populations and of the demand for cooking fuels has resulted in the natural regeneration capacity of forests being exceeded, leading to a loss of trees, the use of animal manure and crop residues for fuel, and, in some cases, a sharp reduction in even subsistence energy consumption levels. The switch to substitute fuels often results in loss of soil fertility, increased soil erosion, and reduced crop yields which, in turn, creates pressures to further clear forested land for agricultural use.

The primary design objectives of improved cookstoves are to: (1) increase the efficiency of cooking and thereby reduce firewood use; and (2) to improve the cooking process and environment which is usually achieved by providing mechanisms for controlling the fire intensity and for smokeless operation. Many different improved designs have been developed during the last thirty years. Successful dissemination of improved cookstove designs could serve to reduce firewood consumption for cooking by rural populations by as much as 50 percent for performance of the same cooking task. This reduction in firewood demand could help slow that part of the deforestation process which is caused by traditionally inefficient firewood use for cooking. It could also reduce the often time-consuming fuel collection efforts of rural villagers and allow them to spend time on more productive activities.

An improved cookstove design, the Lorena stove, developed recently in Guatemala has potential for meeting the need for a better cooking device in rural developing areas. This design has many of the characteristics that are typically found in improved cookstoves. The major design parameters of the Lorena stove are:

- a 50 percent reduction in fuel consumption compared to open-fire cooking practices;
- construction from locally-available materials and with tools and skills common at the village level;
- inexpensive to build ($5 to $15) or to purchase from local entrepreneurs ($30 to $35);
- adaptable in both size and shape to fit the conditions of individual households;
- a large cooking surface and a constant supply of pre-heated water to reduce cooking time;
- compatibility with traditional cooking vessels and utensils;
- high thermal mass to provide a source of heat for space conditioning or firewood drying;
- large horizontal cross section to serve as a work place or kitchen table;
- usable with a variety of fuels including wood, agricultural wastes, charcoal, garbage, and sawdust.

An active dissemination program for Lorena stoves was established at the Choqui Experimental Station in Guatemala in 1976; other groups working with Lorena stoves include CEMAT in Guatemala and Save the Children Federation in Honduras. At the start of 1980, some 800-1,000 Lorena stoves were reported to exist in Guatemala. Some minor problems have been experienced with the design, such as difficulty in determining the proper sand and clay mixture, cracking of the stove structure, and build-up of tars and soot in the chimney. Early surveys indicated that nearly all Lorena cookstove users were pleased with the stove performance, particularly because of the reduced firewood consumption. However, a more recent survey indicates less positive results, with many stoves not being properly utilized. The performance of the Lorena stove design and its early dissemination successes make this stove a likely candidate design for immediate, improved-cookstove extension efforts.

In general, cookstove extension efforts may increase the chances for success by operating through or in connection with persons already involved in other development projects in the region, particularly fuel wood plantation or reforestation projects. Because cookstove designs may still be improved, for example, in durability and operational simplicity, cookstove extension efforts will benefit from establishing links between extension workers and cookstove research, design, and testing centers. Finally, strongly-funded and technically-supported dissemination efforts are now needed to speed the adoption of improved cookstove designs.

Solar Drying Technologies

Grain crop drying

Postharvest food losses are a problem which has become increasingly important to development agencies and developing country governments. Although estimates for postharvest losses vary, at least 10 percent of harvested grain is believed lost in developing countries, of which approximately a third occurs during traditional drying and storing processes. Because the moisture level of the grain is a very important factor in successful grain storage and heavily influences the grain quality, grain drying procedures need to be improved in developing countries.

Most small rural farmers understand the importance of drying grain adequately before storing and devote considerable time and effort to the drying process. In most cases, open-air drying is used in which a thin layer of grain is spread out on dry ground or on concrete or asphalt roads; thus, the grain is exposed to possible contamination by insects or animals. Introduction of new high-yield varieties of grain that permit multiple-cropping practices which require wet-season harvesting have made drying programs so important that the lack of successful drying techniques may jeopardize attempts to improve farm productivity and income. Fossil-fuel fired dryers are generally too expensive and too large for use on small farms in rural areas.

Small-scale solar cabinet dryers using direct-exposure and natural-convection principles are currently under development in India and Southeast Asia, primarily for drying paddy rice. These designs typically have capacities from 20 to 500 kg and drying times from a few hours to a few days. Because they do not have temperature control mechanisms, overheating and damage of the grain may result. As of 1977 or 1978, costs of solar dryers, if constructed by local people from locally-available materials, range from $10 to $20 for a 10-kg design to $100 to $150 for a dryer capacity of 500 kg of grain.

Designs of large-scale dryers, which often serve as storage bins, typically utilize indirect exposure and forced convection, and have capacities of a few tons up to a few hundred tons. These systems usually use the bin roof and walls as solar collectors with the heated air drawn through the grain from below a perforated floor by an electrically-powered fan, although there is some interest in using wind power to circulate the air. In contrast to fossil-fuel dryers, solar dryers use high air-flow rates, low air temperatures, and long drying times which may result in improved grain quality. Supplemental heating by electricity or fossil fuels is normally used but significant fuel savings compared to conventional large-scale dryers may be achieved. Although the capital costs of these dryers are high by developing country standards, fuel savings may offset the capital expenditures in a few years. Research and development on these designs is carried out predominantly in developed countries, particularly the United States and Canada.

Solar grain dryers are neither a technically-proven nor economically-competitive technology in developing countries and, therefore, have had very limited applications. Nevertheless, needs exist in developing countries for low cost, locally-constructed, batch solar dryers for small farm applications as well as for large-scale dryers for use by local entrepreneurs or government installations. We perceive that efforts to develop and use small-scale solar dryers are taking place in many developing countries using inexpensive, locally-available structural materials (e.g., wood, bamboo, mud), and cover materials with less expensive initial costs than glass (e.g., plastic sheeting). However, substantial design innovation and testing appear to be necessary as does a closely-linked research and extension effort.

Timber drying

Open-air drying, which is the most common timber seasoning method utilized in developing countries, is inadequate in several respects: excessive drying times, several months or more, are required; expensive capital is tied up unproductively in drying stock; and drying sequences are incapable of reducing timber moisture contents below about 20 percent. As a result, nearly all wood used in developing countries is not properly dried; exported timber is dried in the importing country; and exported wood products often exhibit unacceptable properties such as warping, cracking, or splitting. Value-added processes in both large and small-scale industries are inhibited by the lack of improved drying technologies because finished wood products for use in temperate climates must be dried to approximately 10 percent moisture content. Improper drying of wood products for indigenous uses in developing countries such as construction may also become an important problem as development efforts proceed. As pressures on indigenous forest reserves in developing countries increase, use of timber species with low shrinkage coefficients may have to be replaced with species that require more careful drying processes.

Small and medium-scale operators in developing countries are at a disadvantage compared to large-scale enterprises in obtaining improved drying technologies because steam-heated or direct-fired kilns have capacities which are an order of magnitude too large and cannot be scaled down economically. Furthermore, if efforts in developing countries to discourage whole-log exports in favor of finished products become successful, or if domestic markets become more discerning, the existence of small and medium-scale operators will be increasingly threatened.

Solar timber kilns may offer an attractive alternative for small-scale operations because moisture contents approaching 10 percent are attainable which will satisfy export criteria for finished products. In addition, although solar kiln drying times are longer than those of conventional kilns, the drying times are typically reduced by 50 percent or more compared to open-air drying. Solar kilns also offer potential for capital and operating cost reductions and require less-technical operating skills.

Extensive solar kiln design development and testing efforts have been carried out, primarily in developed countries. Natural-convection and wind-powered designs have been abandoned as impractical, and all currently known designs utilize electricity for fan power. Most designs employ integral roof and wall air heating solar collectors, although some recent designs use external collector systems. Adequate performance of solar kilns has been demonstrated by past experiments, but preferred operating procedures have not been clearly defined and probably vary widely with location, design, and timber species. Important research directions are apparently not agreed upon because, for example, U.S. efforts have concentrated on improving performance and efficiency while programs in the U.K. have focused on cost reductions.

In spite of past research and development efforts, very few commercial solar kilns appear to be in operation in developing countries. Attractive opportunities for pilot programs to introduce solar kilns in developing countries appear to exist and these implementation efforts should be designed to obtain data on solar kiln performance and drying costs under actual business conditions. Potential design improvements should be prepared and disseminated, and developing country institutions should be supported in efforts to establish drying sequences for important indigenous timber species.

Rice Bran Utilization

Rice bran is presently underutilized for human food in developing countries, despite its substantial content of quality protein and edible oil. Estimates vary, but some indicate that the quantity of protein represented by Asian rice bran is roughly equivalent to the minimum dietary requirements of nearly 200 million people and is present in rice-growing countries where nutritional problems are serious. The oil content of Asian rice bran is also significant, representing the minimum caloric requirements of about 50 million people if the above estimate is used.

The utilization of rice bran for food purposes is hindered by:

- rapid development of rancidity that begins upon separation of the bran from the kernel in the milling process;
- a fiber content that precludes routine consumption in large quantities;
- contamination with inedible hull particles, when milled by handpounding or Engleberg hullers which prevail in many rural areas; and
- lack of acceptance by villagers as a food ingredient.

Cultural, political, and social factors are important. For example, widespread association of rice bran with animal feed tends to limit consumer acceptance of bran as a food component. In some countries, rice bran is a part of an inexpensive feedstuff for animals which is important to rural people. Studies need to take such factors into account and to be approached on a country-by-country basis.

If most of the hull particles were removed by sifting so that hulls amount to only about 20 percent of bran-hull mixtures, and if effective village-scale equipment were available for stabilization against rancidity, the stabilized bran-hull mixtures could be used as feed components in village feed-mill operations. Alternatively, such bran-hull mixtures might be collected for use in regional feed mills. Direct introduction into certain foods, such as bakery products, would be impossible unless the brans were hull-free, and even then must face the aforementioned consumer acceptance obstacle. Collection and transport to regional solvent-extraction operations is likely only if the stabilized bran is essentially hull-free, but oil could be obtained at the village level from hull-free bran, and probably from bran-hull mixtures, by hydraulic pressing. Further, high-protein, low-fiber food products might be obtained from hull-free bran or even bran-hull mixtures, by protein concentration at the village level. These protein products, which are different in appearance from rice bran, might escape the association with animal feed that severely limits the direct utilization of rice bran in foods.

Relatively simple technologies for stabilization against rancidity, for extraction of oil, and concentration of protein appear to be within reach for village application. A substantial amount of work has been done in these areas. However, stabilization and hydraulic pressing studies have been oriented mainly to large-scale applications, and protein concentration work has been largely confined to basic laboratory and pilot-plant studies in developed countries. Work is now needed to develop equipment and process designs appropriate to village-scale industry in developing countries.

Organization and implementation of a program is needed which contains the following elements:

- Studies of key technical processes (sifting, stabilization, protein concentration and fiber removal, oil extraction by pressing, oil refining) that could be combined with rice milling in an integrated village industry for production of food products (milled rice, edible oil, and high protein/low-fiber products); feed (press-cake); and fuel (hulls).
- Experimental research and development activities at several levels, including exploratory laboratory work where needed, pilot-scale activities on village-type equipment, and on-site demonstration tests coordinated with village rice mill operations, plus related cost studies.
- Socio-economic studies related to the above technical development steps (product acceptance and cultural issues; benefit/cost analyses; studies of government incentives and disincentives; financing and policy options; analyses of infrastructure needs; impact studies).

Materials and Products Based on Natural Fibers, Agricultural Residues, and Timber Wastes

Renewable substances such as natural fibers, timber wastes, and agricultural residues represent a developing country resource that may be used to lessen dependence on increasingly costly, imported raw materials. Use of these materials in the building and construction trades is particularly attractive because of the disproportionate share of foreign exchange that is allotted to acquisition of building and construction materials in many developing countries. In rural areas, national and individual cost constraints are likely to require that building and other structural products be based primarily on natural or modified natural materials. On the other hand, in urban areas, traditional natural organic materials are no longer plentiful or inexpensive. These materials often cannot meet stringent urban materials performance requirements (e.g., fire resistance) and are often incompatible with present urban construction methods and skills. The use of agricultural wastes and natural fibers in the building materials industry has been shown to provide an important multiplier effect on both employment and income and, hence, may play a dynamic role in development programs. These opportunities are particularly important because many current national or regional agricultural policies in developing countries recognize the need to improve employment prospects in rural areas, especially during off-season periods, and the need to improve the balance of trade through agricultural exports or the reduction of imports.

A review of available information indicates that less than 20 percent of the current agricultural residue technologies deal with nonfuel, nonfeed/food processes, and a very small fraction of these have been tested and/or utilized on a commercial scale. The lack of information on commercial uses of agricultural residues and organic materials for nonfuel, nonfeed/food uses may not reflect the relative importance of this activity nor the extent to which utilization of these materials occurs at the grass roots level. Even though successful, large-scale commercial production from agricultural wastes and natural fibers appears to have been limited to bagasse, flax, and linseed residues and to European reeds, a number of promising small-scale technologies may be candidates for adoption in developing countries. Rice hulls have been used to produce particleboard fillers for cement blocks, lightweight concrete aggregates, and fillers for plastics in laboratory tests; we have no information on commercial processes. Both natural and carbonized rice and coconut husks have also been demonstrated to be good water filters. A process for producing cement from rice hulls has been patented and developed in the United States to the point of designing and building a commercial-scale processing plant.

Coconut husks and pith, which are traditionally considered as waste products, may be utilized to fabricate particleboard with better rot and insect resistance than wood. Coconut pith has good thermal insulation properties and may be used as an expansion joint filler or as a filler for rubber to yield a high-quality gasket material. However, no commercial-scale processes appear to have been developed for these products. Sugarcane bagasse may be used in small-scale processes to produce particleboard and an insulating wool. Candidates for natural resin adhesives and extenders from agricultural and timber wastes have been identified, but commercial-scale processes for few of these products have been demonstrated. One notable exception is a paper glue produced from cassava starch and plantain which supplies most of the demand for glue in Ghana.

Recommended research and development initiatives to increase the use of local renewable resources for building materials and other products in rural and urban areas include work on: improved preservation techniques; breeding of plants to achieve improved physical properties; composite materials which utilize carbonized plant materials as lightweight, low-cost fillers; and nonpetroleum-based binders from agricultural and timber waste products. The first two of these initiatives are focused primarily toward improving properties of organic materials for rural applications without significantly altering the natural form, whereas the latter two are directed at developing new materials for specific urban uses.

INITIATIVES BASED ON FIVE STUDY AREAS

Tables 6.1 through 6.5 summarize the initiatives we have identified in five technology or product areas selected for study, namely, (1) wind energy for agricultural and other uses, (2) direct biomass utilization for cooking, (3) direct solar energy use for drying, (4) food and feed products from rice bran and other agricultural by-products and wastes, and (5) nonfood/fuel products and materials from natural fibers and agricultural and timber by-products and wastes. These technology or product areas are not meant to be all inclusive. They were chosen jointly by CDT and VITA according to criteria which included potential for helping to meet individual and/or community needs in developing countries at reasonable cost. The initiatives themselves were defined by our CDT study team, using as inputs feedback from the VITA-CDT panels in the first three study areas and some results of the state-of-the-art reviews and analyses in chapters 1 through 5. The choice of technology study areas is discussed in more detail in the introduction.

Table 6.1. Wind Energy Initiatives

1. Support national and international efforts to collect, evaluate, and publish data on village-level energy needs and on wind energy resources.

   Investigate present village-level energy needs and patterns; identify tasks which may be accomplished with wind energy technologies; establish and/or support wind energy monitoring programs; assist in the compilation, interpretation, and publication of existing and new wind resource data useful for project planning and definition.

2. Support information dissemination programs and extension efforts in developing countries for wind energy technologies.

   Support efforts to consolidate, validate, and revise wind machine information for use in implementation programs; assist in the preparation of wind energy system design handbooks that address different levels of technical expertise; initiate demonstration projects to display wind energy technologies; contribute to the organization of in-country wind energy workshops to disseminate hands-on design information.

3. Support research in and among developing countries and collaborative research efforts by U.S. and developing country institutions to improve and adapt wind machine designs for applications in developing countries.

   Expand research efforts on light-capital wind machine designs, especially designs suitable for low wind-speed applications; investigate the potential for developing low-maintenance wind machines; establish wind energy information clearinghouses in developing countries to process and evaluate the results of wind energy projects to identify needed design improvements; investigate techniques to reduce the non-local materials requirements of wind energy technologies; strengthen communication channels between wind energy researchers in the United States and overseas to facilitate exchange of innovative design information and research results.

(continued)

Table 6.1 (continued)

4. Support developing country efforts and cooperative efforts by U.S. and developing country institutions for design testing and performance evaluations of existing and proposed wind machine designs.

   Support the establishment of wind machine testing centers in developing countries; develop standardized wind machine testing procedures and descriptions of test results; identify criteria useful in design selection based on performance data obtainable in developing countries; create personnel exchange programs between U.S. and developing country institutions to facilitate exchange of performance information.

5. Support developing country efforts and cooperative programs between U.S. and developing country institutions to evaluate, assess, and publicize the utilization of wind energy technologies in developing countries.

   Contribute to programs to analyze the environmental, social, and cultural impacts of successful wind machine implementation efforts in developing countries; support efforts to document successful and unsuccessful wind machine installations in developing countries; identify the long-range potential for developing country applications of wind energy technologies; support the establishment of international communication networks for appropriate wind energy activities.

Table 6.2. Improved Cookstove Initiatives

1. Support national and international cooperative efforts to develop effective education and information dissemination programs for improved cookstove designs.

   Create methods of consolidating, validating, and revising cookstove information for extension efforts; ensure proper assessments of variations in regional needs by soliciting direct inputs from village women; train local people to build and operate stoves; support cookstove implementation efforts with monetary, material, and personnel resources.

(continued)

Table 6.2 (continued)

2. Support the development of close cooperation between current cookstove extension programs in developing countries and cookstove research efforts in the United States and developing countries.

    Establish appropriate technology clearinghouses in developing countries to process and evaluate information from local extension efforts to identify needed design improvements; organize communication networks between extension programs and research institutions; create personnel exchange programs between U.S. and developing country institutions to facilitate information transfer.

3. Support forest management expertise exchange programs between U.S. and developing country institutions focused on identification and implementation of actions compatible with deployment of improved cookstove designs.

    Investigate the potential for introducing rapidly growing tree species near villages for firewood use; evaluate and document fuel characteristics of indigenous trees; assess the impacts of conversion of wood to charcoal for cookstove use; analyze the impact of improved cookstoves on local deforestation issues.

4. Support cooperative study programs in U.S. and developing country institutions to identify and assess long-range impacts of successful implementation of improved cookstove designs.

    Analyze environmental, social, and cultural impacts of improved cookstove acceptance; evaluate the impacts of smokeless cookstoves on health issues related to, for example, respiratory and pest control factors; estimate the effects of improved cookstoves on the lifestyles of village women.

5. Support the expansion of centers in developing countries equipped to conduct research on and performance testing of current and emerging cookstove designs.

    Develop research capacities in appropriate institutions in developing countries; analyze the potential of improved cookstove designs from durability, safety, and ease of operation viewpoints; establish standardized testing and reporting procedures for cookstoves.

(continued)

Table 6.2 (continued)

6. Support efforts in developing countries to define and initiate education programs to complement successful improved cookstove implementation projects.

   Develop educational materials highlighting the health benefits of thorough cleaning of utensils and work areas with hot water supplies from improved cookstove designs; generate information packets outlining the potential for improved nutritional plans based on the effective use of improved cookstoves; document the importance of the labor-saving and cost-saving potential of improved cookstove designs.

Table 6.3. Solar Drying Initiatives

1. Support national and international cooperative efforts to collect, evaluate, and publish data on village-level energy needs and on solar energy resources.

   Investigate present village-level energy needs and patterns; identify tasks which are constrained by the lack of energy supplies; establish and/or support solar insolation monitoring programs; provide access to and interpretation of satellite climatic data.

2. Support the establishment and operation of energy information centers in developing countries to facilitate dissemination of appropriate energy-related research results and design developments.

   Strengthen communication channels between solar energy researchers in the United States and overseas; organize communication networks between implementation programs and research institutions; establish appropriate technology clearinghouses in developing countries to process and evaluate information from local extension efforts; support the development of regional energy workshops and personnel exchanges.

3. Support research in and among developing countries and collaborative research efforts by U.S. and developing country institutions to develop improved solar drying designs for applications in developing countries.

(continued)

Table 6.3 (continued)

Expand research efforts on solar drying technologies; support the development and testing of small-scale, natural-convection crop dryers through small grants programs; enlarge solar kiln research and development programs to include work on advanced design concepts; initiate testing and demonstration programs in developing countries to evaluate unproven dryer designs.

4. Support research efforts in developing countries to contribute to the successful deployment of solar dryer technologies.

   Investigate techniques to reduce the cost and nonlocal materials requirements of solar dryers; evaluate the thermal properties of locally-available materials that might be candidates for energy absorption surfaces or insulation; document the performance of simple dryer designs constructed primarily from local materials; evaluate the performance and durability of cover materials under field conditions.

5. Support the establishment of extension programs and rural education, innovation, and training centers in developing countries to assist in the implementation and commercialization of regionally-proven solar drying technologies.

   Initiate demonstration projects to display solar drying technologies; develop user-oriented educational aids and manuals; organize local workshops on dryer construction techniques and operating procedures; and contribute to programs to analyze the impacts of successful solar dryer implementation efforts on farm revenues, postharvest losses, and nutrition.

Table 6.4. Food and Feed Products from Rice Bran and Other Agricultural By-Products and Wastes: Initiatives

1. Support national and international collaborative efforts to assess the availability of agricultural by-products and wastes and the potential alternative uses to which these materials might be put.

(continued)

Table 6.4 (continued)

This effort should include planning, policy-study, and evaluation activity involving economic, social, political, cultural, and environmental as well as technological factors.

2. Support the establishment of centers in developing countries to facilitate the collection, evaluation, and dissemination of information on utilization of agricultural by-products and wastes.

   Some centers may already be in existence or it may be possible to develop such activities at existing agricultural research, development, and extension centers. Ideally, information dissemination, R&D, and extension should be coordinated activities.

3. Support national and collaborative international research and development efforts to develop and utilize appropriate technology for village-level rice bran processing in developing countries.

   Stress retention of protein and caloric values in the villages to improve village-level nutrition. Stress employment retention and generation in villages. Needed are the development of small-scale rice bran stabilization equipment and the development and testing for acceptance of food and feed products of improved nutritional value.

4. Support the development of village-level agri-industries based on agricultural by-products and wastes.

   Needed are research, development, and demonstration efforts in and among developing countries as well as parallel cooperative efforts between developed and developing countries. Stress integration of food processing, food by-product and agricultural waste utilization, and energy generation. Elements include the development of equipment of suitable scale, economy, and efficiency to be viable at the village level.

5. Support efforts to give more attention and prestige to R&D in this area.

   Use mechanisms such as greater involvement of developing and developed country universities working closely with village entrepreneurs and organizations in education, development, demonstration and extension efforts. Support and/or create new rural engineering/innovation centers.

Table 6.5. Nonfood/Fuel Products and Materials From Agricultural and Timber Wastes and Natural Fibers: Initiatives

1. Support national and international collaborative efforts to assess the availability of natural fibers and agricultural and timber by-products and the potential alternative uses to which these materials might be put.

   This effort should include planning, policy-study, and evaluation activity involving economic, social, political, cultural, and environmental as well as technological factors.

2. Support the establishment of centers in developing countries to facilitate the collection, evaluation, and dissemination of information on utilization of natural fibers and agricultural and timber by-products and wastes.

   Some centers may already be in existence or it may be possible to develop such activities at existing agricultural research, development, and extension centers. Ideally, information dissemination, R&D, and extension should be coordinated activities.

3. Support research, development, and demonstration efforts in and among developing countries as well as parallel cooperative efforts between developed and developing countries to produce and gain acceptance for building materials, components and techniques, and other basic nonfood products from agricultural and timber wastes and natural fibers. Key elements of such efforts include:

   a. Development of composite materials which utilize carbonized plant materials and other natural fibers and agricultural and timber by-products and wastes as lightweight, low-cost fillers, together with the development of nonpetroleum-based binders from wood and agricultural products for use in composite materials.

   b. Research and development directed at improved preservation technologies and treatments for organic materials, appropriate for use at rural village scale, so as to improve their physical properties: e.g., fire-retardation for thatch roofs; technologies for combining fire retardation with other preservation requirements.

   c. Research and development directed at the breeding of indigenous plant materials to

(continued)

Table 6.5 (continued)

improve performance for specific end-uses in shelter and other industries. This will require the support of international co-operative efforts to collect information and data on: deficiencies and potential improvements in rural village housing and other materials with respect to durability, strength, performance in fire, and other health and safety-related factors; and the availability, on a regional, geoclimatic basis, of plant materials having superior characteristics for building and other uses, and of indigenous and exotic species with such characteristics that make them candidates for breeding for improved performance.

d. Development of village-level chemical extraction industries based on renewable resources.

4. Support efforts to give more attention and prestige to R&D in this area.

Use mechanisms such as greater involvement of developing and developed country universities working closely with village entrepreneurs and organizations in education, development, demonstration, and extension efforts. Support and/or create new rural engineering/innovation centers.

An examination of these initiatives reveals several common elements. In general, there appears to be a need to:

- collect, evaluate, and publish data on village-level needs and local resources to help meet those needs;
- provide information dissemination programs on appropriate technologies;
- support extension efforts to bring these technologies into productive use. New efforts which differ markedly from "extension" as practiced in the United States may be required;
- perform research within and among developing countries and collaborative research between developing countries and U.S. institutions in order to improve and adapt designs;

- carry out this work to test designs and evaluate performance of the technologies in question; and
- cooperate internationally in the evaluation, assessment, and publicizing of efforts to utilize specific technologies in developing countries.

The degree to which each of these activities is to be emphasized depends upon the technology or product in question. For example, in direct biomass utilization for cooking, the literature and the VITA-CDT Panel Report (see Appendix A) indicate that some improved cookstove designs are already available; therefore, extension efforts to disseminate those designs should receive high priority. However, in the solar crop drying area, few if any appropriate, low-cost designs appear to exist that have been or can be widely utilized; thus, considerable applied research and development work might be needed before extension efforts could go forward.

Table 6.6 provides a cross-matrix of initiatives and problem areas with an indication of which initiatives should receive primary, secondary, or tertiary emphasis. This table should be used as a very rough indication of priorities. Furthermore, the nature of these efforts may be such that the various initiatives in a given topical area are closely coupled and might best be implemented by existing or new centers or other organizations in developing countries which deal with many or all of the initiatives simultaneously. More analysis, including field surveys, is required to define these priorities with greater certainty.

The initiatives share other characteristics in common. All of them, to some degree, address real needs in developing countries. Table 6.7 summarizes the applications and needs for which the technologies are to be utilized. In some areas, for example, cookstoves, the users may be individuals whereas in others, such as windmills, a village or community effort may be required because of cost. Both of these scales are deemed appropriate in this study.

It is our feeling that there is potential in each of the areas for bringing about improvements in living conditions among the poor in developing countries if sufficient resources and effort are expended. However, at this stage, we state this more as a hypothesis to be tested, in view of the complexity and importance of social, cultural, and economic factors and of government policies, than as a well-supported conclusion. There is also, for each technology or project area, considerable uncertainty about the cost and reliability of the technologies. Furthermore, an economic comparison of the technologies which were selected for study with other technologies was beyond the scope of this study and remains to be performed.

Table 6.6. Relative Emphasis for Initiatives Proposed in this Study

Initiatives / Technology or Problem Area	Collect, Evaluate, Publish Data on Village-Level Needs and Local Resources	Provide Information Dissemination Programs in Appropriate Technologies	Support Extension Efforts in Connection with Technologies	Perform Research to Improve and Adapt Designs	Carry Out Testing and Performance Evaluation of Designs	Evaluate, Assess, and Publicize Efforts to Utilize Technologies
Wind Energy for Agricultural Uses	Secondary	Secondary	Secondary	Primary	Primary	Secondary
Direct Biomass Utilization for Cooking	Tertiary	Secondary	Primary	Secondary	Primary	Primary
Direct Solar Energy use for Grain Drying*	Secondary	Tertiary	Secondary	Primary	Primary	Secondary
Utilization of Rice Bran for Food and Feed	Primary	Tertiary	Tertiary	Primary	Secondary	Secondary
Utilization of Agricultural and Timber Wastes for Nonfood/Fuel Uses**	Secondary	Secondary	Secondary	Primary	Primary	Primary

*Different rankings might apply for timber, and fruit and vegetable drying.
**Relative rankings in this technology area are difficult because of the breadth of material covered.

Table 6.7. Needs and Applications for the Technology and Product Areas of this Study

Technology or Product Area	Primary Application	Need Being Met
Wind Energy	Water Lifting	Agriculture, drinking water, watering animals
	Electricity	Household, small industry use
	Rotary Power	Agriculture, small industry
Improved Cookstoves	Cooking Water Heating	Nutrition Health Extend energy resources Reduce time to collect firewood
Direct Solar Energy	Grain Drying Timber Drying Vegetable Drying	Improve agricultural yields Improve income Agriculture Small industry Improved construction materials
Rice Bran Utilization	Food Ingredients Edible Oil	Improve human nutrition
	Animal Feed	Improve animal nutrition Village industry
Agricultural and Timber By-Products and Wastes, Natural Fibers	Improved materials for housing, grain storage, schools, other structures Chemical products	Improve shelter Village industry Waste disposal

Each of the areas provides significant technological challenges. For example, in the area of utilization of agricultural and timber wastes and by-products, there is no reason why some of the techniques and knowledge that were developed in the United States in connection with high performance composite materials for defense and space applications could not be applied to the kinds of materials which affect the lives of poor people in developing countries in a more direct way. It is our impression that there is relatively little activity going on now to support such initiatives. In the area of rice bran utilization, village-level processing could yield considerable benefits to the villager in terms of better nutrition, employment, and income if the technology and organizational structure were scaled to meet his or her needs. The area of greater rice bran utilization has received relatively little attention, although more so of late, and is one which appears to be of considerable importance in Asia and elsewhere.(1)

OTHER INITIATIVES

Village-Level Agri-Industries Based on Renewable Resources

The study of rice bran utilization (chapter 4) and of nonfuel, nonfeed/food products from renewable resources (chapter 5) suggests an initiative which represents an integrated approach to village-level agri-industries based on renewable resources. The primary objectives of such an initiative would be to apply science and technology to:

- generate and retain employment and income in the villages;
- minimize costs and increase the availability and quality of basic products for the village consumer, with emphasis on food, feed, energy, and building materials from food processing by-products and agricultural wastes;
- develop more efficient village-level agri-industries through integration of food processing, food by-product and agricultural waste utilization, and energy generation;
- reduce primary food processing losses; and
- explore the development of village-level chemical extraction industries, based on renewable resources, to supply products to urban centers.

We envision the program as having several elements:

- cooperative research, development, and demonstration between a small number of institutions in developing countries and in the United States;

- field testing of equipment and concepts within the developing countries with back-up technical support in the United States;
- cooperative planning, policy-study, and evaluation activities between U.S. and developing country institutions to generate concepts and to provide economic, social, cultural, and environmental assessments;
- educational involvement of both U.S. and developing country students through thesis projects and courses focused on village-level renewable resource utilization at U.S. and at developing country institutions.

Among the village industries to be considered would be those involving rice, oilseeds, and other agricultural products. Village industries might include, for example:

- protein for food and/or feed from rice bran, other cereal brans, oilseeds (cottonseeds, groundnuts), certain leaves and algae, and from rice straw and other straws;
- chemicals from rice hulls (e.g., sodium silicate, furfural, and activated carbon);
- building materials from rice hulls (e.g., hull-resin particleboard, cement-hull insulation, cement-hull blocks for nonload-bearing walls, and insulating bricks made from hull ash and soil);
- low-Btu fuel gas from agricultural wastes and food processing by-products via pyrolysis;
- edible vegetable oils from oilseeds and rice bran by extraction followed by alkaline refining of the crude oil.

In this effort, light-capital, labor-intensive approaches would be sought and identified. For example, such an approach is known for village-level production of sodium silicate from rice hulls, utilizing simple batch operations typical of early chemical process industries in developed countries. If sodium silicate can be made successfully from rice hulls in a village-level industry, and this may have already been demonstrated in India, it seems likely that other chemical and building material production processes suited for village-level activities can be identified and developed.

Full consideration of village renewable-resource industry needs might include development and/or adaptation of key items of process equipment, such as:

- multipurpose dryer-combustors for agricultural wastes and food processing by-products;
- low-pressure steam generators for paddy parboiling and rice bran stabilization;
- rice milling equipment that will not contaminate the bran with hull fragments and that improves upon the single rubber-roll huller;

- rice bran stabilization equipment, with emphasis on approaches that seem to be suitable for village application (e.g., simple extruder-cookers, thin stationary beds, and rotating drums);
- multipurpose pyrolysis units and digesters for generation of low-Btu fuel gas from agricultural wastes;
- oil-extraction equipment including crushing devices to support revival of village oilseed extraction industries, and hydraulic pressing equipment to stimulate new industries for extraction of oil from rice bran in rice-growing, oil-short regions;
- multipurpose equipment for refining oils from oilseeds and rice bran by alkaline treatment to establish edible oil industries in villages;
- multipurpose equipment for protein extraction from cereal brans;
- fermenters for animal feed production;
- choppers and composters for conversion of wastes to fertilizers and soil amendments;
- compressing/briquetting equipment for agricultural wastes;
- equipment for pelletized feed production including drying, chopping/grinding, mixing, and pelletizing components;
- evaporating equipment for sugar processing;
- sugarcane presses.

Information is available from diverse sources on most of these individual technologies, but information on concepts for village industry integration is often lacking. The following integrative schemes are examples of potentially valuable study areas:

- integrated village rice industries which would incorporate, for example, parboiling, milling, and bran stabilization and processing to produce crude oil; edible oil production by alkaline refining; feed production based on press-cake and other feed components; protein extraction from bran; building materials and chemicals from rice hulls;
- integrated village farming/milling/manufacturing industries which would incorporate, for example, paddy, sugarcane, and fish farming in rotation; rice milling and sugarcane processing; chickens, pigs, and fish based on millfeed; slaughtering and fish processing; bran oil extraction and refining; brick manufacturing using black ash from rice hulls.

Renewable Energy Resource Centers

Chapters 1, 2, and 3 examine a small number of topics within the broader field of renewable energy resources. It is not likely that these topics would be worked on in isolation. More likely, they would form a part of a larger program centered on either the "appropriate technology" or the "renewable energy resources" theme. Table 6.8 indicates the range of technologies and applications which might fall within the rubric of a renewable energy resource center. Table 6.9 contains a list by developing country of solar energy R&D activity underway as of 1977.

One initiative we feel is worthy of consideration is the establishment and/or support of renewable energy resource centers in developing areas which could become involved in the full range of initiatives illustrated for wind energy, improved cookstoves, and solar drying (see tables 6.1, 6.2, and 6.3). Thus, the centers would have capability for needs surveys, R&D, design and testing, extension and information dissemination over a broad spectrum of renewable energy technologies for rural areas. They would work with and involve rural communities and individuals. Some organizations in developing countries may be doing some or all of these activities for a limited number of topics. The Asian Institute of Technology has established a Renewable Energy Resources Information Center. FUNDAEC in Colombia is pioneering in new forms of rural engineering and education for development. Appropriate technology organizations in several countries have the potential for serving as renewable energy centers. However, a larger commitment of resources would be required than is now available. We also stress the need to integrate fully the R&D activity with the outreach aspects so that they reinforce and build upon each other.

Other Areas for Future Study

There were several areas which we were not able to examine due to time and resource limitations which we believe are worthy of future study. Although we were not able to examine these areas in detail, it seems likely to us that such an examination would identify possible initiatives in the following areas:

- Biogas generation from wastes is the area we most frequently received information on in response to our requests for information from developing country institutions. This area is potentially very important because an energy source, methane gas, can be produced without destroying the fertilizer values of the dung and other wastes. Between 60,000 and 70,000 units have reportedly been built in India and much larger numbers of smaller units are apparently widespread in China. Problems to be overcome are technical, economic, and sociocultural. In India, according to Hoda, "It [biogas generation] has

Table 6.8. Applications of Renewable Energy Resource Technologies

Renewable technology	Water pumping	Lighting	Cooling	Communi-cations	Water desalting	Spinning	Sawing	Heating: Cooking	Heating: Space	Heating: Domestic water	Grinding	Drying	Trans-port	Fertil-izing
Solar cells	+	+	+	++	-	+	-	-	-	-	-	-	-	-
Flat-plate solar collectors[a]	+	-	+	-	++	-	-	+	++	++	-	++	-	-
Concentrating solar collectors	+	+	+	-	+	-	-	+	-	-	-	-	-	-
Solar, Stirling cycles (small scale)	+	+	+	+	-	+	+	-	-	-	+	-	-	-
Solar, Rankine cycles	+	+	+	-	-	+	+	-	-	-	+	-	-	-
Wind (mechanical)	++	-	-	-	-	+	+	-	-	-	++	-	-	-
Wind (electrical)	++	++	++	++	-	+	+	-	-	-	++	-	-	-
Water (mechanical)	++	-	-	-	-	++	++	-	-	-	++	-	-	-
Hydroelectric	++	++	++	++	-	++	++	-	-	-	++	-	-	-
Bioconversion wood/charcoal	-	+	-	-	-	-	-	++	++	++	-	++	-	+
Biogas	+	++	+	-	-	-	-	++	-	+	-	+	+	++
Draft animals[b]	++	-	-	-	-	-	-	+	-	+	++	-	++	++

Symbols: ++, applicable; +, potentially applicable; -, not applicable.

[a]Solar stills are included in this category.

[b]Draft animals are included as a renewable technology based on photosynthesis. For many villages in developing countries, the use of animals would represent a modernizing step. Use of dung for burning and fertilizer is included in this category.

Source: Tanzania National Scientific Research Council, Workshop on Solar Energy for the Villages in Tanzania, Dar es Salaam, Tanzania, 1978, p. 27.

Table 6.9. Solar Energy R&D in Selected Developing Countries in 1977

Country	Solar heating	Solar cooling of buildings	Crop drying	Water pumping	Solar electricity; thermal	Solar electricity; photovoltaic	Wind energy	Biological energy
Argentina	x		x			x	x	
Barbados			x				x	
Bolivia			x			x	x	
Brazil	x		x			x		
Chad		x			x		x	
Chile	x				x			
China	x		x					
Costa Rica	x		x				x	
Cuba		x					x	
Ecuador	x							
Egypt	x			x	x		x	
El Salvador								
Guatemala	x		x					
India	x			x	x	x	x	
Iran	x		x		x	x		
Iraq	x							
Israel	x	x		x	x	x		x
Jamaica	x		x					
Jordan	x							
Kuwait	x				x	x		
Malawi	x		x					
Malaysia		x	x		x	x		x
Mali		x			x			
Mauritius		x						
Niger			x	x	x			
Nigeria			x					
Oman					x		x	
Pakistan	x	x	x			x		x
Papua New Guinea			x	x				
Peru	x		x					
Philippines	x		x				x	
Qatar					x			
Saudi Arabia	x		x	x	x	x		
Senegal	x				x			
Singapore					x			x
Sri Lanka	x		x	x			x	
Sudan			x					
Thailand			x	x			x	
Total number of countries engaged in R&D activities	22	7	21	8	15	10	12	4
Percentage of countries engaged in R&D activities	58	18	55	21	39	26	32	11

Source: United Nations, "Research in Non-Conventional Sources of Energy: Report to the Secretary-General," E/C.8/56, New York, January 1978, Annex.

not yet been designed in such a way as could be within the reach of poor men."(2)

- Direct solar energy applications other than drying. Table 6.10 contains a list of rural energy needs prepared by the Working Group on Energy for Rural Requirements at the 1978 UNIDO Forum on Appropriate Industrial Technology. Priorities identified by the Working Group for direct solar energy using flat-plate collectors include hot water for process heat and domestic uses, cold storage for agricultural products, and mechanical shaft power from solar engines, particularly for water pumping. The latter two topics are solar energy technologies that have not yet been demonstrated to be technically feasible or economically attractive for most applications in rural areas of developing countries.

  Another priority area is the solar cooker. Much work has been done on solar cookers in the past but utilization is not widespread. The Appropriate Technology Development Association in India appears to be embarking on a renewed effort to develop and utilize solar cookers.

  Finally, the Working Group recommended that consideration should be given to the possible use of solar generated electricity (either mechanical or photovoltaic) for village electrification and possibly to produce mechanical power. These applications would require programs to fabricate low-cost photovoltaic cells.

- Low-head hydroelectric power technologies are apparently being produced commercially in several European countries and China.(3) Over 1,000 units of less than 100 kw capacity have reportedly been constructed and installed in rural areas in China.(4) Low-head, low-power hydropower, which may be used to perform mechanical work directly or used to drive a generator to produce electricity, is a mature energy technology that has recently received renewed interest in many parts of the world. A number of hydraulic rotating prime movers have been developed and field tested, many of which may be suitable for applications in developing countries. Mid-1970s costs of commercial hydroelectric units range from about \$1,500/kw_e to \$7,500/kw_e, across a capacity range of 20 to 0.5 kw. Widespread implementation of this technology will probably require research and development efforts to design simpler and cheaper equipment which may be constructed from locally-available materials and labor, as well as active extension programs.
- Liquid fuels from biomass, which are produced by anaerobic processes, may be important future sources of energy in many developed and developing countries.

Table 6.10. Rural Energy Needs

Sector or task	Needs
Domestic	Lighting Cooking Hot Water Heating
Water Pumping	Drinking and Domestic Cattle Watering Irrigation
Agriculture	Land Preparation Planting Cultivation Harvesting
Agriculture Processes	Grinding Drying Storage Refrigeration
Rural Industry	Pottery Spinning and Weaving Furniture Metal Work
Transport	Land Water
Social and Community	Health Services Drinking Water Treatment Education Communication

Source: Draft Report of Working Group XI, "Appropriate Technology for Energy Rural Requirements," UNIDO Forum on Appropriate Industrial Technology, New Delhi, India, November 1978.

Many commercial-scale processes were developed and implemented in developed countries during World War II; however, reducing the scale of these processes to village-level needs in developing countries may present a number of difficult technological challenges.(5) A variety of raw materials, which are commonly available in developing countries, such as acetone, ethanol, and butanol, may be used to produce fuels in fermentation processes. However, the most easily fermentable substrates (e.g., molasses, sugar, starches) are likely to be less abundant and more important to the economy of developing countries than less attractive, from a technological point of view, waste products (e.g., wood, hulls, citrus wastes).(6) A notable example of a program to produce liquid fuels from biomass is the multi-million dollar, long-term commitment to ethanol production from sugarcane and manioc in Brazil.(7)

Oilseeds that can grow on marginal lands unsuitable for growing food crops are a relatively unexplored possibility which might prove a valuable source of energy in countries that have firewood shortages or are petroleum importers. Although some attention has been given to a limited number of hydrocarbon bearing plants, high-yielding oilseeds of astonishing variety and with a diversity of growth characteristics have been neglected until now. Sunflower farmers in North Dakota have begun to experiment with sunflower oil as a substitute for diesel fuel in tractors. The Center for Development Technology at Washington University is currently exploring possibilities for cooperative international R&D in the broad area of fuels and chemicals from oilseeds.

ORGANIZATIONS

Introduction

For each of the technology or project areas, there are a variety of U.S. and developing country organizations that could contribute to cooperative efforts in appropriate technology for renewable resource utilization. Tables 6.11 and 6.12 list some of these organizations which have been active in the designated subject areas. An expanded list of organizations is included in Appendix B.

The organizations listed in these tables were selected from those we received information from or that we felt were well known and active in the specific topic. They represent organizations which, based upon the information available to us, appear to be very active and prominent in the specified renewable resource activity. Some but not all of the U.S. organizations have had international involvements.

The lists in this section and in Appendix B are by no means all inclusive. There are undoubtedly other organizations which should have been included. We view these lists as a first effort, to be further improved and refined; we welcome additions and corrections. We have generally not included organizations which could make contributions but which have had no prior involvement with the subject area; however, in some cases, new involvements of such organizations is desirable.

U.S. Organizations: Analysis

The organizations in table 6.11 may be classified as follows.

Universities

The university is a logical focal point for cooperative international R&D efforts that are called for in this study. Universities are in evidence in all of the subject areas in table 6.11; however, there is a much larger number of universities that could develop capability and participate. Many have had extensive international involvement(8) and have resources and personnel to bring to such efforts. However, as is the case with most of the U.S. organizations in this section, their participation generally requires external financial support.

Appropriate technology (A.T.) organizations

These organizations participate in a wide spectrum of activity. For example, they include Volunteers in Technical Assistance (VITA) which stresses information dissemination, and the New Alchemy Institute which is heavily involved in science and technology applications and development. Some of the potential contributions of U.S. A.T. organizations have been summarized in a report edited by Smith.(9) Appropriate technology organizations are probably more in touch with the kinds of problems at the village and grass roots level that need to be faced than some other institutions. Some U.S.-based A.T. organizations have a strong orientation toward international activity but many focus predominantly on local domestic problems.

Table 6.11. Partial List of U.S Organizations Involved in Appropriate Technology for Renewable Resource Utilization

Wind Energy	Improved Cookstoves	Solar Drying	Rice Bran Processing*	Non-Food Products from Agricultural Wastes
American Wind Energy Assn.	A.T. International	Center for Rural Affairs (Small Farm Energy Project, NE)	Colorado State University (Dept. of Agricultural Eng'g)	Beagle Corporation (CA)
Appropriate Technology Group (Oskaloosa, KS)	Aprovecho Institute	Georgia Institute of Technology (Eng'g Experiment Station)	Denver Research Institute	North Carolina State University (Dept. of Biol. and Agri. Eng'g)
New Alchemy Institute	Arca Foundation	Iowa State University (Agri. and Home Economics Station)	Koehring Farm Division, Brady Operations (IA)	Oregon State University (Dept. of Agronomic Crop. Sci.)
Oklahoma State University	Community Services Assn.	Michigan State University (Agri. Engineering Dept.)	Meals for Millions Foundation	University of Illinois (Dept. of Forestry)
Princeton University (Flight Concepts Lab)	Georgia Institute of Technology (Eng'g Experiment Station)	South Dakota University (Dept. of Agri. Eng'g)	Ralston Purina Company (MO)	URS Research Co. (CA)
Rockwell International Wind Systems Program	Princeton University (Center for Environmental Studies)	U.S. Department of Agriculture (Agri. Research Service, TX) (Forest Products Lab, WI)	Riviana Foods (TX)	U.S. Department of Agriculture (Eastern Regional Research Center: PA) (Richard B. Russell Research Center: GA)
Univ. of Massachusetts (Dept. of Civil Eng'g)	Save the Children Federation	University of Hawaii (Hawaii Agri. Expt. Sta.) (Hawaii Natural Energy Inst.)	Texas A&M University (Food and Protein R&D Center)	
Volunteers in Technical Assistance (VITA)	Volunteers in Asia (VIA)	University of Maryland (Agri. Experiment Station)	U.S. Department of Agriculture (Western Regional Research Center: CA) (Southern Regional Research Center: LA)	
Windworks (WI)	Volunteers in Technical Assistance (VITA)		Washington University (MO) (Center for Development Technology)	
	World Visions International			

*Some of these organizations are not currently involved in small-scale, rice bran processing work. They are listed for their potential.

Table 6.12. Partial List of Developing Country Organizations Involved in Appropriate Technology for Renewable Resource Utilization

Wind Energy	Improved Cookstoves	Solar Drying	Rice Bran Processing	Non-Food Products from Agricultural Wastes
Centro Internacional de Agricultura Tropical (CIAT) COLOMBIA	University of Sao Paulo BRAZIL	Grupo de Conversao de Energia Solar, Universidad Estudual de Campinas (UNICAMP) BRAZIL	Food Corporation of India INDIA	ICIDA CUBA
Las Gaviotas COLOMBIA	Nasinu Forestry Station FIJI	Forest Products Research Institute GHANA	Central Food Technology Research Institute INDIA	Universidad del Valle COLOMBIA
Indian Institute of Science (Cell for the Application of Science and Technology to Rural Areas) INDIA	University of the South Pacific (Center for Applied Studies in Development) FIJI	Central American Institute for Industrial Research (ICAITI) GUATEMALA	Indian Institute of Technology, Kharagpur INDIA	Ministry of Forests FIJI
National Aeronautical Laboratory INDIA	Department of Social Welfare and Community Development GHANA	Forest Products Research Institute, Dehra Dun INDIA	Paddy Processing Research Center INDIA	University of Science and Technology (Technology Consultancy Center) GHANA
Bandung Institute of Technology (Aero and Hydrodynamic Laboratory) INDONESIA	Mesoamerican Center for the Study of Appropriate Technologies (CEMAT) GUATEMALA	Indian Institute of Technology, Kharagpur INDIA	Appropriate Technology Development Association INDIA	ICAITI GUATEMALA
University of Nairobi (Dept. of Mech. Eng'g) KENYA	Alliance for Community Development GUATEMALA	Shri A.M.M. Murugappa Chettiar Research Centre INDIA	Jadavpur University INDIA	Council of Scientific and Industrial Research INDIA
Malaysian Agricultural Research and Development Institute MALAYSIA	Choqui Experimental Station GUATEMALA	Caribbean Food and Nutrition Institute JAMAICA	Appropriate Technology Development Organization PAKISTAN	Indian Polywood Industries Research Institute INDIA
International Rice Research Institute (IRRI) PHILIPPINES	Investigaciones Cientificas Asociadas Del Altiplano GUATEMALA	UNICEF Village Technology Unit KENYA	International Rice Research Institute PHILIPPINES	National Autonomous University of Mexico (UNAM) (Materials Research Institute) MEXICO
Korean Advanced Institute of Science (Dept. of Mech. Eng'g) KOREA	Central Fuel Research Institute, Dhanbad INDIA	National Autonomous University of Mexico (UNAM) (Materials Research Institute) MEXICO	Kasetsart University THAILAND	Pakistan CSIR Labs PAKISTAN
Arusha Appropriate Technology Project TANZANIA	Ungra Extension Center INDIA	Centro Internacional de la Papa PERU	Applied Scientific Research Corporation THAILAND	FORPRIDECOM PHILIPPINES
University of Dar es Salaam (Faculties of Eng'g and of Agri. and Forestry) TANZANIA	Indian Institute of Sciences (Cell for the Application of Science and Technology to Development) INDIA	International Rice Research Institute (IRRI) PHILIPPINES	Dept. of Science, Ministry of Industry THAILAND	Caribbean Industrial Research Institute TRINIDAD and TOBAGO
Ministry of Agriculture and Cooperatives THAILAND		Asian Institute of Technology THAILAND		Catedras de Proyectode Fabrica y de Quimica Industrial y Control Analytica URUGUAY

Federal laboratories

Federal laboratories are active in some areas of interest. For example, two U.S. Department of Agriculture (USDA) Research Laboratories are working on rice research; other USDA labs have solar drying and agri-waste materials experience. In the energy field, Brookhaven National Laboratory and the Solar Energy Research Institute (SERI) have performed planning studies related to energy needs and utilization in developing countries.(10) There is capability in the federal laboratories but, in general, there has been no strong commitment as yet, to become heavily involved internationally in renewable resource activity.

Research institutes

Organizations like the Denver Research Institute (DRI) and the Battelle Memorial Institute have the capability for participating in these areas. DRI has been heavily involved in village-level food processing and other AID-supported programs.

Peace Corps

Peace Corps Volunteers represent a group that has the necessary outreach and in some cases the necessary skills to work in these areas. There is an active Peace Corps Energy Program.(11) The connection of Peace Corps Volunteers with universities might help to improve the output of such efforts.(12)

Private industry

We have devoted little effort to examining the role of private industry. Small and medium private industries, in particular, are often mentioned as possible contributors in international technology activity. A workshop supported by AID as part of UNCSTD preparations examined the role of industrial research in science and technology for development.(13) Some companies that we feel can contribute to research and development on rice bran processing are listed in table 6.11.

Private voluntary organizations

Organizations like CARE and Save the Children Federation support field projects in the areas in question. They can provide both monetary and, in some cases, technical resources.

Funding organizations

Organizations which provide financial support in the area of renewable resources include a foundation (Arca Foundation), a government organization (Community Services Administration - CSA), and a private non-profit organization (A.T. International) established in response to a congressional mandate. In the case of CSA, the mission is primarily a domestic one, but the expertise in the agency might be applicable elsewhere.

Developing Country Organizations: Analysis

The success of initiatives in appropriate technology for renewable resource utilization will depend very heavily upon the developing countries themselves and the effectiveness of their policies and organizations. The documentation associated with a recent FORUM on Appropriate Industrial Technology sponsored by UNIDO indicates that, in many countries, a considerable amount of activity is on-going that falls within the topical areas of this study.(14)

Table 6.12 lists some of the developing country organizations that could participate in such efforts. Again, we emphasize that these lists and those of Appendix B are partial lists which are not likely to be comprehensive. Much A.T. and other developing country work goes on locally and does not make its way into international information channels that we have had access to in this study. For example, we came across an article in the May 11, 1979 edition of the New York Times: the Indian government has set aside $1 billion for solar research. The Indian Institute of Technology at Madras is one of 30 centers in India where solar energy research is being conducted. This article illustrates to us the difficulty in keeping track of worldwide activity in the renewable resources field. It also indicates that renewable resource activity in some developing countries may be more extensive than we have indicated. Furthermore, the emphasis on India and Southeast Asia in some of the subject areas may be reflective of the information sources available to us rather than what is actually going on in geographic regions not represented. Developing country participation can be categorized as follows.

Universities

Universities in some developing countries are active in several of the topical areas. For example, there are universities involved in solar energy research in India and Mexico at a level similar to that in universities in the United States. Other institutions, such as the Allahabad Polytechnic Institute, India, are directly involved in A.T. rural extension efforts.

Still other universities contain organizations which appear to possess the capability for both applied R&D and extension, such as the Technology Consultancy Center of the University of Science and Technology, Kumasi, Ghana. It is doubtful that these involvements of universities extend to all countries; however, there are a small number of institutions which have an international reputation for working in these areas and could serve as regional resources (e.g., Asian Institute of Technology, Bangkok). In addition, there are interesting new forms of activity emerging, such as the Cell for the Application of Science and Technology to Rural Areas (ASTRA), associated with the Indian Institute of Science, Bangalore; and FUNDAEC which was started by personnel from the University del Valle in Colombia. Both of these organizations are involved with the people in the villages and provide a new framework for applying science and technology to rural areas. Universities have often been accused of being aloof from development problems; however, there are some that clearly are providing a framework for making significant contributions in the field of appropriate technology for renewable resource utilization.

Government organizations

There is a broad range of government organizations - laboratories, research institutes, ministries - in a wide variety of subject areas. These organizations are an important presence in most if not all developing countries; some are very active in specific areas and are making important contributions. For example, the Central Building Research Institute in Roorkee, India, has been active in applied research on the use of local materials in building. ICAITI in Guatemala has had much experience in a variety of projects concerned with renewable resources. Local institutions in developing countries with expertise in renewable resources need more support and international recognition for their efforts. In addition, the international agricultural research institutes such as IITA in Nigeria, CIAT in Colombia, and IRRI in the Philippines represent important resources.

Appropriate technology organizations

There is a small but growing number of A.T. organizations that are active in the areas of this study. For example, the Appropriate Technology Development Association (ATDA) in Lucknow, India, is very interested and active in many areas of appropriate technology for renewable resource utilization. ATDA represents an international model and resource for work in this area, as do other organizations such as Las Gaviotas in Colombia, CEMAT in Guatemala, and the Arusha project in Tanzania. A.T. organizations are sometimes government related, sometimes private; some are connected through informal or formal networks such as TRANET with A.T. groups in the United States.

Other developing country organizations

Other organizations or individuals in developing countries which are active or could contribute to the areas of interest include small industries, cooperatives, entrepreneurs, missionaries, and private voluntary organizations. Many of these organizations and individuals need information which they can obtain through organizations like VITA.

Other Organizations: National and International

Another important organizational resource in appropriate technology for renewable resource utilization consists of national organizations in countries other than the United States and the developing world, as well as international organizations which are active in the areas of interest. Names and addresses of some of these organizations are included in Appendix B.

Appropriate technology organizations in these countries include ITDG in the United Kingdom, GRET in France, Brace Research Institute in Canada, and TOOL in the Netherlands. Government research labs include the Tropical Products Institute in the United Kingdom. Government development organizations, such as the Swedish International Development Authority (SIDA), the Swiss Association for Technical Assistance (SATA), and the Canadian International Development Research Centre (IDRC), as well as similar organizations in other countries support innovative programs in renewable resources. In The Netherlands, there is a great deal of appropriate technology activity getting underway at universities, including Eindhoven and Delft.(15)

The World Bank and the regional development banks have, within the last year or two, become more active in support of appropriate technology activity. United Nations organizations involved in subject matter areas of interest to this study include UNICEF, UNIDO, UNEP, FAO, and the UN University.

There exist significant national and international organizational resources outside of the United States which can contribute to and/or cooperate with organizations in developing countries in appropriate technology for renewable resource utilization. Some specific activities are summarized by Ashworth.(16) Such resources need to be taken into account in designing U.S. program initiatives.

MECHANISMS FOR IMPLEMENTATION

In order for the initiatives defined previously to be implemented, a variety of mechanisms need to be in place. We use the term "mechanisms" very broadly to include programs, organizational forms, and sources of support created by legislation as well as arrangements between U.S. and developing country organizations and/or individuals.(17) The questions that need to be answered are: (1) Does legislation and organizational infrastructure exist in the United States to support U.S. initiatives in appropriate technology for renewable resource utilization of direct benefit to developing countries? (2) Do adequate arrangements and linkages exist for cooperative efforts between U.S. and developing country organizations? (3) Is the financial support base adequate? (4) If the answers to one or more of these questions is no, what needs to be done?

U.S. Legislation and Organizational Infrastructure for Supporting the Initiatives

In a previous study of the international role of U.S. universities in science, engineering, and agriculture, we reviewed the more general legislative mandate for U.S. involvement in science and technology for development.(18) Based upon that review and subsequent developments, we believe that the legislative mandate and U.S. government organizational infrastructure provide a basis, albeit a somewhat shaky one, for supporting many of the initiatives. As the 1970s drew to a close, renewable energy sources, in particular, received increased funding and increased attention in connection with the 1981 UN Conference on New and Renewable Sources of Energy (UNCNRSE). We deal later in this section with the adequacy of financial support for the initiatives.

In the field of appropriate technology for renewable resource utilization, the following U.S. government or U.S. government-supported organizations represent the principal sources of support for the initiatives defined in this study.

The Agency for International Development (AID)

AID has long been active in the field of appropriate technology for renewable resource utilization, although some of its activities may have gone by different names. AID has supported efforts to gather and analyze information on its programs in connection with UNCSTD preparations. A study by The International Science and Technology Institute, Inc. (ISTI) lists a variety of projects in food storage and processing, renewable energy, and nonconventional sources of energy.(19) A study by the Solar Energy Research Institute (SERI) contains brief descriptions of selected AID-funded and other projects involving renewable energy sources as well as more comprehensive lists of bilateral and multilateral assistance projects.(20)

It is likely that AID will continue to be a major focal point for U.S. financial support of programs in appropriate technology for renewable resource utilization. Efforts will be directed toward the approximately 40 countries which are the focus of AID programs. Within AID, the Office of Energy has grown significantly. Total AID obligations in new and renewable energy have grown from $19 million in FY 78 to $33.5 million in FY 79 to $36 million in FY 80. AID expects to spend about $77 million in this energy field in FY 81. AID's Office of Science and Technology, the Regional Bureaus, and the Agriculture and Nutrition Programs also carry out activities which relate to one or more of the initiatives. An appropriate technology program exists within AID; in addition, efforts to infuse appropriate technology activity into all aspects of AID's program have gone forward.

The Institute for Scientific and Technological Cooperation (ISTC)

Had ISTC been approved and adequately funded, it would have represented a major new source of support for the more research and development oriented aspects of the initiatives not only in AID countries but in countries not eligible for AID support, that is, "non-AID" countries. ISTC plans called for spending 75 percent of its funds directly in developing countries, in both "problem-solving" and "capacity-building" activities. Among the areas planned for new ISTC program initiatives were Nutritional Improvement, Energy Planning and New Energy Supplies, and Non-Agricultural Employment.(21) In addition, ISTC was to take over a variety of activities currently located in AID, including activities concerned with increasing agricultural productivity. The latter included support for the international agricultural research centers and a collaborative research support planning grant under Title XII for reducing postharvest food losses.(22)

In late 1979, ISTC failed to win final congressional approval. An effort began to have AID assume some of the functions and programs envisaged for ISTC but on a greatly reduced scale.

AT International (ATI)

ATI is a private, nonprofit corporation, funded primarily by the U.S. government. It provides direct support to entrepreneurs, small businesses, cooperatives, and other organizations to promote the development and dissemination of appropriate technologies.(23) ATI has provided a grant to CEMAT in Guatemala to disseminate the construction of the Lorena cookstove (see chapter 2) and is also supporting related work in Honduras. Although its budget is very small relative to AID's total budget, ATI could play a significant role in supporting the information dissemination and extension aspects of the initiatives.

Primarily domestic government organizations: Department of Energy (DOE); Department of Agriculture (USDA); National Science Foundation (NSF)

There are at least three U.S. government organizations with primarily domestic missions that could contribute to implementing selected initiatives. The Department of Energy's domestic program in renewable energy sources has expanded greatly in recent years. Furthermore, DOE is authorized, in accordance with the Nuclear Non-Proliferation Act of 1978, to cooperate with the State Department in helping developing countries to develop their renewable energy resources in order to lessen their need for nuclear power. In 1980, joint planning between DOE, State, and AID/IDCA was underway to define new program initiatives in renewable energy. The Department of Agriculture has an Office of International Cooperation and Development. Its nutrition and agri-business program has been involved in rice bran stabilization work using pass-through funds from AID. The U.S. Forest Service of USDA has been involved in studies of solar drying applications in the Philippines (see chapter 3). Both DOE and USDA support research within the United States which might be relevant to developing countries, such as DOE's current A.T. small-grants program, and employ personnel with interest and experience in international activity.

The National Science Foundation is supporting a domestic program in appropriate technology on an experimental basis in the United States.(24) Major topics for the program include: appropriate technology and urban innovation; small-scale industrial technology; recycling, resource recovery, and conservation; appropriate technology and rural revitalization; and food and nutrition. Some of these activities could help build the base for further collaborative international efforts in A.T. Although NSF has a mandate for collaborative international research, developing country activities have not been a major focus in the past. Some work is being supported through NSF's Division of International Programs in the Directorate of Scientific and Technological Affairs,(25) including a small Science for Developing Countries Program which was created in 1980 from previously existing program elements.

The U.S. domestic, mission-oriented, government agencies represent an underutilized resource in the field of appropriate technology for renewable resource utilization directed toward developing countries. Recent years have been a time of tight budgets; and, in spite of requests by agencies such as NSF to increase their developing country program involvements, the outlook for this occurring has not been encouraging. In the spring of 1979, the Office of Management and Budget (OMB) appeared to have taken a dim view of expanded international involvements by domestic agencies and, hence, the budgetary support for such expansion may not be forthcoming. Perhaps the OMB stance reflected not only budgetary limitations but also the expectation that the ISTC would provide the leadership and financial support for new international initiatives. With the demise of ISTC, both AID and the domestic agencies are likely to renew their efforts to expand their programs, with renewable energy being a likely growth area. Whether the Reagan administration and/or the Congress concur with such expansion remains to be seen. Early (February 1981) indications are that budgets for both international development activity and renewable energy resources may be substantially reduced.

Other organizations

There are other sources or potential sources of support for activity in appropriate technology for renewable resource utilization. They include private foundations (e.g., Ford, Rockefeller, Clark, Kettering, Arca); contributions by individual citizens to private, voluntary organizations (e.g., Save the Children Federation, Oxfam, Unitarian-Universalist Service Committee); foreign governments and/or organizations (e.g., al Diŕiyyah Institute); and international organizations (e.g., UNIDO, UN University, UNICEF, World Bank). The report by Ashworth should be consulted for a description of activities of many of these organizations in renewable energy sources for international development.(26)

Concluding remarks

Much has happened in the United States since 1978 which makes it difficult to provide a simple answer to the question: are the policy framework and legislative mandate adequate within the U.S. government to support the initiatives we have proposed in appropriate technology for renewable resource utilization? Furthermore, policy pronouncements and legislation do not always ensure action. On the negative side, efforts failed to establish an Institute for Scientific and Technological Cooperation to provide a focal point for international cooperative R&D activity. The United States has been slow in responding to requests for financing the new fund for science and technology within UNDP called for at UNCSTD. On the positive side, funding for international activity in renewable energy sources has been increasing. The 1981 UNCNRSE could result in still greater levels of support and cooperation.

Some of our more integrated initiatives focus on "renewable resources" or "village-level development" as opposed to "food" or "energy," but we do not believe that new support organizations are necessarily needed; existing organizations might establish new program categories if and when desirable. The label "appropriate technology" sometimes causes problems, but relevant work has been going on, under both that label and other labels. The deficiencies seem not to stem so much from infrastructure as they do from lack of priority, commitment, and financial support.

Financial Support Base

Adequate financial and human resources are needed to support the initiatives identified in this study. We have not done an analysis of how much money is being spent currently by the United States in the study areas; such data are difficult to acquire although it might be possible now to do so, using Ashworth's data(27) and information being compiled for UNCNRSE.

The initiatives we have defined are generally quite broad (see tables 6.1 through 6.5), involving R&D, design and testing, extension, dissemination, and needs analyses. We would guess that U.S. spending for the R&D segment of the initiatives geared toward developing countries in the specific topical areas we have examined is of the order of $1 million annually. It is our opinion that U.S. resources devoted to R&D in these areas could be increased by an order of magnitude with good effect, without causing severe strain on budgets.

The range of activities encompassed under the rubric of renewable resources is much broader than the topical areas considered in our study. Ashworth and Neuendorffer indicate that over $225 million is currently committed by foreign assistance organizations worldwide to the development, installation, field testing, adaptation, and manufacture of renewable energy systems in developing countries.(28) This figure does not include the substantial funds provided by the developing countries themselves. The U.S. contribution to the $225 million runs in the tens of millions of dollars. This level of resources represents a substantial increase over previous activity ten or even five years ago. Careful documentation and evaluation of experiences is required.

Mechanisms for United States-Developing Country Interaction

A key element in using appropriate technology for renewable resource utilization to improve the life of poor people in developing countries is to strengthen the ability of those countries to carry out the activities described in the initiatives. The developing country organizations previously listed plus numerous others - both existing and new - as well as developing country individuals are the means for achieving this goal. Because financial and human resources are limited, a key element in the process is to identify developing country organizations and individuals that are truly effective and for which U.S. assistance and cooperation will make a difference. The process by which such identification takes place is not well defined and likely to be an ongoing, challenging element of program implementation. The organizations we have listed in table 6.12 are those that seemed to us to be good candidates during the course of this study; they may or may not prove to be the ones to focus on. There are, undoubtedly, other organizations worthy of support which we were unaware of or for which we had no information.

There is a point of view held strongly by many A.T. organizations that, unless the poor themselves are involved in the process of defining needs and choosing or developing the technologies to help meet those needs, neither the technology nor the development which takes place will be truly appropriate. Whereas many of the organizations we have previously listed operate in accordance with this philosophy, we have not imposed this condition in compiling the lists.

U.S. government action should either support developing country organizations directly or use other mechanisms to provide such support. Some such mechanisms are described below.

Support developing country organizations directly

The U.S. government through AID supports developing country organizations directly. The level of support is determined to some extent by political considerations; for example, Egypt is currently receiving a considerable amount of aid from the United States. Sometimes there are counterpart funds available which can be used for such support. The direct support mechanism is one that will probably continue to be heavily emphasized in the future. Furthermore, involvement of developing country personnel in the decision processes of the U.S. initiatives, although difficult, would seem to be very desirable. Such involvement is encouraged by development agencies in other countries, e.g., Canada.

Sometimes funds are provided to private U.S. organizations which in turn seek out developing country groups to support. A.T. International is active in a range of appropriate technology activity. VITA now has a small grants program in renewable energy.

Support bilateral agreements between governments

Mexico and the United States have been conducting research on the guayule rubber plant through a Scientific and Technological Cooperation Agreement administered by the U.S. National Science Foundation and the Mexican Science and Technology Council (CONACYT).(29) Such agreements, implemented through government agencies, usually involve commitment of support from both governments. The U.S.-India Subcommission on Science and Technology is another mechanism which serves to implement such bilateral agreements. These bilateral agreements are responsive to political considerations and are likely to be continued to be used in the future.

Support U.S. organizations to collaborate with developing country organizations

There are a variety of U.S. organizations which collaborate with developing country organizations. Universities are a prime example. Their involvements have been extensive and are documented elsewhere.(30) Appropriate technology groups such as VITA serve an important role in this area. Research and development organizations such as the Denver Research Institute and the Georgia Institute of Technology Engineering Experiment Station are also active.

Concern sometimes arises in the Congress and elsewhere that the U.S. organization may get a disproportionately large share of the funds. ISTC attempted to respond to this concern by targeting 75 percent of its funds for spending directly overseas. On the other hand, if international collaborative R&D activity is to be carried out, as has been repeatedly called for by various national and international policy statements and studies, it is essential that some support for the U.S. end of the collaboration be made available.

Support international organizations which in turn support developing country organizations

International organizations potentially loom as important in this area. UNIDO has shown strong interest in appropriate industrial technology in some of the topical areas described in this study. The UN University is focusing on several areas relevant to the topics under consideration.(31) FAO, UNICEF, and UNEP are also pertinent. There is considerable developing country interest in having developing country aid funds given to international organizations because, in some cases, it may give them more influence and control over direction of the program than might be the case in a bilateral arrangement. This mechanism received major emphasis by developing countries at UNCSTD. A new fund for scientific and technological development was established within the UN Development Programme (UNDP).(32) However, financial support for this fund has been slow in coming.

Support regional centers for work in topical areas

The International Agricultural Research Centers provide a model for work in other areas. For example, the Asian Institute of Technology (AIT) in Bangkok, Thailand, has a variety of activities underway in appropriate technology for renewable resource utilization. AIT has students from a wide range of countries and could represent the kind of institution that should be supported. National centers serving regional needs deserve consideration.

Create and/or support international appropriate technology mechanisms

There was some emphasis in 1978 and 1979 on creating and obtaining support for international mechanisms for appropriate technology which bypass governments. A recommendation supporting such activity was made at the Winrock Conference on Appropriate Technology(33) and a proposal for an International Mechanism for Appropriate Technology (IMAT) has been discussed at several UN forums.(34) The emphasis here was on grass roots involvement accompanied by integration of R&D with other A.T. activities. IMAT advocates generally felt that such an organization would be more effective if it were not part of the UN system, but cooperated with it. Whether something like IMAT does, in fact, get established and is effective remains to be determined. However, international A.T. mechanisms like SATIS and TRANET as well as informal links between VITA and other A.T. groups represent important communications channels for dissemination of ideas and information in the areas of interest.

Create and support international renewable energy resource networks

The renewable energy resource theme may be one whose time has come. There is rapidly growing interest in the United States, if the proliferation of conferences on solar energy and biomass utilization is any indication. The 1981 UN Conference on New and Renewable Sources of Energy can stimulate important new initiatives. All indications are that renewable energy resources will not prove to be a passing fad. It may be that existing professional societies or international scientific organizations can be persuaded to take part in creating a resource network. However, it is also possible that new organizations will be needed.

NEXT STEPS

A logical follow-on to this particular study would be to assemble working groups of individuals from several countries to structure work programs in each of the five topical areas. This might be done in a relatively short time, using this study as a basis for making contacts and organizing such an effort. Working groups would be in a position to help define long-range programs with R&D, extension, analysis, information dissemination, and education components in each of these areas. Organizations to help support such efforts include AID, DOE, and USDA; in fact, we would encourage inter-agency cooperation, given the broad range of activities required to carry out the initiatives (see tables 6.1 through 6.5).

We also would emphasize the necessity of bringing together representatives from various kinds of organizations, including A.T. groups, universities, research institutes, private industry, and government agencies. There is much that these groups have to contribute to each other as well as to the initiatives proposed in this study. The collaboration between Washington University and VITA in this study reaffirms this belief and lends support to this recommendation.

Appendix A
VITA/CDT Panel Reports

This Appendix consists of the Reports of the VITA/CDT expert panels held on three of the topical areas of our study: wind energy, improved cookstoves, and solar drying. These reports were drafted initially by L. Icerman and were reviewed by the panelists prior to final preparation. Each of these reports contains findings that supplement the discussions on wind energy in Chapter 1, improved cookstoves in Chapter 2, and solar drying in Chapter 3.

The panels were assembled for full day sessions as follows: Wind Energy Panel on January 12, 1979 and Improved Cookstoves Panel on February 9, 1979 at VITA Headquarters, Mt. Rainier, Maryland; Solar Drying Panel on March 9, 1979 at the Center for Development Technology, Washington University, St. Louis, Missouri. Panel members were selected based on their interest, expertise, and experience in the subject of the panel as related to developing countries. The goal of the panels was for our study project personnel to extract useful inputs for analysis from the practical insights and experiences of panel members regarding applications and programs in developing countries for the topics under consideration.

REPORT OF THE VITA-CDT WIND ENERGY PANEL

Panel Participants

<u>Invited members</u>

Ron Alward
National Center for Appropriate Technology

Carl Aspliden
Federal Wind Energy Program
Department of Energy

Steve Blake
Appropriate Technology Group

Charles (Sandy) Butterfield
Rockwell International
Rocky Flats Plant

Fred Perkins
Solar Energy Research Institute

Priscilla Reining
International Office
American Association for the Advancement of Science

Marcus Sherman
Wind Energy Consultant

VITA and CDT staff

Elizabeth Andrews
Center for Development Technology
Washington University, St. Louis

Larry Icerman
Center for Development Technology
Washington University, St. Louis

Eduardo Perez
Volunteers in Technical Assistance

Additional contributors

Forrest (Woody) Stoddard, U.S. Wind Power Associates, submitted written comments on the materials that were discussed by the panel.

Review of Panel Discussions

Morning session

A number of topics were discussed at the morning session, including: energy needs in developing countries which could be satisfied by windmills; preferred applications for windmills in developing countries; wind machine designs; commercially-available windmills and their potential use in developing countries; design modifications for improved task matching in developing countries; categorization of windmill designs with respect to materials and construction, including special emphasis on the design potential for component production or fabrication by light-capital indigenous industries. Two particularly important items were emphasized in the discussions.

First, the need for a worldwide data base that includes wind resources data, developing country end-use energy needs, and conventional energy supplies and prices was agreed upon by the panelists. Data bases of this type would be useful in project planning and definition since, without compilations of these data, even the most elementary attempts to evaluate or define desirable development efforts for windmill technologies are extremely difficult. This is not to imply, however, that these data would be sufficient for making decisions related to individual installations. Many important parameters surrounding a specific windmill application are too site-specific for making detailed design decisions based on a rather sparse data network of this type. On the other hand, the importance of site-specific resource issues is reduced as the level of technical sophistication of appropriate windmill designs declines.

Secondly, information dissemination appears to pose a significant barrier to the widespread utilization of wind energy systems in developing countries. Examples of data in need of more effective dissemination are tabulations of available designs, relevant research results, evaluations of technical parameters that influence wind machine designs, and maintenance instructions and/or manuals. Two general problems impede information dissemination programs: (1) how to get the information to the people who need it, when many of them may not understand the descriptions or, in some cases, even be able to read effectively; and (2) how can the information be packaged to facilitate the extraction of data useful in a particular situation from the general material that would be disseminated? The following are conclusions which resulted from the discussions:

1. Small-scale water pumping windmills could satisfy major needs for power commonly found on subsistence farms in developing countries with land holdings of one to two hectares. Cattle watering and selected irrigation are examples of these needs.
2. Light-capital wind machine designs that are appropriate for use in developing countries are best focused toward water pumping applications rather than the production of electricity.
3. Windmill designs for low wind speed applications (i.e., below 3 m/sec) should be developed and tested.

4. Maps should be developed that display mean wind energy density contours adjusted to a specific height for the world based on compilations of existing wind data.
5. End-use needs maps should be developed on a worldwide scale. An example of data that would be particularly useful for windmill applications is the average power requirements for water lifting. In some cases, end-use needs may exhibit significant seasonal variations.
6. Case studies of the success and/or failure of present and past wind machine applications in developing countries should be prepared. Critical reasons for failures should be identified and evaluated.
7. Wind energy system design handbooks should be prepared. These handbooks should include, for example, discussions of important windmill design parameters, materials, and skills requirements, construction and fabrication instructions that are widely applicable, operation instructions, troubleshooting guidelines, and maintenance directions. A number of these handbooks are needed that address different levels of technical expertise (e.g., local entrepreneurs, windmill owners and operators, engineers in developing country institutions).
8. Skills and manufacturing techniques which may be in use for applications other than windmills (e.g., precision machining, welding, fabrication processes) in developing countries but could contribute to implementation of windmills should be identified, assessed, and documented.
9. In-country workshops to disseminate hands-on design information may be an effective way to transfer information and should utilize the appropriate-level design handbooks as a resource. These workshops are probably best coordinated by competent local personnel, possibly in connection with local or regional educational institutions.

Afternoon session

The basis of the afternoon discussion was a list of questions which was distributed in advance for consideration by the panelists (see table A.1). Answers to each specific question were presented by the panelists and discussed. The following remarks summarize the responses by the panelists.

1. Desirable actions that could be taken by the United States to facilitate the deployment of light-capital wind energy technologies in developing countries include:
   - technical and financial support of wind resource and application surveys;
   - incentives for foreign students and especially technicians to study in the United States and to return to their home country;

Table A.1. Questions for Advance Consideration: Wind Energy Panel

1) What is the single best action the United States could take to assist in the deployment of wind energy technologies in developing countries?

2) What is the single best action that could be taken by developing country institutions and/or governments to support the introduction of wind energy technologies?

3) What are the realistic opportunities for U.S. and LDC cooperative efforts in wind energy?

4) What U.S. institutions have capability and know-how to work with LDC institutions in wind energy applications? List.

5) What LDC institutions have capability and know-how to work in wind energy applications? List.

6) What is the largest current barrier to the implementation of wind energy technologies in developing countries? Is it highly site- and/or country-specific? Is this barrier also the most difficult one to overcome?

7) Where should the most emphasis be placed in order to expand the use of wind energy systems? Resource assessment? Design improvements? Maintenance reductions? Cost reductions? Social or cultural acceptability? Evaluation of needs? Performance evaluation and/or certification?

8) What are the preferred financing arrangements for rapid introduction of wind energy technologies?

9) To what extent are the preferred programs for implementation assistance for wind energy systems country-specific?

10) Are information dissemination problems a major impediment to widespread use of wind energy systems?

- support program for developing country institutions to operate hands-on wind energy workshops;
- utilization of wind energy systems by U.S. facilities in developing countries for demonstration purposes;
- establishment of central or regional warehouses in developing countries for wind machine maintenance needs;
- market guarantees for products which have wind energy as a factor of production; and
- publication of U.S. and other developments in wind power in developing countries in native languages.

2. Important programs for possible adoption by developing country institutions and/or governments to support the introduction of wind energy technologies include:

    - a variety of financial incentives (e.g., loans, tax benefits, credits, financing arrangements);
    - educational programs on alternative energy technologies in developing country schools and universities;
    - establishment and deployment of professional societies to facilitate information and personnel changes;
    - utilization of wind energy systems by local governments; and
    - participation by developing countries rich in conventional resources in the development of alternative energy sources in countries lacking conventional resources.

3. The consensus of the panel was that realistic opportunities for U.S. and developing country cooperative efforts in wind energy are essentially unlimited.

4. U.S. institutions with the capability and know-how to work with developing country institutions in wind energy applications include:

    - private businesses including multinational companies,
    - government institutions,
    - universities,
    - appropriate technology groups,
    - trade organizations,
    - churches,
    - professional societies, and
    - trade associations.

    Potentially, any organization involved in wind energy activities in the United States could participate and contribute.

5. Developing country institutions with the capability and know-how to work in wind applications are similar to those listed for the United States. Agricultural engineering universities and personnel were judged to play a particularly important role in this area.
6. Many barriers impede the implementation of wind energy technologies in developing countries. For example:

   - lack of both general and site-specific siting and design information;
   - lack of local skilled construction labor;
   - lack of personnel qualified to assess nondesign-related, site-specific issues (e.g., site choices, social or cultural impacts);
   - lack of indigenous and/or on-site manufacturing capabilities;
   - lack of project continuity;
   - lack of a variety of adequately proven and commercially available wind machines;
   - lack of widespread information on windmill design, performance, and costs;
   - unavailability of maintenance materials and/or experienced personnel;
   - inadequate financing opportunities;
   - current energy prices and high capital costs of even light-capital wind energy systems;
   - indirect subsidies for fossil fuels;
   - unproven techniques of introduction;
   - shortages of suitable wind machines;
   - unclear definition of ownership responsibilities; and
   - vandalism.

7. The panel was unable to isolate a single area on which the most emphasis should be placed in order to effectively expand the use of wind energy technologies in developing countries. Resource assessments, improved access to maintenance materials and availability of qualified maintenance personnel, improved financing arrangements, design improvements and modifications, evaluation of end-use needs, social and cultural acceptability, etc. are all important areas for emphasis. Moreover, the most important of these may be determined largely by country-specific, site-specific, design-specific, or application-specific parameters. Since successful implementation programs are difficult to initiate without, at least, the minimum required resource assessment, this area may represent the most basic effort. Development of reliable and/or maintenance-free wind machines as well as the establishment of effective maintenance programs are certainly areas of extreme importance.

8. Desirable financing arrangements are necessary for the successful widespread implementation of wind energy systems in developing countries. However, general descriptions of favorable programs are extremely difficult to formulate.

Additional remarks

Several general items surfaced repeatedly during the panel discussions. For example, small-scale (i.e., less than a few hundred watts of shaft power) water pumping windmill designs should be the technology targeted for early deployment in developing countries.

Wind energy technology transfer programs must possess an element of continuity in order to be successful. Short-term or crash programs will fail because the necessary institutions and/or infrastructures will not have time to mature within a developing country to support the project after the initial transfer process is completed.

Technical problems or design improvements, related only to performance considerations and not necessarily to reliability and/or capital cost reductions, are not, in general, a major concern or need in the case of small-scale water pumping windmills for export to developing countries. On the other hand, only a few reliable small-scale systems for electricity production are currently being manufactured in developed countries and probably no reliable home-built designs are available. However, for both water pumping and electric designs, the complete package necessary for successful transfer (e.g., hardware, support materials and information, space or replacement parts, competent personnel) is yet to be assembled.

Deployment programs for indigenous wind machine designs will probably be more successful if the focus is toward assisting the growth of a group of experienced local entrepreneurs rather than direct contact with the user. This concept is compatible with the establishment of hands-on training workshops initiated by U.S. and developing country cooperative efforts.

REPORT OF THE VITA-CDT COOKSTOVE PANEL

Panel Participants

Invited members

Sydney Draper
World Bank

G. S. Dutt
Center for Environmental Studies
Princeton University

Ianto Evans
The Aprovecho Institute

John Fry
Agency for International Development

Paul Jankura
Overseas Development Council

Irene Tinker
Equity Policy Center

VITA and CDT staff

Laurel Druben
Volunteers in Technical Assistance

Larry Icerman
Center for Development Technology
Washington University, St. Louis

Bill Magrath
Volunteers in Technical Assistance

Eduardo Perez
Volunteers in Technical Assistance

Holly Winger
Center for Development Technology
Washington University, St. Louis

Review of Panel Discussions

Morning session

Three principal areas related to the implementation of improved cookstove designs were discussed at the morning session: cookstove needs and performance; firewood use and deforestation issues; and impacts on women and village life resulting from the deployment of improved cookstove designs. A number of issues were addressed within each of these three subject areas based in part upon responses to advance questions (see table A.2).

Table A.2. Questions for Advance Consideration: Issues and Impacts of Improved Cookstoves

Cookstove needs and performance

1. What is the major need that may be satisfied by improved cookstove designs? Is this need country-specific?

2. Are village cooking needs being effectively met by the proposed improved cookstove designs? If not, could this be the major barrier to widespread acceptance?

3. Have villagers' own perceived cooking needs been investigated to determine how the benefits resulting from use of improved cookstove designs will be valued?

4. Have inputs from village women been solicited in developing the range of more efficient stove designs?

5. How important is the deployment of improved cookstove designs? Are specific design improvements needed?

6. How much more efficient are the new cookstove designs than the traditional cooking methods?

7. Do current methods of measuring cookstove efficiencies reflect properly the actual day-to-day conditions of stove use? Are these measurements reliable and valid?

8. How easy are more efficient stove designs to build, operate, and maintain? Is durability a major problem?

9. Do present improved stove designs show a need for further refinements in the building materials mix? Have past attempts to introduce improved stoves been hampered by difficulties with poor performance of various clay mixtures?

10. Are there ongoing efforts to improve designs of cookstove utensils to accompany cookstove improvements? Are these efforts subject to significantly different cultural acceptability problems?

11. What is the major use for the hot water that is provided by most of the improved cookstove designs? Could this hot water be supplied more effectively by other means?

12. Are chimney fires resulting from creosote build-up a major potential hazard from use of "smokeless" stoves? Has this issue been given enough consideration?

Table A.2. (continued)

Deforestation Issues

1. Is the documentation of the role played by woodburning for cooking in deforestation problems adequate? If not, what additional information is needed?

2. Will implementation of improved cookstoves have measurable positive impacts on deforestation problems? If so, how quickly can the benefits be realized?

3. Is there a lack of village-level perception of the impacts of deforestation? If so, is this an impediment to recognizing the need for more efficient use of wood for cooking?

4. Is it conceivable that having a firewood supply nearby would encourage villagers to stockpile wood for drying before burning?

5. Will plans for managed village woodlots interfere with traditional land use patterns?

6. Has a coordinated effort to document traditionally preferred sources of cooking fuel by region been completed?

7. Is the use of charcoal, with the associated conversion loss, a reasonable procedure since even the improved stove efficiencies may not fully compensate for the loss?

8. Have the trade-offs between using wood, cow dung, and agricultural waste mixtures for fuel been evaluated adequately?

Impacts on women and village life

1. Who at the village-level stands to gain the most from acceptance of improved stoves? Will women's positions be improved? Will children benefit? How?

2. Will use of more efficient cookstoves produce a measurable lightening of the burden of cooking and related activities performed by women?

3. If more efficient stoves significantly reduce the amount of time and/or money spent procuring firewood, who stands to benefit the most?

Table A.2. (continued)

4. Is exposure to the smoke from traditional cooking methods viewed by village women as a major unpleasantness?

5. Has the role of smoke from traditional cooking methods been evaluated? Have the preservative and pest-control properties of smoke been weighed against the problems of eye and respiratory irritation?

6. What will the potential effects be on daily living of having hot water available on a continual basis?

7. Will improved cookstoves encourage and/or permit preparation of more nutritional foods?

8. What will the effects of more efficient stove acceptance be on village-industry activities such as: making of brick and other fireclay products; soapmaking, cloth weaving and other domestic products made by women; construction of improved cookstoves?

First, efforts to deploy improved cookstove designs have not always been successful in the past (e.g., programs in India in the 1950s), which may be explained, in part, by the fact that apparently some needs of users were not met effectively. This situation may still be an impediment to implementation of current programs. Needs of users have still not been adequately assessed and inputs from women in the identification of needs are conspicuously absent. Basic data on village-level and household energy end-uses and supplies are presently inadequate in most countries to support the identification of cookstove needs; however, the Peace Corps program to survey energy uses in villages may be an important step toward improving the required data base.

Second, improved cookstove designs must be capable of demonstrating a substantial improvement in performance compared to traditional cooking methods to ensure that a sufficient incentive exists for women to adopt the new designs. This requirement can certainly be met in terms of technical performance under closely controlled laboratory construction and operation practices. However, imperfect construction and/or individual operational preferences may easily offset the expected thermal efficiency improvements of a new design. Accordingly, poor off-design performance characteristics (i.e., significant declines in stove efficiency resulting from only small deviations from recommended operational procedures) may result in disturbing and, in some cases, insolvable barriers to cookstove deployment programs. On the other hand, deployment efforts for Lorena cookstove in Guatemala are reportedly becoming well established.

Third, in some regions (e.g., India), firewood consumption for cooking may not be contributing to deforestation problems as much as commercial harvesting of forests for use as fuel in small-scale industries (e.g., firing of clay bricks). In many villages, women and children gather twigs for cooking fuel while men often participate in commercial firewood collection activities. Although national deforestation and the receding of local forest boundaries are related issues, the development of programs to deal with these problems will require substantially different strategies. Villagers are not, in general, concerned with national deforestation problems; however, they are affected by the ever-increasing time required to collect firewood as nearby trees are cut and not replanted. The fact that inefficient cooking devices and practices require more firewood than more efficient designs and practices is apparently not widely recognized by villagers.

Finally, emphasis on cookstove improvements should be directed toward both high-mass, stationary designs and low-mass, portable stoves. High-mass stoves are typically constructed from mud and/or measured clay and sand mixtures. Low-mass stoves usually have metal structures or are built from waste metal products such as oil drums or paint cans. Since, in urban areas, cooking is often performed outdoors to reduce the impacts on the congested space within the household, urban dwellers in temperate climates often prefer portable cookstove designs whereas rural villagers typically select large high-mass cookstoves. The following are additional points that were made during the discussions:

1. Deployment strategies for improved cookstoves should be focused toward in-country development of improved designs. Region-based design modifications may be important to successful implementation programs.
2. Cookstove extension programs should involve women and should attempt to transfer basic technical principles of cooking and cookstove construction rather than just recipes for cookstove construction.
3. Some practice is required to become proficient in cookstove construction techniques. Durability is an important consideration, and improvements in durability of high-mass cookstoves can apparently be realized by utilizing properly chosen mixtures of sand and clay in place of mud.
4. Cookstoves should be designed to require a minimum of tools for construction, and designs should be customized to utilize traditional cooking pots and utensils.

5. In some areas (e.g., Guatemala), personal washing appears to be the principal use of hot water available from improved cookstove designs. However, since hot water use varies widely with region and social class, an assessment of how effectively improved cookstoves meet hot water needs would be helpful in evaluating this aspect of improved cookstove designs. In some areas, alternative hot water supply technologies (e.g., simple solar water heaters) may be preferred to supplies from cookstoves.
6. Laboratory cookstove efficiency measurements and comparisons have little utility without accounting for operational practices and, hence, are not particularly useful measures of the efficiency of day-to-day cooking schedules. Villagers are more likely to evaluate cookstove designs by daily firewood requirements rather than by relative efficiency data.
7. Improved cookstove efficiencies lead to increased potential for chimney fires since tar build-up increases with reductions in flue gas temperatures.
8. Assessments of the availability of combustible waste materials in villages and evaluations of technologies capable of improving the utilization of these materials to replace firewood for cooking should be performed.
9. Managed village woodlots may be a realistic option for expanded firewood supplies in some special cases. Stable villages and communal land may be prerequisites for establishing successful programs.
10. Utilization of charcoal for cooking is not, in general, an attractive option from an energy viewpoint; however, comparisons between direct firewood and charcoal use must account for transportation energy carefully. If transportation distances are large enough, transport energy costs may make charcoal an economically competitive alternative to firewood. Nonsmoking characteristics of charcoal are attractive to many middle class urban dwellers. Charcoal use is probably desirable only in special circumstances such as in areas where charcoal is a way of life, high temperatures are required, or when charcoal is obtained as a by-product from materials that would otherwise be discarded.
11. Trade-offs between using wood, cow dung, and agricultural wastes for cooking fuels should be more thoroughly evaluated. Deforestation, declines in land fertility and productivity, and use of chemical fertilizers are related issues to be assessed.
12. Household budgeting practices (i.e., the apparently common occurrence in parts of Africa of separate budgets for men and women, with women often responsible for firewood purchases) may be an important aspect of the

successful implementation of improved cookstove designs. Assessments of household budgeting practices should be performed since the understanding of household budgets may be an important component in the design of successful cookstove extension programs.

13. Evaluations of the importance of smoke from traditional cooking methods for pest control should be performed. Potential positive and negative impacts of smokeless stoves should be documented (e.g., reductions in respiratory ailments versus health hazards from insect infestation).
14. Improved cookstove designs may contribute to the adoption of new foods with long cooking times and/or to larger quantities of a particular food being prepared during a single cooking period. Impacts of these modified cooking practices should be evaluated from health, nutritional, and firewood consumption viewpoints.
15. Acceptance of improved cookstove designs may expand economic contributions from women to their families by reducing the expenditures required for firewood (either monetary or time) and by increasing the opportunities for income generation activities by women during periods which otherwise would have been spent cooking and/or collecting firewood.

In general, the panelists expressed concern that information acquisition efforts and/or assessment programs should be closely linked with specific cookstove extension programs. This approach helps to ensure that desirable communication networks are established between cookstove users, promoters, and data collection efforts; and is compatible with the operation of central cookstove information clearinghouses located in developing countries where cookstove extension programs are planned.

Afternoon session

The basis of the afternoon discussion was a list of questions which was prepared to direct the responses of the panelists toward topics that would contribute to the refinement of a list of preliminary initiatives that the United States might undertake to support the adoption of improved cookstove designs in developing countries (see table A.3). The following remarks summarize the responses by the panelists.

1. The panel was unable to isolate a single area on which the most emphasis should be placed in order to effectively expand the use of improved cookstove designs in developing countries. Designs improvements, evaluation of end-use needs, matching of designs with durability requirements, improved financing arrangements, social

Table A.3. Questions for Advance Consideration: Initiatives and Mechanisms for Implementation of Improved Cookstoves

1. Where should the most emphasis be placed in order to expand the use of improved cookstove designs? Design improvements? Maintenance reductions? Cost reductions? Improvement of social or cultural acceptability? Evaluation of needs? Performance evaluations?
2. Are information dissemination problems a major impediment to widespread use of improved cookstove designs?
3. Is there a need to emphasize improvements in cookstove durability and ease of maintenance and operation before implementation will be successful? What kinds of inputs would this effort require?
4. Are special financing arrangements needed for effective introduction of improved cookstove designs?
5. What is the single best action the United States could take to assist in the deployment of improved cookstove designs in developing countries?
6. Should providing funding sources be a main focus of U.S. contributions to cookstove implementation efforts? Should supply of special materials such as small metal parts be emphasized? What U.S. organizations would be best able to participate?
7. What developing country institutions have the capability and know-how to work in cookstove applications? List.
8. What is the single best action that could be taken by developing country institutions and/or governments to support the introduction of improved cookstove designs?
9. What U.S. institutions currently have the capability and know-how to work with developing country institutions in cookstove applications? List.
10. What are the realistic opportunities for U.S. and developing country cooperative efforts in cookstove development and implementation?
11. Is the use of cooperative U.S. and developing country institutional capabilities the best vehicle to organize, disseminate, and update cookstove design information?
12. What are the best institutional vehicles to carry out extension efforts in developing countries for cookstove implementation?
13. Can U.S. institutions contribute effectively to the development of an international cooperative effort to evaluate cookstove extension activities?
14. What mechanisms could be established to handle feedback-response systems between cookstove extension and design improvement efforts?
15. Are universities the best place to do research to improve stove designs in response to feedback from extension projects? Can U.S. universities or other institutions contribute effectively to further design refinements?

Table A.3. (continued)

16. Can U.S. expertise in wood burning technologies contribute to cookstove technology improvements for applications in developing countries?
17. Are there possibilities for cooperative efforts between U.S. forestry expertise and developing country institutions in transfer or exchange of silvicultural techniques directed toward accelerated firewood production?
18. Are additional inputs needed to effectively document what trees in each region are best suited to intensive management for firewood use?
19. Is implementation of an improved stove extension program only a partial effort without accompanying attempts to facilitate access to dry firewood supplies?
20. How much infrastructure building is needed in developing countries to handle extensive improved stove extension efforts? How would development of accompanying forest management efforts affect this need?
21. How can extension programs best be developed to incorporate the inputs of village women who must ultimately accept or reject the improved stoves?
22. What contributions can the United States make to aid in a coordinated effort to measure the costs and benefits of "smokeless" stoves?
23. Will successful cookstove implementation efforts assist in the establishment of complementary hygiene/nutrition education programs?

---

and cultural acceptability, effective training and extension programs, etc. are all important areas for emphasis. The relative importance of these issues will be determined primarily by site-specific factors. Effective training and extension programs covering the construction and operation of improved cookstoves may be the one area that is essential if cookstove implementation efforts are to be successful.

2. Cost constraints may be an important factor influencing cookstove implementation efforts in some locations. Improved cookstoves typically cost from $5 to $35 depending upon the design and whether or not the stove is home-built or purchased. Novel financing strategies such as village lotteries in which each participant has no chance of winning twice until all participants have won once may be a useful capital generation technique. Similar programs are reported to have been used in rural villages for the purchase of other durable goods. Credit programs for women may be necessary to provide capital

for cookstove construction materials. Price subsidies for competing cookstove designs and fuels (e.g., kerosene) may be an effective way to buy time to allow woodlots time to mature before implementing improved woodburning cookstoves in areas where widespread deforestation is already common and fuel switching has occurred.

3. Desirable actions that could be taken by the U.S. to facilitate the deployment of improved cookstove designs in developing countries include:

    - establish mechanisms to disseminate small quantities of money to organizations in developing countries to initiate large numbers of cookstove extension programs;
    - support programs to provide access to or the capability to produce tools required for cookstove construction;
    - facilitate the development of small businesses and/or entrepreneurs to provide cookstove construction capability; and
    - reinforce the appeal of appropriate technology designs by inviting personnel from developing countries to discuss and demonstrate desirable technologies in the United States.

4. Information exchange on improved cookstoves is currently a serious implementation problem since an effective worldwide information network does not exist. Educational television may be a useful mechanism for information exchange, in addition to exhibitions of improved cookstove designs at local fairs or in public areas. A network of information brokers (e.g., organizations similar to VITA) should be established for transfer of cookstove information. Information dissemination programs should be focused toward women and could possibly be linked with or follow patterns established by family planning and/or preventive health care programs.
5. Institutions which provide outreach capabilities rather than research expertise should be the principal target groups in both the United States and in developing countries to contribute to the implementation of improved cookstove designs. Research institutions may play an important supportive role but, in many cases, are too far removed from the rural setting to take the lead in cookstove extension efforts.
6. Effective cookstove extension programs should utilize local entrepreneurs or promoters in developing countries as the backbone of, the extension efforts and should tap the resources of, and/or the communication channels already established by, currently existing extension programs in

other areas (e.g., family planning and preventive health care programs) which have been specifically designed to reach women in developing countries.

7. Efficient communication channels should be developed between cookstove extension programs and design improvement efforts. U.S. institutions should contribute to the coordination and management of an effort to establish effective communication networks. Centers should be organized in developing countries to act as clearinghouses for extension programs. U.S. universities and appropriate technology groups should be able to contribute to the establishment and maintenance of communication between research programs and extension efforts and should be able to provide inputs into needed design improvement and modification efforts identified by extension programs.

Additional remarks

Several additional items came up a number of times during the discussions. The following remarks summarize these issues.

Forestry projects selected to expand food production (e.g., fruit and nut trees) as well as to provide firewood supplies may be a useful technique to allow forests to mature prior to being harvested for use as firewood. Successful utilization of this strategy would require careful forest management practices and identification of forestry experience applicable to the needs of developing countries.

Counterpart institutions in the United States and in developing countries should be identified and/or established to provide support for improved cookstove extension programs. Industrial research institutions in developing countries may be suitable target organizations to couple with U.S. university and appropriate technology groups. There was some difference of opinion among the panelists on the amount of desirable inputs that could be made, in general, by industrial research institutes in developing countries.

The United States should enhance the credibility of appropriate technology programs by demonstrating commitments to these technologies in U.S. institutions and by attempting to raise the status of the concept of appropriate technology by adopting suitable technologies into U.S. life-styles. For example, some interest has been shown in the United States for applications of improved cookstove designs in private outdoor patios, remote cabins, and mountain chalets. U.S. educational institutions may contribute in this area by providing educational opportunities in appropriate technologies to students from developing countries.

REPORT OF THE VITA-CDT SOLAR DRYING PANEL

Panel Participants

Invited members

Merle Esmay
Michigan State University

Carl Lindblad
Private Consultant

Richard Meunier
Solar Energy Research Institute

John Tschernitz
Forest Products Laboratory

Larry Van Fossen
Iowa State University

VITA and CDT staff

James Chapman
Center for Development Technology
Washington University, St. Louis

Larry Icerman
Center for Development Technology
Washington University, St. Louis

Robert Morgan
Center for Development Technology
Washington University, St. Louis

Eduardo Perez
Volunteers in Technical Assitance

Review of Panel Discussions

Morning session

Three principal application areas of solar drying technologies were discussed at the morning session: crop drying; drying of fruits, vegetables, and fish; and timber drying (see table A.4). A number of issues were addressed within each of these three subject areas.

Table A.4. Questions for Advance Consideration: Issues and Impacts of Solar Drying

Crop Drying

1. What are the major needs for and applications of crop drying technologies? How region-specific are these needs and applications?
2. To what extent are heated crop dryers currently being used in developing countries? What are the typical designs, materials and skills requirements, fuels, and costs?
3. Are there natural-convection solar crop dryer designs that are applicable to the needs in developing countries? How large an advantage do dryers possess over open-air drying? Do forced-air solar crop dryers have applications in developing countries?
4. What economic advantages do crop dryers offer in developing countries? Are price differentials for dried and undried crops sufficient to induce adoption of solar crop dryer designs?
5. Are there significant numbers of farmers in developing countries that have the financial capability of withholding a substantial fraction of their crops from markets to permit drying operations? Are there differences between the socioeconomic groups that can afford to utilize natural-convection solar dryers and those that employ other drying methods?
6. How dependent are programs for crop storage improvements on the deployment of crop drying technologies?
7. Are marketing conventions (e.g., sales by weight) important barriers to the use of crop drying? Is the extent of drying accounted for in the terms of payment for in-kind loans?

Drying of Fruits, Vegetables, and Fish

1. What are the major drying needs and opportunities for food products? How region-specific are these needs and opportunities?
2. What food products are currently dried in significant quantities in developing countries? What is the main objective of the drying efforts? Home storage? Delayed marketing? Ease of transport? Product upgrading?
3. What proportion of drying activities is open-air efforts? To what extent are heated dryers currently being used?
4. Are there solar cabinet dryer designs that are applicable to the needs in developing countries? How large of an advantage do cabinet dryers have over open-air drying processes? What is the typical cost of cabinet dryers?

Table A.4. (continued)

5. Is the lack of storage and/or preservative treatments a significant barrier to the expansion of perishable food production? Do seasonal surpluses commonly occur that could be dried and stored for off-season marketing? Are transportation costs an important factor?
6. What economic advantages do cabinet dryers offer in developing countries? Do food products exhibit significant seasonal variations in price and availability? Are dried food products likely to face consumer resistance?

<u>Timber Drying</u>

1. What are the major needs for and applications of timber drying technologies? How region-specific are these needs and applications?
2. What fraction of timber production in developing countries is currently open-air dried? What other kinds of timber drying technologies are commonly found in developing countries?
3. Are there nonsolar designs for timber kilns that are applicable to the needs in developing countries? What are typical capacities, fuels, and costs? Can these designs be constructed easily in developing countries?
4. Are there solar designs for timber kilns that can compete effectively with conventional kilns and/or air drying? Can these designs be constructed easily? What are the typical capacities and costs?
5. What economic advantages do timber kilns offer in developing countries? Are there substantial price differentials favoring more completely dried timber, wood products, or firewood in developing country markets?
6. Are solar timber kilns only applicable for industries with access to export market channels?

<u>Implementation Programs</u>

1. What efforts have been made in the past to disseminate information about and to promote the use of solar drying technologies? Are there any examples of successful programs in the past?
2. What factors have been significant barriers to the widespread adoption of solar drying technologies in developing countries? Capital costs? Financing or ownership issues? Information dissemination? Lack of construction materials? Performance problems? Lack of markets for dried products? Consumer resistance? Lack of adequate storage and/or packaging facilities? How technology-specific are these factors?

Table A.4. (continued)

3. Could the introduction of construction kits facilitate the adoption and use of solar drying technologies?
4. Are solar drying technologies subject to cultural and social acceptability problems?
5. What impacts are likely to occur following the successful implementation of solar drying technologies?
6. Have requests been received in the United States for assistance in the deployment of solar drying technologies in developing countries?

First, applications of solar drying technologies, excluding open-air sun drying, in each of the three principal end-uses are not currently commonly found in developing countries. This situation appears to be the result of the immature status of the development of appropriate technical designs and the lack of proven dryer performance data, as well as the unavailability of economically-attractive systems. Accordingly, extension programs for the deployment of solar dryers have rarely been established in the past and were judged by the panelists to be premature unless preceded by local performance trials and adaptive research.

Second, solar drying technologies, especially solar crop dryers, must be capable of demonstrating a substantial improvement in performance compared to sun or other current drying practices to ensure that a sufficient incentive exists for farmers and villagers to adopt the solar drying designs. For solar crop drying applications, initial target areas for adoption should be regions where significant postharvest losses occur from processes (e.g., rotting, pests) that result in large quantity losses rather than less serious quality losses. Quality losses (e.g., cracked rice from sun drying processes) are ordinarily not perceived as losses by small farmers and villagers, but these losses are important for participants in export markets. Therefore, improved drying technologies to be implemented in the small farmer sector must be coupled with suitable crop storage programs to effectively reduce quantity losses.

Third, crop losses from production to consumption are often substantial in many regions of the world. Although site-specific losses as high as 40 percent have been reported for maize, the panelists felt that realistic average loss levels are much lower (e.g., grain losses of 10 to 15 percent in West Africa, rice losses in Asia below 25 percent with occasional site-specific losses of 33 percent). The principal motivations for adoption of improved drying technologies are the reduction of postharvest losses, the ability to harvest in inclement weather, the possibility of adopting two-season growing patterns, and the maintenance of consistent product quality.

Fourth, the use of cabinet dryers for drying fruits, vegetables, and fish had not been observed personally by the panelists in developing countries. In fact, with the exceptions of limited canning operations and some smoking and/or drying of fish, the preserving of fresh foods was not observed at any level by the panelists, even though price fluctuations for fresh produce as high as a factor of ten or more occurred regularly during off-season periods. In any case, expanded domestic and/or export markets for dried produce appear to be the main application areas for solar cabinet dryers. However, of the three uses for solar drying technologies discussed (i.e., crop, produce, and timber), cabinet dryers appear to have the largest potential for the development of small-scale, inexpensive designs which could be appropriate for use at the single-household or small farmer level.

Finally, solar timber drying appears to be potentially applicable in developing countries only to meet the needs for export markets. These needs occur, on a large scale, for timber drying to improve quality prior to making bulk export shipments and, on a small scale, for cottage industries to dry indigenous wood materials for use in the fabrication of products for export, thereby improving the quality and value of wood export products. Utilization of solar timber kilns supports the current trends in many developing countries toward exporting processed goods rather than bulk raw materials. The following are additional points that were made during the discussions:

1. Crop drying needs in developing countries occur at three levels: individual farmers, farmer organizations, and industrial or government-sponsored installations. Solar crop drying systems are not, in general, easily scaled down to the size appropriate for individual farmers in developing countries; however, potential applications may occur at the level of farmer organizations formed by 10 to 15 individual farmers.
2. Heated crop dryers in developing countries are generally owned and operated by large entrepreneurs or by government organizations. Forced-air, fossil-fuel fired dryers are typically introduced by government subsidy programs and are imported or built by local, experienced craftsmen, with the dominant cost item ordinarily being an internal combustion engine.
3. Operators of fossil-fuel fired dryers have the tendency to try to accelerate the drying process and often damage grain from exposure to excessive temperatures.
4. Most batch-type crop dryer designs could be built in many developing countries, but nearly all continuous-flow

dryer designs would have to be imported. Nevertheless, nearly all forced-air and/or heated-air crop dryers operating in developing countries have been imported as a complete unit from developed countries.

5. Very low-cost crop dryer designs are most easily obtained by eliminating the need for perforated floors, usually constructed of metal, which are a very expensive component of the designs. One low-cost design observed in India is a natural-convection dryer constructed of plastic stretched over a wood or bamboo frame which is apparently placed over grain spread out to dry.
6. Most grain in developing countries is sold at the time of harvest rather than during or at the end of a storage season. This situation creates a buyer's market and often results in high moisture content grain being sold.
7. Price fluctuations for grain in developing countries commonly reach three- to seven-fold increases from harvest times to periods shortly before the following harvest. Hence, economic potential for drying and storage capabilities exists, but most farmers do not have the means to capitalize on these market changes.
8. In order to realize significant improvements over open-air sun drying, crop dryer designs must have temperature control mechanisms. Designs employing automatic temperature controls to reduce the potential for overheating the crops to be dried are the most attractive for use in developing countries.
9. The quality of crop products is not rewarded normally in developing country domestic markets which effectively creates an economic disincentive for the adoption of solar crop drying technologies.
10. Timber kilns are being installed in a number of wood exporting countries. These systems range in size upward from capacities of 30,000 board feet, which is well beyond the cottage industry scale in developing countries.
11. Timber for export is commonly air dried to a 20 to 25 percent moisture content in developing countries and then kiln dried upon receipt at the importing country. Timber for domestic consumption in developing countries is ordinarily not dried and, if drying is utilized, the required moisture levels in tropical countries are reachable by air drying technologies.
12. Solar timber drying technologies reduce air drying times by about a factor of two on the average. However, stress relief of the dried timber is more difficult in solar drying systems because elevated temperatures (i.e., 130°F as a minimum) and increased humidity levels (90 percent) which are the most effective method for stress relief are less easily obtainable in solar drying designs than in fuel-heated drying kilns.

13. Solar timber drying should be technically and economically feasible in many developing countries even though temperature and humidity control is very important for timber drying and the preferred drying sequence varies considerably with timber species.

In general, solar drying technologies for applications in developing countries are not well developed. The technologies typically suffer from difficulties in meeting the end-use needs in developing countries, especially in terms of system size. In addition, solar drying technologies have not penetrated potential markets in the United States, which are not as scale limited, as the result of being technically (i.e., performance) and/or economically unattractive. These technologies may not be deployed successfully in developing countries on a widespread basis until after the technologies are demonstrated and adopted to a greater extent in the United States. However, alternative low-cost designs that would not, in general, be acceptable for U.S. applications may be deployed succesfully in developing countries without the support of developed markets in the United States.

Afternoon session

The afternoon discussion focused on implementation programs for solar drying technologies and on potential U.S. initiatives to support the implementation of these technologies in developing countries (see table A.5). The following remarks summarize the response by the panelists.

1. Solar drying technologies have generally not matured to the stage where extension efforts in developing countries are likely to be successful, except in limited areas where the technologies have been tested and proven locally.
2. Since solar drying technologies are still largely unproven, demonstration programs in developing countries may be the most effective way to gain operational experience and to expose individuals in developing countries to these technologies.
3. Cooperative efforts between U.S. institutions and developing country research organizations should be established to further develop appropriate solar drying designs for use in developing countries and to form a foundation for a closely linked research/extension effort when the technologies have been adequately developed. U.S. universities should play an important role in these activities.
4. Cost reduction and performance improvements should be the primary objectives of research and development programs on solar drying technologies.

Table A.5. Questions for Advance Consideration: Initiatives and Mechanisms for Implementation of Solar Drying

1. Where should the most emphasis be placed in order to expand the use of solar dryer technologies? Design improvements? Cost reductions? Improvement of social or cultural acceptability? Evaluation of needs? Performance evaluations?
2. Are information dissemination problems a major impediment to widespread use of solar drying technologies?
3. What is the single best action the United States could take to assist in the deployment of solar dryers in developing countries?
4. What developing country institutions have the capability and know-how to work on solar drying applications? List.
5. What is the single best action that could be taken by developing country institutions and/or governments to support the introduction of solar drying technologies?
6. What U.S. institutions currently have the capability and know-how to work with developing country institutions on solar drying applications? List.
7. What are the realistic opportunities for U.S. and developing country cooperative efforts in solar dryer development and implementation?
8. Is the use of cooperative U.S. and developing country institutional capabilities the best vehicle to organize, disseminate, and update solar dryer design and operational information?
9. What are the best institutional vehicles to carry out extension efforts in developing countries for implementation of solar dryers?
10. How much infrastructure building is needed in developing countries to handle extensive solar drying extension efforts?
11. Can U.S. institutions contribute effectively to the development of an international cooperative effort to evaluate solar drying extension activities?
12. Can U.S. expertise in solar drying technologies contribute to technical improvements in designs for applications in developing countries? What U.S. institutions could contribute effectively in this area?

5. Capital availability is likely to remain a significant barrier to the deployment of solar drying technologies, especially for small-scale systems appropriate for most potential applications in developing countries.
6. Solar drying technologies for developing countries must be designed so that all or nearly all of the required hardware can be manufactured in the developing country where the extension efforts are being directed.
7. Additional effort needs to be focused toward the identification and definition of needs that solar drying technologies may be able to fulfill in developing countries.
8. Desirable actions that could be taken by the U.S. to facilitate the deployment of solar drying technologies in developing countries include:

   - establish programs to place U.S. experts on-site in developing countries to assist in implementation and research efforts at the request of developing country institutions;
   - encourage U.S. universities to contribute to the education of foreign students in solar technologies that are appropriate for applications in their own countries;
   - provide assistance in the development of solar drying technology demonstration programs in developing countries.

9. Examples of developing country institutions that have or could easily acquire the capability and know-how to contribute to the development and implementation of solar drying technologies are (see Appendix B for a more comprehensive tabulation):

   - International Institute for Tropical Agriculture, Nigeria
   - Asian Institute of Technology, Thailand
   - Forest Research Institute and Colleges, Dehra Dun, India
   - International Rice Research Institute, Philippines
   - National Forestry Laboratory of India
   - Indian Institute of Technology, Kharaghpur, India
   - UNICEF Eastern Africa Regional Office, Nairobi, Kenya
   - University of Sri Lanka
   - Centro Internacional de Agricultura Tropical, Cali, Columbia

Additional remarks

In general, solar drying technologies may have greater potential for applications in developing countries than in the United States because of special situations found in developing countries such as (1) the lack of power for forced-air dryers which may make natural-convection designs attractive, (2) the relatively high availability of solar radiation, (3) the relatively high levels of humidity which make ambient air drying difficult, and (4) less stringent product quality requirements.

Solar crop drying research activities have been one of the most heavily funded areas in agricultural systems research in the United States during the past several years. One of the results of the research efforts has been the finding that, in most U.S. applications, heated-air dryers are not required because drying can be achieved equally effectively by passing high volumes of ambient air through the dryer.

The U.S. model of agricultural extension services linked with research activities in universities may be an important approach to be utilized in the deployment of solar drying technologies in developing countries. Arrangements of this type facilitate the important linking of extension and research efforts and help provide the necessary communication channels for transmitting feedback from extension efforts to the research institutions responsible for design development, modification, testing, and demonstration. However, experience indicates that the U.S. extension service approach cannot be transferred successfully without modification and must be adapted to meet the local situation in the target developing country.

Appendix B Organizations Involved in Appropriate Technology for Renewable Resource Utilization

This Appendix consists of listings of names and addresses of organizations involved in the topical areas of our study: wind energy, improved cookstoves, solar drying, and rice bran processing. Tables B.1 through B.4 list U.S. organizations; tables B.5 through B.8 list developing country organizations; and tables B.9 through B.12 list other national and international organizations. As was indicated in Chapter 6, these lists are by no means complete. We welcome additions and corrections.

Because of the information contained on raw materials and products, the lists of organizations dealing with research on manufacturing of building materials from agricultural wastes are included in Chapter 5. Chapter 3 contains lists of windmill manufacturers.

Table B.1. U.S. Organizations Involved in Wind Energy

Alternative Sources of Energy
Route 2, Box 90-A
Milaca, MN 56353

American Wind Energy Association
1621 Connecticut Avenue, N.W.
Washington, D.C. 20009

Appropriate Technology Group
Rte. 1, Box 93-2
Oskaloosa, KS 66066

Appropriate Technology Research
1938 Hano Road
Santa Fe, NM 87501

Brown University
Prospect Street
Providence, RI 02912

Earthmind
5246 Boyer Road
Mariposa, CA 95338

Georgia Institute of Technology
Dept. of Aerospace Engineering
Atlanta, GA 30332

New Alchemy Institute - East
Box 432
Woods Hole, MA 02543

Oklahoma State University
School of Electrical Engineering
South Stillwater, OK 74074

Princeton University
Flight Concepts Laboratory
Princeton, NJ 08540

Rockwell International Wind Systems
Program
P.O. Box 464
Golden, CO 80401

Small Farm Energy Project
Center for Rural Affairs
P.O. Box 736
Hartington, NE 68739

University of Massachusetts
Dept. of Civil Engineering
Amherst, MA 01002

Volunteers in Asia, Inc.
Appropriate Technology Project
Box 4543
Stanford, CA 94305

Volunteers in Technical Assistance, Inc.
3706 Rhode Island Avenue
Mt. Rainier, MD 20822

Windworks
Box 329, Rte. 3
Mukwonago, WI 53149

Table B.2. U.S. Organizations Involved in Improved Cookstoves

Appropriate Technology
International (ATI)
1709 N Street, N.W.
Washington, DC 20036

Aprovecho Institute
359 Polk Street
Eugene, OR 97402

Arca Foundation
New York, NY

CARE
New York, NY

Center for Environmental Studies
Princeton University
Engineering Quadrangle
Princeton, NJ 08540

Community Services Administration
Region II
26 Federal Place
New York, NY 10007

Georgia Institute of Technology
Small Industry Development Network
Atlanta, GA 30332

National Center for Appropriate
Technology (NCAT)
P. O. Box 3838
Butte, MT 59701

Northeastern Forest Experiment
Station
Forest Products Marketing
Lab.
Princeton, WV 24740

Northwest Appropriate Technology
Institute
Box 1211
Friday Harbor, WA 98250

Peace Corps.
806 Connecticut Avenue, N.W.
Washington, DC 20525

Save the Children Federation (SCF)
48 Wilton Road
Westport, CT 06880

Technical Assistance Information
Clearinghouse
200 Park Avenue, South
New York, NY 10003

Transitional Network for Appropriate
Alternative Technology (TRANET)
P.O. Box 567
Rangeley, ME 04970

Volunteers in Asia (VIA)
Appropriate Technology Project
Box 4543
Stanford, CA 94305

Volunteers in Technical Assistance
(VITA)
3706 Rhode Island Avenue
Mt. Rainier, MD 20822

Wood Energy Institute
P. O. Box 800
Camden, ME 04843

World Vision International
919 West Huntington Dr.
Monrovia, CA 91016

Table B.3. U.S. Organizations Involved in Solar Drying

ACTION
806 Connecticut Avenue, N.W.
Washington, DC 20525

Agricultural Research and
Education Center
University of Florida
P. O. Box 1088
Lake Alfred, FL 33850

Colorado State Agricultural Station
Colorado State University
Fort Collins, CO 80521

Community Environmental Council
Solar Research Group
109E De La Guerra
Santa Barbara, CA 93101

Domestic Technology Institute
12520 W. Cedar Drive
Lakewood, CO 80215

Florida Agricultural Experiment
Station
University of Florida
1022 McCarty Hall
Gainesville, FL 32611

Florida Solar Energy Center
Florida Technological University
Cape Canaveral, FL 32920

Hawaii Agricultural Experiment
Station
University of Hawaii
Honolulu, HI 96822

Hawaii Natural Energy Institute
University of Hawaii
Honolulu, HI 96822

Helio Associates, Inc.
P. O. Box 17960
Tucson, AZ 85731

Illinois Agricultural Experiment
Station
University of Illinois
109 Munford Hall
Urbana, IL 61801

Iowa Agricultural and Home Economics
Experiment Station
Iowa State University
Ames, IA 50010

Kansas Agricultural Experiment Station
Kansas State University
Manhattan, KS 66506

Kentucky Agricultural Experiment
Station
University of Kentucky
Agricultural Science Building
Lexington, KY 40506

Living Food Dehydrators
P. O. Box 546
Fall City, WA 98025

Michigan State University
Agricultural Engineering Dept.
East Lansing, MI 48824

Nebraska Agricultural Experiment
Station
University of Nebraska
Lincoln, NE 68583

North Carolina Agricultural Experiment
Station
North Carolina State University
Box 5847, College Station
Raleigh, NC 27650

(continued)

Table B.3. (continued)

Office of International Programs
Engineering Experiment Station
Georgia Institute of Technology
Atlanta, GA 30332

Ohio Agricultural Research and Development Center
Ohio State University
Wooster, OH 44691

Oklahoma Agricultural Experiment Station
Oklahoma State University
Stillwater, OK 74074

Small Farm Energy Project
Center for Rural Affairs
P. O. Box 736
Hartington, NE 68739

South Dakota State University
Dept. of Agricultural Engineering
College Station
Brookings, SD 57006

Tennessee Agricultural Experiment Station
University of Tennessee
P. O. Box 1071
Knoxville, TN 37901

U.S. Department of Agriculture
Agricultural Research Service
Beaumont, TX 75215

U.S. Department of Agriculture
Citrus and Subtropical Products Lab.
Winter Haven, FL 33880

U.S. Department of Agriculture
Coastal Plain Experimentation Station
Tifton, GA 31794

U.S. Department of Agriculture
Forest Service
Forest Products Laboratory
P. O. Box 5130
Madison, WI 53705

U.S. Department of Agriculture
Forest Service
Northeastern Forest Experiment Station
Princeton, WV 24740

University of Maryland
Agricultural Experiment Station
College Park, MD 20742

Volunteers in Technical Assistance
3706 Rhode Island Avenue
Mt. Rainier, MD 20822

Table B.4. U.S. Organizations Involved in Rice Bran R&D*

Center for Development Technology
Washington University
St. Louis, Missouri

Colorado State University
Department of Agricultural Engineering
Ft. Collins, Colorado

Denver Research Institute
Denver, Colorado

Insta-Pro Division
Triple "F" Company
Des Moines, Iowa

Koehring Farm Division
Brady Operations
Des Moines, Iowa

Meals for Millions Foundation
Santa Monica, California

Ralston-Purina Company
St. Louis, Missouri

Riviana Foods, Inc.
Houston, Texas

Southern Regional Research Center
U.S. Department of Agriculture
New Orleans, Louisiana

Texas A&M University
Food Protein R&D Center
College Station, Texas

Western Regional Research Center
U.S. Department of Agriculture
Albany, California

---

*Few of these are actually involved, but they have the potential to contribute to such efforts.

Table B.5. Developing Country Organizations Involved in Wind Energy

Centro Internacional de
Agricultura Tropical (CIAT)
Apartado Aereo No. 6713
Cali
COLOMBIA

Christian Relief and Development
Association
P. O. Box 5674
Addis Ababa
ETHIOPIA

Aligarh Muslim University
Aligarh, Uttar Pradesh
INDIA

Appropriate Technology Unit
Allahabad Polytechnic
Allahabad U.P. 211003
INDIA

Auroville Centre for Environmental
Studies
Pondicherry - 2, 605001
INDIA

Bharat Heavy Electricals
Hyderabad
INDIA

Central Arid Zone Research Institute
Wind and Solar Division
Jodhpur
INDIA

Indian Agricultural Research Institute
Agricultural Engineering Division
Dairy Road
New Delhi 110012
INDIA

Indian Institute of Science
Cell for the Application of
Science and Technology to
Rural Areas
Bangalore, Karnataka 560012
INDIA

Indian Institute of Technology
Appropriate Technology Unit
Powai, Bombay 400076
INDIA

Ministry of Industry and Civil Supplies
Appropriate Technology Cell
268 Udyog Bharan
New Delhi - 11011
INDIA

National Aeronautical Laboratory
Post Bag 1779
Bangalore, Karnataka 560017
INDIA

National Council of Science and Technology
Technology Bharan
New Meharavley Rd.
New Delhi
INDIA

Organization of the Rural Poor
Kosumihkalan
P. O. Maduahi Urf Pahonehi
Ghazipur District
Uttar Pradesh
INDIA

Social Work and Research Centre
Tiloria
Admer District
Rajasthan
INDIA

Bandung Institute of Technology
Aero and Hydrodynamic Laboratory
Bandung
INDONESIA

National Christian Council of Kenya
P. O. Box 45009
Nairobi
KENYA

(continued)

Table B.5. (continued)

University of Nairobi
Dept. of Mechanical Engineering
P. O. Box 30197
Nairobi
KENYA

Malaysian Agricultural Research and Development Institute
P. O. Box 208
Sungai Besi,
Sedang, Salangor
MALAYSIA

Tribhuvan University Research
Centre for Applied Science and Technology
Tripureswor, Kathmandu
NEPAL

International Institute of Tropical Agriculture
P.M.B. 5320
Ibadan
NIGERIA

National Institute for Development Technology (ITINTEC)
Av. Salaverry 2461
Lima 27
PERU

International Rice Research Institute
Agricultural Engineering Dept.
P. O. Box 933
Los Baños, Manila
PHILIPPINES

Korea Advanced Institute of Science
Dept. of Mechanical Engineering
P. O. Box 150
Cheong-Ryang-Ri, Seoul
REPUBLIC OF KOREA

Ministry of Irrigation, Power, and Highways
Secretariat Bldg.
Colombo 1
SRI LANKA

Sarvodaya Educational Development Institute
Appropriate Technology Project
77 De Soysa Road
Moratuwa
SRI LANKA

Arusha Appropriate Technology Project (AATP)
P. O. Box 764
Arusha
TANZANIA

Christian Council of Tanzania
Dept. for Development Services
P. O. Box 2537
Dar es Salaam
TANZANIA

Ministry of Lands, Settlement, and Water Development
Water Development and Irrigation Division
P. O. Box 3030
Arusha
TANZANIA

University of Dar es Salaam
Engineering Faculty
P. O. Box 35091
Dar es Salaam
TANZANIA

King Mongkut's Institute of Technology
Sukswas 48, Bangmod
Rasburana, Bangkok 14
THAILAND

Ministry of Agriculture and Cooperative
Agricultural Engineering Division
Kaselsart University Grounds
Bangkhen, Bangkok 9
THAILAND

National Energy Administration
Design and Energy Research Section
Pibultham Villa
Ksastsuke Bridge
Bangkok
THAILAND

Thai USA Industrial Factory
2 Pracharaj II Road
Duist, Bangkok
THAILAND

L'Ecole Inter-Etats d'Ingéneurs de l'Equipement Rural
B.P. 7023
Ougadougou
UPPER VOLTA

Table B.6. Developing Country Organizations Involved in Improved Cookstoves

BRIDEC
Ministry of Education
Private Bag 005
Gaborone
BOTSWANA

Rural Industries
Innovation Centre
BOTSWANA

University of Sao Paulo
Sao Paulo
BRAZIL

Arab States Fundamental Education Center (ASFEC)
Sirs el Layyan
Menoufia Province
EGYPT

African Training and Research Centre For Women
ECA, Box 3001
Addis Ababa
ETHIOPIA

Center for Applied Studies in Development (CASD)
University of the South Pacific
Box 1168
Suva
FIJI

Nasinu Forestry Station
Timber Utilization Office
Suva
FIJI

School of Engineering
University of the South Pacific
P. O. Box 1168
Suva
FIJI

South Pacific Commission Community Leadership Training Center
Samabula, Suva
FIJI

Department of Social Welfare and Community Development
Accra
GHANA

Technology Consultancy Center (TCC)
The University of Science and Technology
Kumasi
GHANA

Alliance for Community Development
El Quiche
GUATEMALA

Estación Experimental Choqui
Choqui Experimental Station
Apartado Postal 159
Quezaltenango
GUATEMALA

Investigaciones Cientificas Asociadas Del Altiplano (ICADA)
Apartado Postal 159
Quezaltenango
QUATEMALA

Centro Mesoamericano de Estudios Sobre Tecnologia Apropiada (CEMAT)
Apartado Postal 1160
GUATEMALA

Ministry of Education
Extramural Department
GUATEMALA

Agricultural Tools Research Center
Suruchi Vasahat
P. O. Box 4
Bardoli - 1
District Surat Gujarat
INDIA

Appropriate Technology Development Association
P. O. Box 311
Gandhi Bhawan
Lucknow 226001, U.P.
INDIA

Central Fuel Research Institute
Dhanbad
INDIA

Farm Information Unit
Ministry of Food and Agriculture
New Delhi
INDIA

Indian Agricultural Institute
New Delhi
INDIA

Indian Institute of Science
Application of Science and Technology to Rural Areas (ASTRA)
Bangalore
INDIA

National Building Organization
Nirman Bhavan
G Wing
Maulara Azad Road
New Delhi 11
INDIA

Saghan Kshetra Vikas
Samiti, Sewapuri
Varnasi, Uttar Pradesh
INDIA

Ungra Extension Center
Karnataka State
INDIA

Village Polytechnics
Youth Development
Dept. of the Ministry of Housing and Social Services
KENYA

Community Development Foundation
Honduras
PHILIPPINES

Table B.7. Developing Country Organizations Involved in Solar Drying*

Christian Organization for Relief
and Rehabilitation
P. O. Box 994
Dacca - 2
BANGLADESH

Bangladesh University of
Engineering and Technology
Ramna, Dacca
BANGLADESH

Centro de Pesquises do Cacav (CEPEC)
KM 26, Rodivia Ilheus - Itabung
Caixa Postal 7
Itabuna, B.A.
BRAZIL

Grupo de Conversao de Energia Solar
Universidade Estadual de Campinas
(UNICAMP)
Cidade Universitaria
Barao Geraldo
Caixa Postal 1170
Campinas, Sao Paulo
BRAZIL

Laboratorio de Energia Solar de U.F.Pb.
Centro de Technologia
Cidade Universitaria
Joao Pessoa
Paraiba 58000
BRAZIL

Centro Nacional de Investigaciones
De Café (CENICAFE)
Federación Nacional de Cafeteros de
Colombia
Chinchina, Caldas
COLOMBIA

Universidad del Valle
Department de Ingeniera Agricola
Ciuded Universitaria
Melendez
Apartado: Aereo 21-88, Nal. 439
Cali
COLOMBIA

Forestry Utilization Division
Nasinu
FIJI

University of Science and Technology
Department of Agricultural Engineering
Kumasi
GHANA

Forest Products Research Institute
Univ. P. O. Box 63
Kumasi
GHANA

Estación Experimental Choqui
Apartado Postal 159
Quezaltenango
GUATEMALA

All India Solar Energy Working Group
Central Salt & Marine Chemical
Research Institute
Waghawadi Road
Bhavnagar 2, Gujarat
INDIA

Annamalai University
Annamalalainagar P. O.
Tamil Nadu 608101
INDIA

---

*Includes only institutions which currently are involved or have been involved. A much larger list could be constructed if potential institutions were included.

(continued)

Table B.7. (continued)

Appropriate Technology Unit
Allahabad Polytechnic
Allahabad 211002
Uttar Pradesh
INDIA

Auroville Centre for Environmental Studies
Freternity Auroville
Pondicherry, Tamil Nadu
INDIA

Central Arid Zone Research Institute
Solar Energy Studies Section
Jodhpur, Rajasthan
INDIA

Central Building Research Institute
Roorkee, Uttar Pradesh
INDIA

Forest Products Research Institute
Wood Seasoning Branch
P. O. New Forest
Dehra Dun, Uttar Pradesh
INDIA

Indian Agricultural Research Institute
Agricultural Engineering Division
New Delhi 110012
INDIA

Indian Institute of Technology-Kanpur
Dept. of Mechanical Engineering
Kanpur 208016, Uttar Pradesh
INDIA

Indian Institute of Technology-Kharagpur
Kharagpur 721302, Dist. Midnapore
West Bengal
INDIA

Indian Institute of Technology-Madras
Dept. of Mechanical Engineering
Madras 600036, Tamil Nadu
INDIA

National Industrial Development Corp. Ltd.
Puri, New Delhi 110012
INDIA

National Physical Laboratory
Hillside Road
New Delhi 110012
INDIA

Shri A.M.M. Murugappa Chettiar Research Centre
Tiam House
11/12 North Beach Road
Madras, Tamil Nadu 600001
INDIA

University of Agricultural Sciences
Hebbal, Bangalore
Kornataka 560024
INDIA

Development Technology Center
Institute of Technology
P. O. Box 276
Jalan Ganeska 10, Bandung
INDONESIA

Caribbean Food and Nutrition Institute
Jamaica Centre
P. O. B. 140
Mona, Kingston 7
JAMAICA

Village Technology Unit
UNICEF
Box 44145
Nairobi
KENYA

(continued)

Table B.7. (continued)

Centre Technique Forestier Tropical
B. P. 904
Tananarive
MADAGASCAR

Universiti Malaya
Mechanical Engineering Department
Lembah Pantai
Kuala Lumpur 22-11
MALAYSIA

Laboratoire de l'Energie Solaire de Mali
(Mali Solar Energy Laboratory)
B. P. 134 Bamako
MALI

Research Centre for Applied Science and Technology
Tribhuvan University
Tripureswar, Kathmandu
NEPAL

University of Nigeria
Dept. of Mechanical Engineering
Nsukka, Anambra State
NIGERIA

Atomic Energy Centre
Lahore
PAKISTAN

Pakistan Council of Scientific and Industrial Research Labs
Shahrah-e-Kemal Ataturk Road
PPI Building
Karachi
PAKISTAN

Centro Internacional de la Papa (CIP)
(International Potato Center)
Apartado 5969
Lima
PERU

Instituto de Investigaciones de Applicaciones de Energias
Naturales de Ayacucho
PERU

Bureau of Energy Development
PHILIPPINES

Forest Products Research Institute
College, Laguna
PHILIPPINES

International Rice Research Institute (IRRI)
P. O. Box 933
Manila
PHILIPPINES

University of Dakar
Dakar - Fann
SENEGAL

Forest Department Utilization Section
Moshi
TANZANIA

Asian Institute of Technology
P. O. Box 2754
Bangkok
THAILAND

University of the West Indies
Dept. of Mechanical Engineering
St. Augustine, Trinidad
TRINIDAD and TOBAGO

Ankara Universitesi Ziraat
Facultesi Zirai
Kuvvet Makilanair Kursusu
Aydinlikevler, Ankara
TURKEY

Table B.8. Developing Country Organizations Involved in Rice Bran R&D

Appropriate Technology Development Association
INDIA

Central Food Technology Research Institute
Mysore
INDIA

Food Corporation of India
Mysore
INDIA

Indian Institute of Technology
Kharagpur
INDIA

Jadavpur University
Calcutta
INDIA

Paddy Processing Research Center
INDIA

International Rice Research Institute
PHILIPPINES

Applied Scientific Research Corporation
THAILAND

Kasetsart University
THAILAND

Ministry of Industry
Dept. of Science
Bangkok
THAILAND

Table B.9. Other National and International Organizations Involved in Wind Energy

Flinders University
Bedford Park
AUSTRALIA 5042

Brace Research Institute
MacDonald College
McGill University
P. O. Box 900
Ste. Anne de Bellevue 800
Quebec
CANADA

International Development Research Center (IDRC)
Box 8500
Ottawa K1G 3H9
CANADA

National Research Council
Low-Speed Aerodynamics Laboratory
Montreal Road
Ottawa K1H 6R6
CANADA

Folketeknik
Christianmindeus 11
Copenhagan 2100
DENMARK

Center of Wind Energy Research
Forschungsprojekt Windenergie
University of Regensburg
D-8400 Regensburg
FEDERAL REPUBLIC OF GERMANY

Technische Universitat Berlin
Interdisziplinare Projektgruppe for Angepasste Technologie (IPAT)
Lentzealle 86
D-1000 Berlin 33
FEDERAL REPUBLIC OF GERMANY

Wind Energy Research Institute
University of Stuttgart, E.U.
Pfaffenwaldring 31
D-7000 Stuttgart 80
FEDERAL REPUBLIC OF GERMANY

Arya-Mehr University
Mechanical Engineering Dept.
P. O. Box 3406
Teheran
IRAN

Ecodevelopment Cluster
Bu-Ali Sina University
P. O. Box 302
Hamadan
IRAN

Pahlavi University
Dept. of Mechanical Engineering
Shiraz
IRAN

Low Energy Systems
3 Larkfield Garden
Dublin 6
IRELAND

Eindhoven University of Technology
Technische Hogeschool Eindhoven
Den Dolech 2
Posthus 513, 5600 MB
Eindhoven
THE NETHERLANDS

Steering Committee for Wind Energy in Developing Countries (SWD)
P. O. Box 85
Amersfoort
THE NETHERLANDS

(continued)

Table B.9. (continued)

TOOL Foundation
Mauritskade 61a
1092 AD
Amsterdam
THE NETHERLANDS

Auckland University
Dept. of Mechanical Engineering
Auckland
NEW ZEALAND

Lincoln College
Dept. of Mechanical Engineering
Canterbury
NEW ZEALAND

Royal Institute of Technology
Wind Tunnel Dept.
Stockholm 70
SWEDEN

World Council of Churches
Commission of the Church's Participation in Development
150 Route de Ferney
P. O. Box 66
1211 Geneva 20
SWITZERLAND

World Meteorological Organization (WMO)
Case Postale No. 5,
CH-1211 Genève 20
SWITZERLAND

Brynoyre Farm
Talybonto-on-Usk
Brecon, Powys, LD3 7Y5
Wales
UNITED KINGDOM

Cranfield Institute of Technology
Cranfield, Bedfordshire
UNITED KINGDOM

Electrical Research Association
Power Engineering Division
Cleeve Road
Leatherhead, Surrey KT 227 SA
UNITED KINGDOM

Intermediate Technology Development Group, Ltd. (ITDG)
9 King Street
Covent Garden
London WC 2
UNITED KINGDOM

National Centre for Alternative Technology (NCAT)
Llwyngwern Quarry
Machnylleth, Dyfed
Wales
UNITED KINGDOM

Oxford Committee for Famine Relief
274 Banbury Road
Oxford OX2 70Z
UNITED KINGDOM

Royal College of Art
School of Industrial Design
Kensington
London SW7 2EV
UNITED KINGDOM

University of Liverpool
Dept. of Mechanical Engineering
Liverpool 3
UNITED KINGDOM

University of Oxford
Dept. of Engineering Science
Parks Road
Oxford OX1 3PJ
UNITED KINGDOM

Table B.10. Other National and International Organizations Involved in Improved Cookstoves

Brace Memorial Research Institute
MacDonald College
CANADA

Canadian Freedom from Hunger Foundation
CANADA

Canadian International Development Agency
CANADA

International Development Research Council (IDRC)
CANADA

Group de Recherches Sur Les Techniques Rurales (GRET)
34, ru Dumont d'Urville
75116, Paris
FRANCE

Organization for Economic Cooperation and Development (OECD)
Development Centre
94, ru Chardon-Lagache
75016 Paris
FRANCE

Home Economics Section of the Human Institutions and Agrarian Reform Division
Food and Agricultural Organization (FAO)
Rome
ITALY

Japanese Ministry of Agriculture and Forestry
JAPAN

TOOL Foundation
P. O. Box 525
Eindhoven
THE NETHERLANDS

Swedish International Development Authority (SIDA)
SWEDEN

World Employment Program of the International Labour Organization
Geneva
SWITZERLAND

Intermediate Technology Development Group (ITDG)
9 King Street
London WC2E 9HN
UNITED KINGDOM

Rural Communications Services
South Petherton, Somerset
UNITED KINGDOM

al Dir'iyyah Institute of Geneva Switzerland
1925 North Lynn Street
Arlington, Virginia
UNITED STATES

Table B.11. Other National and International Organization Involved in Solar Drying

Commonwealth Scientific and Industrial Research Organization (CSIRO)
Solar Energy Studies Unit
Box 225
Dickson, A.C.T. 2602
AUSTRALIA

CSIRO
Division of Mechanical Engineering
P. O. Box 26
Highett, Victoria 3190
AUSTRALIA

University of Melbourne
Parkville, Victoria 3052
AUSTRALIA

Brace Research Institute
MacDonald College of McGill University
Ste. Anne de Bellevue
Quebec, H0A LC0
CANADA

Lakehead University
Oliver Road
Thunder Bay "P"
Ontario, P7B 5E1
CANADA

University of Saskatchewan
Dept. of Mechanical Engineering
Saskatoon, Saskatchewan, S7N 0W0
CANADA

Centre Technique du Bois
10 Avenue St. Mande
75012 Paris
FRANCE

Centre Technique Forestier Tropical
45 Bis Avenue de la Belle Grabrielle
94000 Nogent sur Marne
FRANCE

Institute Francais du Cafe du Cacao
34 Rue des Renaudes
75017 Paris
FRANCE

Orta Dogu Teknik Universitsi
(Middle East Technical University)
Mechanical Engineering Department
Ismet Inonu Bulvari
Ankara
TURKEY

Commonwealth Forestry Institute
University of Oxford
South Parks Road
Oxford
UNITED KINGDOM

Jak Forest Fuels Ltd.
21 Biggar Road
Silverburn, Penicuik
Midlothian EH26 9LG
UNITED KINGDOM

National Institute of Agricultural Engineering
Scottish Station
Bush Estate
Penicuik
Midlothian KH26 OPH
UNITED KINGDOM

Tropical Products Institute
Ministry of Overseas Development
Industrial Development Department
Culham, Abingdon
Oxfordshire OX14 3DA
UNITED KINGDOM

Table B.12. Other National and International Organizations Involved in Rice Bran Processing

UNIDO
Vienna
AUSTRIA

Rice Bran Oil Technical Research Institute
Tokyo
JAPAN

Instituto de Agroquimica y Tecnologia de
Alimentos (IATA)
Valencia
SPAIN

Ron Hawkey Partners, Ltd.
Rice Consultants
Heve, Sussex
UNITED KINGDOM

Appendix C
Monetary Exchange Rates

Approximate monetary exchange rates for selected currencies and time periods are given in table C.1. There are limitations to the use of these data because of fluctuations in monetary exchange rates. In addition, many of the data in the report were taken from sources in which the dates of the cost estimates were not cited explicitly or the period of reference was such that these approximate exchange rates should be applied with caution. Consequently, the data in this table should be used only to provide rough estimates of U.S. dollar equivalents of cost estimates quoted in this report, where those U.S. dollar equivalents are not indicated.

Table C.1. Approximate Monetary Exchange Rates

Country	Currency	Approximate U.S. $ equivalent (current)					
		1955	1960	1965	1970	1975[a]	1978[b]
Australia	A$	2.24[c]	2.24[c]	2.24[c]	1.12	1.26	1.16
Egypt	pound (£E)	2.87	2.87	2.87	2.87	2.56	2.56
Ethiopia	Eth ($)	0.4	0.4	0.4	0.4	0.48	0.48
India	rupee (Rs)	0.21	0.21	0.21	0.13	0.11	0.13
Philippines	peso (P)	0.5	0.5	0.27	0.27	0.13	0.14
Tanzania	shilling (Sh)	-	-	-	0.14	0.12	0.13
Thailand	baht	-	-	0.05	0.05	0.05	0.05
United Kingdom	pound (£)	2.8	2.8	2.8	2.4	2.02	1.97

[a]The values refer to end of period rates in December 1975.

[b]The values refer to end of period rates in September 1978.

[c]These values represent Australian pounds, which were later decimalized to A $2.00 per pound.

Source:

International Monetary Fund, International Financial Statistics.

Notes

Introduction

1. R. Diwan and D. Livingston, *Alternative Development Strategies and Appropriate Technology* (New York: Pergamon Press, 1979); *Preliminary Report from a Workshop on Appropriate Technology*, December 1-5, 1978, Winrock International Conference Center, Morrilton, Arkansas, 1979.

2. *Methane Generation from Human, Agricultural and Animal Wastes* (Washington, D.C.: National Academy of Sciences - National Research Council, 1977).

3. *Preliminary Report from a Workshop on Appropriate Technology*, pp. 9-10.

4. United Nations Industrial Development Organization (UNIDO), *Draft Report of the Technical/Official Level Meeting to the Ministerial Level Meeting, International Forum on Appropriate Industrial Technology, New Delhi/Anand, India, 20-30 November 1978*. p. 4.

5. R. P. Morgan, *Science and Technology for Development: The Role of U.S. Universities* (New York: Pergamon Press, 1979), pp. 225-28 (with E. E. Irons, E. A. Perez, T. N. Soule, and A. K. Fried).

6. Donald D. Evans and L. N. Adler, eds., *Appropriate Technology for Development: A Discussion and Case Histories* (Boulder, Colorado: Westview Press, 1979).

Chapter 1

1. Board on Science and Technology for International Development, Energy for Rural Development: Renewable Resources and Alternative Technologies for Developing Countries (Washington, D.C.: National Academy of Sciences, 1976) (hereafter referred to as Energy for Development); Peter Fraenkel, "An International Development Programme to Produce a Wind-Powered Water-Pumping System Suitable for Small-Scale Economic Manufacture," Proceedings of the Second International Symposium on Wind Energy Systems, Amsterdam, The Netherlands, October 3-6, 1978; and Economic and Social Commission for Asia and the Pacific, Proceedings of the Meeting of the Expert Working Group on the Use of Solar and Wind Energy, Energy Resources Development Series No. 16, New York, United Nations, 1976 (hereafter referred to as ESCAP Proceedings).

2. VITA/CDT Wind Energy Panel, January 12, 1979.

3. E.W. Golding, The Generation of Electricity by Wind Power (New York: Philosophical Library, 1955).

4. P.D. Henderson, India: The Energy Sector (London: Oxford University Press, 1975), pp. 76, 165.

5. Arjun Makhijani, "Solar Energy and Rural Development for the Third World," Bulletin of the Atomic Scientists 32 (June 1976): 14-24; and ESCAP Proceedings, p. 8.

6. R.S.P. Govinda and R.N. Narasimha, "Windmills for Rural Use," Fluid Mechanics Report No. 77 FM 14, Indian Institute of Science, Bangalore, December 1977, p. 1.

7. Roland Reichel, "Rural Electrification in Tanzania," Research Report No. EP5, University of Dar es Salaam, Faculty of Electrical Engineering, Dar es Salaam, Tanzania, October 1977.

8. Peter Johnston and James Reierson, "Wind-Generated Electricity for Rural Areas of Fiji," August 1974.

9. ESCAP Proceedings, p. 16.

10. Marcus M. Sherman, "Practical Application of Wind Powered Water Pumps," Economic and Social Commission for Asia and the Pacific, Bangkok, Thailand, February 1977.

11. Ibid.

12. D.F. Warne and P.G. Calnan, "Generation of Electicity from the Wind," Proc. IEE 124 (November 1977): 963-85.

13. Fraenkel, "An International Development Programme."

14. W.A.M. Jansen and P.T. Smulders, Rotor Design for Horizontal Axis Windmills (Amersfoort, The Netherlands: Steering Committee for Wind Energy in Developing Countries, May 1977).

15. Sharat K. Tewari, "Economics of Wind Energy Use for Irrigation in India," Science 202 (1978): 481-86.

16. Personal communication from P.T. Smulders, November 23, 1979.

17. Frank R. Eldridge, Wind Machines (New York: Van Nostrand Reinhold, 1980), p. 110.

18. Steve Blake, "Wind Driven Water Pumps: Economics, Technology, Current Activities," Appropriate Technology Group, Oskaloosa, Kansas, December 1978; and personal communication from Dr. James Spain, March 2, 1979.

19. J. Kozlowski, "Investigation of Whether a Savonius Rotor is a Suitable Prime Mover," Intermediate Technology Development Group, London, U.K., 1976.

20. J. Needham, Science and Civilization in China, Vol. 4: Physics and Physical Technology, Part II - Mechanical Engineering (London: Cambridge University Press, 1965), p. 558.

21. ESCAP Proceedings.

22. Sherman, "Practical Application of Wind Powered Water Pumps."

23. M.M. Sherman, "6,000 Hand-crafted Sailwing Windmills of Lassithiou, Greece and Their Relevance to Windmill Development in Rural India," International Conference, Appropriate Technologies for Semiarid Areas: Wind and Solar Energy for Water Supply, West Berlin, German Foundation for International Development, September 15-20, 1975.

24. For a description of Cretan-type Windmill implementation efforts in Ethiopia, see Peter Fraenkel, Food for Windmills (London: Intermediate Technology Publications, 1975). See also M.M. Sherman, "An Interim Report: The Design and Construction of an Appropriate Water-Pumping Windmill for Indian Agriculture," International Conference, Appropriate Technologies for Semiarid Areas: Wind and Solar Energy for Water Supply, West Berlin, German Foundation for International Development, September 15-20, 1975.

25. ESCAP Proceedings, p. 68.

26. E. Barbour, The Homemade Windmills of Nebraska (Occidental, Calif.: Farallones Institute, 1976); and Energy for Development, p. 117.

27. Govinda and Narasimha, "Windmills for Rural Use"; and H. Djojodihardjo, "Research and Prospects of Utilization of Wind Energy in Indonesia," Proceedings of the Meeting of the Expert Working Group on the Use of Solar and Wind Energy, Energy Resources Development Series No, 16 (New York: United Nations, 1976), pp. 125-29.

28. Peter Fraenkel, "Vertical Axis Wind-Machines," Appropriate Technology 3 (August 1976): 17-19.

29. ESCAP Proceedings, p. 68.

30. Personal Communication from Dr. James Spain, March 2, 1979.

31. Ibid.

32. Blake, "Wind Driven Water Pumps," p. 21.

33. Personal Communication from Dr. James Spain, March 2, 1979.

34. Blake, "Wind Driven Water Pumps," p. 21.

35. Ibid., p. 23.

36. Govinda and Narasimha, "Windmills for Rural Use."

37. Ibid., pp. 8-9.

38. ESCAP Proceedings, pp. 65-67.

39. National Energy Administration, "Research, Development, and Use of Wind Energy in Thailand," Proceedings of the Meeting of the Expert Working Group on the Use of Solar and Wind Energy, Energy Resources Development Series No. 16 (New York: United Nations, 1976) pp. 108-14.

40. ESCAP Proceedings, p. 9.

41. N. van de Ven, Construction Manual for a Cretan Windmill (Amersfoort, The Netherlands: Steering Committee for Wind Energy in Developing Countries, October 1977).

42. Windworks, "25 Foot Diameter Sail Windmill: Design Manual," Windworks, Mukwonago, Wisconsin, June 1974.

43. National Aeronautical Laboratory, "A Programme for the Development and Utilization of Windmills in India," National Aeronautical Laboratory, Bangalore, India, September 1978, pp. 11-12; and Tewari, "Economics of Wind Energy."

44. Fraenkel, Food from Windmills.

45. T.E. Sweeney, "The Bishop's Windmill," Princeton University, Department of Aerospace and Mechanical Sciences, Princeton, New Jersey, July 1976.

46. Marcus M. Sherman, "A Water Pumping Windmill that Works," Journal of the New Alchemists 2 (1974): 22-27.

47. Hartmut Bossel, Low-Cost Windmill for Developing Nations: Multivane Fan Type (Mt. Rainier, Md.: Volunteers in Technical Assistance, 1970).

48. ESCAP Proceedings, p. 68.

49. NAL, "A Programme for the Development and Utilization of Windmills in India," p. 2.

50. Ibid., pp. 12-14, Table 5.

51. Fraenkel, "An International Development Programme"; and personal communication from Peter Fraenkel, February 4, 1979.

52. Sherman, "A Water Pumping Windmill that Works."

53. "Peruvian Windmills - Spinning at Last!" Unitarian Universalist Service Newsletter from Headquarters 7 (Winter 1978-79).

54. Volunteers in Technical Assistance, "Co-op To Produce Windmills Developed by Arusha Center," VITA News (August 1978): 6; Dick Stanley and Ken Darrow, "The Arusha Windmill: A Construction Manual," Volunteers in Technical Assistance, Mt. Rainier, Maryland, 1978.

55. Personal communication from Chris Ahrens, undated.

56. Personal communication from Jorge Zapp Glauser, December 12, 1980. For further information about the Gaviotas program, see J. Zapp Glauser, "The Gaviotas Program", in Fundamental Aspects of Appropriate Technology, edited by J.

de Schutter and G. Bemer (Delft, The Netherlands: Delft University Press, 1980), pp. 9-13.

57. National Energy Adminstration, "Research, Development"; and William E. Heronemus, "A Survey of the Possible Uses of Windpower in Thailand and the Philippines," University of Massachusetts, Amherst, Mass., November 1974.

58. Chung-Oh Lee, "Windpower Studies in Korea," Proceedings of the Meeting of the Expert Working Group on the Use of Solar and Wind Energy, Energy Resources Development Series No. 16 (New York: United Nations, 1976), pp. 121-24.

59. Peter Bade, "Flapping-Vane Wind Machine and Rod-Piston Pump, An Integrated Delivery System for Large Well Depths and Small Flow Rates," and "Flapping-Vane Wind Machine," International Conference, Appropriate Technologies for Semiarid Areas: Wind and Solar Energy for Water Supply, West Berlin, German Foundation for International Development, September 15-20, 1975, pp. 71-82, 83-88.

Chapter 2

1. Erik Eckholm, "Planting for the Future: Forestry for Human Needs," Worldwatch Paper 26, Worldwatch Institute, Washington, D.C., February 1979, p. 19.

2. Keith Openshaw, "Wood Fuels the Developing World," New Scientist 61 (January 1974): 271-72.

3. Arjun Makhijani and Alan Poole, Energy and Agriculture in the Third World (Cambridge, Mass.: Ballinger, 1975), p. 68.

4. Ianto Evans, Lorena Owner-Built Stoves, (Stanford, Calif: The Appropriate Technology Project of Volunteers in Asia, 1979).

5. E.G.K. Rao, "Improving the Domestic Chula," Indian Farming, (January 1962).

6. Whitney Garberson, "Preliminary Testing of Wood Consumption in Four 'Smokeless Cooker' Models and the Traditional Three-Stone Fire," Peace Corps, Ouagadougou, Upper Volta, January 1979.

7. J. Goldemberg and R.I. Brown, "Cooking Stoves: The State of the Art," Institute of Physics, University of Sao Paulo, Sao Paulo, Brazil, January 1979, p. 11.

8. Rao, "Improving the Domestic Chula."

9. Openshaw, "Wood Fuels the Developing World."

10. Goldemberg and Brown, "Cooking Stoves," p. 13.

11. J.E.M. Arnold and Jules Jongma, "Fuelwood and Charcoal in Developing Countries," FAO, Rome, 1978, p. 7.

12. Ibid., p. 5.

13. Roger Revelle, "Requirements for Energy in the Rural Areas of Developing Countries," in Renewable Energy Resources and Rural Applications in the Developing World,

edited by Norman L. Brown (Boulder, Colo.: Westview Press, 1978), p. 24.

14. Erik Eckholm, "The Other Energy Crisis: Firewood," Worldwatch Paper 1, Worldwatch Institute, Washington, D.C., September 1975, p. 8.

15. Arnold and Jongma, "Fuelwood and Charcoal," p. 6.

16. Eckholm, "The Other Energy Crisis," p. 10.

17. Arnold and Jongma, "Fuelwood and Charcoal," p. 6.

18. Eckholm, "Planting for the Future," p. 21.

19. Walter Parham, "LDC Deforestation Problem - A Preliminary Data Base," draft (Washington, D.C.: Dept. of State, Office of Science and Technology, March 1978).

20. Ibid.

21. Ianto Evans, "The Lorena Fuel Conserving Stove: A Construction Manual," Appropriate Technology Project of Volunteers of Asia, Stanford, Calif., 1978; and VITA/CDT Cookstove Panel, held in Washington, D.C., February 9, 1979.

22. S.P. Raju, Smokeless Kitchens for the Millions (Madras, India: The Christian Literature Society, 1953), p. 1.

23. Rao, "Improving the Domestic Chula."

24. Institute of Natural Resources, University of the South Pacific, and Department of Forestry, Fiji Government, "Wood as an Alternative Energy Resource." (Report of the seminar held at the University of Bure, Laucala Bay Campus, University of the South Pacific, Suva, Fiji, July 3-4, 1978), pp. 127-29.

25. Evans, "The Lorena Fuel Conserving Stove," p. 7.

26. David French, "Firewood in Africa" (Discussion paper for the African Bureau Firewood Workshop, June 12-14, 1978); U.S. Agency for International Development, Washington, D.C., June 1978, p. 10.

27. Rao, "Improving the Domestic Chula."

28. H. Singer, "Report to the Government of Indonesia on Improvements of Fuelwood Cooking Stoves and Economy in Fuelwood Consumption," FAO Report #1315, Rome, 1961.

29. T. Kalluppatti, "Magan Choola," Gandhiniketan Ashram, Madurai District, India, 1955.

30. Rao, "Improving the Domestic Chula."

31. Raju, "Smokeless Kitchens for the Millions," p. 37.

32. B. Acherjee, S.K. Chanda, and R. Chatterjee, "Efficiency of Masonry Domestic Chullah," Indian Standards Institution Bulletin 26 (September 1974): 344-49.

33. Evans, "The Lorena Fuel Conserving Stove."

34. Rao, "Improving the Domestic Chula."

35. B. Theodorovic, "Experiments with the Improved Egyptian Rural Stove," Arab States Fundamental Education Centre, Sirs-el-Layyan, Menoufia, Egypt, 1953.

36. Kalluppatti, "Magan Choola."

37. Ibid.

38. Ibid.

39. Farm Information Unit, "A Smokeless Chula for Every Home," Home Science Series No. 2, Ministry of Food and Agriculture, New Delhi, India, September 1961, p. 6.
40. Singer, "Report to the Government."
41. Ibid., p. 10.
42. Ibid., p. 16.
43. "Wood as an Alternative Energy Resource," p. 7.
44. Ibid., p. 8.
45. Ibid., pp. 8 and 103.
46. Canadian Freedom from Hunger Foundation and Brace Research Institute, "The Smokeless Stove - Ghana," Case Study #7, Appropriate Technology Handbook, Ottawa, Canada, 1976.
47. Evans, "The Lorena Fuel Conserving Stove."
48. Estación Experimental Choqui, "The Lorena Mudstove," Quezaltenango, Guatemala, 1977.
49. Evans, "The Lorena Fuel Conserving Stove."
50. Ibid.
51. Appropriate Technology International, "Dissemination of Home Cooking Stoves Made of Sand and Clay," Washington, D.C., 1978.
52. "Smokeless Fogones," unpublished note, March 26, 1965.
53. Gwendolyn Peters, "Construction of a Smokeless Fireplace," unpublished, undated.
54. Appropriate Technology Development Association, "Improved 'Chula' for Kitchen," Appropriate Technology Directory of Tools, Equipments, Machines, Plants, Processes and Industries, Vol. I, undated.
55. Singer, "Report to the Government," p. 7.
56. Georgia Institute of Technology, "Recent Efforts to Develop Simple Wood Burning Stoves," A State of the Art Survey of Solar Powered Irrigation Pumps, Solar Cookers and Wood Burning Stoves for Use in Sub-Sahara Africa (Atlanta, Ga.: Small Industry Development Network, 1978), p. 68.
57. Eric Simon and Pedro Solis, "Economic Stove that Burns Sawdust as Fuel," Appropriate Technology 4 (May 1977): 23-24.
58. Ibid.
59. Jeffrey Wartluft, "Double-Drum Sawdust Stove," VITA Technical Bulletin 29, Mt. Rainier, Md., undated.
60. Ken Darrow, "Rice Hull Stove," A.T. Sourcebook (Stanford, Calif.: Volunteers in Asia, 1976), pp. 177-78.
61. Mario P. Chanco, "The Rice Husk Stove," Appropriate Technology 5 (November 1978): 18-19.
62. VITA, "Outdoor Oven," Village Technology Handbook, 1975, p. 339.
63. Rev. Bertrand Saubolle, "Wood-Burning Oven," VITA Technical Bulletin 11, Mt. Rainier, Md., undated.

64. Rural Communication Services, "The Village Bakery," Leaflet No. 5, South Petherton, Somerset, England, undated.

65. G.S. Dutt, "Improved Wood Burning Cooking Stoves for LDC's," unpublished memo, undated; and Stephen Joseph and Yvonne Shanahan, "Designing a Test Procedure for Domestic Woodburning Stoves," Interim Report No. 1, Intermediate Technology Development Group, London, November 1980.

66. See for example, "Wood as an Alternative Energy Resource"; Singer, "Report to the Government"; Acherjee, Chanda, and Chatterjee, "Efficiency of Masonry Domestic Chullah"; and B.D. Ahuja and O.N. Gupta, "A Study on the Efficiency of Chulahs," Technical and Research Reports: 17, UDC 643.332, Government of India National Buildings Organisation, New Delhi, 1964.

67. Garberson, "Preliminary Testing"; and "Wood as an Alternative Energy Resource."

68. G.S. Dutt, "Reduced Cooking Energy Use in Rural India," Report PU/CES 74, Princeton University Center for Environmental Studies, Princeton, N.J., November 1978, p. 4.

69. Ibid.

70. Kalluppatti, "Magan Choola"; Acherjee, Chanda, and Chatterjee, "Efficiency of Masonry Domestic Chullah"; and Ahuja and Gupta, "A Study on the Efficiency of Chulahs."

71. Acherjee, Chanda, and Chatterjee, "Efficiency of Masonry Domestic Chullah."

72. G.S. Dutt, "Efficient Wood Burning Cooking Stove Literature," draft report, (Princeton, N.J.: Princeton University Center for Environmental Studies,undated).

73. Ibid.

74. Ibid.

75. Dutt, "Reduced Cooking Energy Use in Rural India," p. 7.

76. Garberson, "Preliminary Testing"; and "Wood as an Alternative Energy Resource."

77. VITA/CDT Cookstove Panel; Kalluppatti, "Magan Choola"; Ahuja and Gupta, "A Study on the Efficiency of Chulahs"; Dutt, "Efficient Wood Burning Cooking Stove Literature"; and B. Acherjee, S.K. Chanda, and R. Chatterjee, "Efficiency of Domestic Chullah," Indian Standards Institution Bulletin 26 (January 1974): 17-22.

78. Arnold and Jongma, "Fuelwood and Charcoal," p. 5.

79. Ibid.

80. Ibid., p. 3.

81. French, "Firewood in Africa," p. 3.

82. Goldemberg and Brown, "Cooking Stoves," p. 2.

83. Eckholm, "Planting for the Future," p. 29.

84. Eckholm, "The Other Energy Crisis," p. 7.

85. Arnold and Jongma, "Fuelwood and Charcoal," p. 6.

86. Eckholm, "The Other Energy Crisis," p. 6.

87. Ianto Evans, "Using Firewood More Efficiently," Aprovecho Institute, Eugene, Ore. (Special paper submitted to Eighth World Forestry Congress, Jakarta, Indonesia, 1979), p. 2.

88. S.A. Draper, "Wood Processing and Utilization at the Village Level," Third FAO/SIDA Expert Consultation on Forestry for Local Community Development, held in Semarang, Indonesia, December 1977, p. 6.

89. Synthetic Fuels Office, Development Division, California Energy Commission, "Woodburning Stoves," Staff draft report, November 1978.

90. Eckholm, "Planting for the Future," p. 29.

91. Arnold and Jongma, "Fuelwood and Charcoal," p. 2.

92. Eckholm, "The Other Energy Crisis," p. 7.

93. Revelle, "Requirements for Energy," p. 13.

94. Makhijani and Poole, Energy and Agriculture in the Third World, p. 109.

95. Eckholm, "The Other Energy Crisis," p. 7.

96. Ibid., p. 10.

97. Theodorovic, "Experiments with the Improved Egyptian Rural Stove."

98. J.E.M. Arnold, "Wood Energy and Rural Communities," Position paper for the Eighth World Forestry Congress, Jakarta, Indonesia, October 16-28, 1978. (Rome: FAO 1978), p. 8.

99. Rao, "Improving the Domestic Chula."

100. Arnold, "Wood Energy and Rural Communities," p. 14.

101. "Wood as an Alternative Energy Resource," p. 7.

102. Goldemberg and Brown, "Cooking Stoves," p. 8.

103. Arnold, "Wood Energy and Rural Communities," p. 16.

104. Ibid., p. 14.

105. French, "Firewood in Africa," p. 24.

106. "Wood as an Alternative Energy Resource," p. 30.

107. Evans, "Using Firewood More Efficiently," p. 3.

108. Arnold, "Wood Energy and Rural Communities," p. 14.

109. Chanco, "The Rice Husk Stove."

110. Arnold, "Wood Energy and Rural Communities," p. 7.

111. French, "Firewood in Africa," p. 11.

112. John W. Powell, "Wood Waste as an Energy Source in Ghana," in Renewable Energy Resources and Rural Applications in the Developing World, edited by Norman L. Brown (Boulder, Colo.: Westview Press, 1978), p. 23.

113. Arnold, "Wood Energy and Rural Communities," p. 9.

114. Ibid., p. 14.

115. VITA/CDT Cookstove Panel.

116. Arnold, "Wood Energy and Rural Communities," p. 14.

117. "Wood as an Alternative Energy Resource," p. 45; and Powell, "Wood Waste as an Energy Source in Ghana," p. 119.

118. Powell, "Wood Waste as an Energy Source in Ghana," p. 124.

119. Amulya Kumar N. Reddy, "Energy Options for the Third World," Bulletin of the Atomic Scientists 34 (May 1978): 28-33.

120. French, "Firewood in Africa," p. 5.

121. Ibid., p. 9.

122. Arnold, "Wood Energy and Rural Communities," p. 24.

123. French, "Firewood in Africa," p. 10.

124. Arnold, "Wood Energy and Rural Communities," p. 24.

125. Ibid., p. 25.

126. National Academy of Sciences, "Leucaena: Promising Forage and Tree Crop for the Tropics," Report of a study conducted jointly by the Philippine Council for Agriculture and Resource Research and the United States National Academy of Sciences, Washington, D.C., July 1977, pp. 3 and 5.

127. Arnold, "Wood Energy and Rural Communities," p. 25.

128. Mario P. Chanco, "Asia Has a Miracle Tree," Alternative Sources of Energy, Special Edition: "Wood Heat" 35 (1978): 28-31.

129. National Academy of Sciences, "Leucaena," p. 40.

130. Ibid., pp. 6 and 49.

131. Ibid., p. 55.

132. Ibid.

133. "Wood as an Alternative Energy Resource," p. 46.

134. Arnold and Jongma, "Fuelwood and Charcoal," p. 3.

135. Draper, "Wood Processing and Utilization," p. 4.

136. Reddy, "Energy Options for the Third World."

137. Makhijani and Poole, Energy and Agriculture in the Third World, p. 69.

138. Ibid.

139. French, "Firewood in Africa."

140. Ibid., p. 18.

141. Ibid., p. 9.

142. Revelle, "Requirements for Energy," p. 24.

143. Makhijani and Poole, Energy and Agriculture in the Third World.

144. Eckholm, "The Other Energy Crisis," p. 15.

145. Ibid., p. 16.

146. Eckholm, "Planting for the Future," p. 36.

147. John S. Spears, "The Changing Emphasis in World Bank Forestry Lending." Paper presented to the Eighth World

Forestry Congress, Jakarta, Indonesia, October 16-28, 1978, p. 4.

148. Eckholm, "Planting for the Future," p. 40.

149. Arnold, "Wood Energy and Rural Communities," p. 27.

150. Eckholm, "Planting for the Future," p. 44.

151. Arnold, "Wood Energy and Rural Communities," p. 27.

152. Eckholm, "Planting for the Future," p. 47.

153. Raju, "Smokeless Kitchens for the Millions."

154. Ibid., p. 1.

155. Goldemberg and Brown, "Cooking Stoves," p. 13; and Evans, "Using Firewood More Efficiently," p. 5.

156. Evans, "Using Firewood More Efficiently," p. 5.

157. Eckholm, "The Other Energy Crisis," p. 7.

158. Eckholm, "Planting For the Future," p. 1.

159. Ibid., p. 56.

160. Wm. S. Wiser and Charlotte V. Wiser, Behind Mud Walls 1930-1960 (Berkeley, Calif.: University of California Press, 1963), p. 203.

161. Rao, "Improving the Domestic Chula."

162. "Wood as an Alternative Energy Resource," p. 12 and 133-4.

163. Evans, "The Lorena Fuel Conserving Stove," p. 6.

164. VITA/CDT Cookstove Panel.

165. Ibid.; and Choqui Experimental Station, "Preliminary Progress Report of the Lorena Stove Program," Quezaltenango, Guatemala, February 1979.

166. Ibid., p. 4; and S.S. Edwards, "Central America: The Lorena Cookstove," in Appropriate Technology for Development: A Discussion and Case Histories, edited by D.D. Evans and L.N. Adler (Boulder, Colorado: Westview Press, 1979), p. 188.

167. Choqui Experimental Station, "Preliminary Progress Report," p. 7.

168. ATI, "Dissemination of Home Cooking Stoves."

169. Appropriate Technology International, "Sand and Clay Stove Dissemination as Catalyst in Community Development," Washington, D.C., 1978.

170. Choqui Experimental Station, "Preliminary Progress Report."

171. Ibid., p. 4.

172. Evans, "The Lorena Fuel Conserving Stove," p. 5.

173. Choqui Experimental Station, "Preliminary Progress Report," p. 5.

174. Ibid., p. 7.

175. Dale V. Shaller, "A Sociocultural Assessment of the Lorena Stove and Its Diffusion in Highland Guatemala," unpublished report, March 1979, p. 2.

176. Ibid., p. 4.

177. Ibid., p. 6.

178. Ibid., p. 14.

179. John S. Spears, "The Changing Emphasis in World Bank Forestry Lending," p. 11.

180. Appropriate Technology International, "A Report," Washington, D.C., January 1979.

181. Volunteers in Technical Assistance, VITA News, No. 2, (January 1980): 11; and personal communication from Margaret Crouch, March 25, 1980.

182. Volunteers in Technical Assistance, "Wood Conserving Cook Stoves: A Design Guide," Mt. Rainier, Maryland 1980.

183. Ianto Evans, Lorena Owner-Built Stoves.

184. "Report of the First Meeting of the Technical Panel on Fuelwood and Charcoal," held at the Headquarters of the Food and Agricultural Organization of the United Nations, Rome, January 21-25, 1980, in preparation for the 1981 United Nations Conference on New and Renewable Sources of Energy.

185. VITA/CDT Cookstove Panel.

186. Dutt, "Reducing Cooking Energy Use in India," p. 12.

187. Volunteers in Technical Assistance, "Wood Conserving Cook Stoves," Energy Fact Sheet No. 3, Mt. Rainier, Maryland, April 1980, p. 2.

188. Dutt, "Improved Wood Burning Cooking Stoves," and VITA/CDT Cookstove Panel.

Chapter 3

1. "Draft Report of Working Group XI: Appropriate Technology for Energy for Rural Requirements," UNIDO Forum on Appropriate Industrial Technology, N.L. Brown, Chairman; P.D. Dunn, Rapporteur. New Delhi, India, November 24, 1978, p. 28.

2. Personal communication from T.R. Satish Chandran to R.P. Morgan, New Delhi, India, November 1978.

3. "Reports and Recommendations of Workshop on Direct Solar Energy, Symposium on Energy and Development in the Americas," Gustavo Best Brown, Chairman; Jaime Moragues, Secretary. Interciencia 3 (May-June 1978): 183-84.

4. "U.S. Science and Technology for Development: A Contribution to the 1979 U.N. Conference," National Research Council, Washington, D.C., 1978, p. 116.

5. Ibid.

6. "Postharvest Food Losses in Developing Countries," National Academy of Sciences, Washington, D.C., 1978, Chapter 3, p. 50.

7. Histamitsu Takai, Leonarda Ebron, and Bart Duff, "Nature and Characteristics of Farm Level Paddy Storage in

Luzon, Philippines," mimeograph report, International Rice Research Institute (IRRI), Philippines, June 3, 1978, pp. 16-17.

8. "Postharvest Food Losses," p. 51.

9. Report of the VITA/CDT Solar Drying Panel (see Appendix A).

10. D.W. Hall, "Handling and Storage of Food Grains in Tropical and Subtropical Areas," FAO, Rome, 1970, p. 70.

11. Report of the VITA/CDT Solar Drying Panel.

12. Ibid.

13. Hall, "Handling and Storage of Food Grains."

14. George H. Foster and Robert M. Peart, "Solar Grain Drying: Progress and Potential," Agricultural Information Bulletin No. 401, U.S. Government Printing Office, Washington, D.C., 1976. Also, in March 1979, a Conference on Solar Grain Drying was co-sponsored by DOE, USDA, and the Agricultural Engineering Department of Iowa State University at the ISU Campus. See also "Soil, Water and Air Sciences Research Annual Report," U.S. Department of Agriculture, 1978.

15. "Postharvest Food Losses," p. 70.

16. Report of VITA/CDT Solar Drying Panel.

17. "Postharvest Food Losses," p. 70.

18. Ibid.

19. T.A. Lawand, "The Potential of Solar Agricultural Dryers in Developing Areas," presented to UNIDO Conference, Vienna, February 14-18, 1977.

20. Ibid., p. 7.

21. Ibid.

22. Ibid., p. 16.

23. "A Handbook on Appropriate Technology," Section B5, Canadian Hunger Foundation/Brace Research Institute, Canada, 1976.

24. Ibid.

25. Ibid.

26. Ibid.

27. "Small Farm Grain Storage, Vol. 1: Preparing Grain for Storage," VITA Publication No. 35E, Volunteers in Technical Assistance, Mt. Rainier, Md., 1977.

28. Ibid.

29. Algal Division, "Solar Dryers: Periodical Technical Notes No. 3," Shri A.M.M. Murugappa Chettiar Research Centre, 11/12 North Beach Rd., Madras 600-001,India, March 1978, mimeo.

30. "Small Farm Grain Storage," pp. 95-126.

31. Robert Exell and Sommai Kornsakoo, "Solar Rice Dryer," Asian Institute of Technology, Bangkok, Thailand, 1978; and Robert Exell and Sommai Kornsakoo, "A Low Cost Solar Dryer," Appropriate Technology 5 (May 1978): 23-24.

32. Exell and Kornsakoo, "Solar Rice Dryer."

33. V.R. Muthuveerappan et al., "Utilization of Solar Energy for Paddy Drying," Sixth Meeting, All Indian Solar Energy Working Group, Allahabad, December 23-24, 1974.

34. S.K. Tandon and D.K. Biswas, "Studies on Drying of Paddy in Solar Drier," Journal of IASAE (Indian Agricultural Research Institute, New Delhi), February 1974.

35. O.P. Singhal and C.P. Gupta, "Solar Drying," Seminar Proceedings: Industrial Applications of Solar Energy, Madras, June 4-6, 1975, National Productivity Council, New Delhi, 1976.

36. I.C. Macedo and C.A.C. Altemani, "Experimental Evaluation of Natural Convection Solar Air Heaters," Solar Energy 20 (1978): 367-69.

37. Lawand, "The Potential of Solar Agricultural Dryers," p. 16.

38. Ibid., p. 13.

39. D.K. Biswas, "A Solar Dryer for Small Farmers," Indian Farming 24 (August 1974): 21-22.

40. Ibid.

41. Muthuveerappan et al., "Utilization of Solar Energy."

42. Ibid.

43. Lawand, "The Potential of Solar Agricultural Dryers," p. 16.

44. Ibid.

45. Allan L. Phillips, "Drying Coffee with Solar Heated Air," Solar Energy 9 (1965): 213-16.

46. Foster and Peart, "Solar Grain Drying."

47. Don Marier, "Solar Crop Drier," DISE No. 20; and Eugene E. Williams, Robert M. Peart, John R. Barrett, Jr., and George H. Foster, "Solar Grain Drying," Agricultural Experiment Station, Purdue University, Lafayette, Ind., No. AED 128, September 1977.

48. William Peterson, "Solar Heat for Drying Shelled Corn." (Paper presented to 1973 Annual Meeting, North Central Region ASAE, Regina, Canada. Paper No. NC73-302); "Harness the Sun to Dry Grain," World Farming 10 (June 1968): 36-37.

49. Frederick Buelow, "Drying Crops with Solar Energy," Annuals of Arid Zones 15 (1976): 172-76.

50. "Postharvest Food Losses," p. 15.

51. Ibid., pp. 173-78.

52. Ibid., p. 22.

53. John L. Tschernitz and William T. Simpson, "Solar Kilns: Feasibility of Utilizing Energy for Drying Lumber in Developing Countries," U.S. Forest Products Laboratory Final Report to U.S. Agency for International Development, FPL-AID-PASA TA (AG) 02-75, January 1977, p. 18.

54. M. Chudnoff, E.D. Maldonado, and E. Goytia, "Solar Drying of Tropical Hardwoods," Forest Service Research Paper ITF-2, Institute of Tropical Forestry, Rio Piedras, Puerto Rico, April 1966.

55. Tschernitz and Simpson, "Solar Kilns," p. 17.

56. R.A. Plumptre, "The Design and Operation of a Small Solar Seasoning Kiln on the Equator in Uganda," Commonwealth Forestry Review 46 (1967): 298-309.

57. Ibid.

58. E.D. Maldonado and E.C. Peck, "Drying Lumber by Solar Radiation in Puerto Rico," Forest Products Journal 12 (October 1962); and S.N. Sharma, Prem Nath, and B.I. Bali, "A Solar Timber Seasoning Kiln," J. Timber Development Association of India 17 (April 1972).

59. Ibid.

60. Tschernitz and Simpson, "Solar Kilns," p. 17.

61. Ibid.

62. "Kendel Village Restoration," The International Rural Housing Journal (English edition) 1 (January-April 1973).

63. Sharma, Nath, and Bali, "A Solar Timber Seasoning Kiln," p. 10.

64. U.S. Forest Products Laboratory, Papers for Conference on Improved Utilization of Tropical Forests, Madison, Wisc., 1978.

65. Jack Ewel and Louis Conde, "Environmental Implication of Any-Species Utilization in the Moist Tropics" (Papers for Conference on Improved Utilization of Tropical Forests, Madison, Wisconsin, May 21-26, 1978), pp. 65-81.

66. Tschernitz and Simpson, "Solar Kilns," p. 20.

67. Ibid., p. 28.

68. John L. Tschernitz, VITA/CDT Solar Drying Panel, Washington University, St. Louis, Mo., March 9, 1979.

69. Tschernitz and Simpson, "Solar Kilns," p. 23.

70. Plumptre, "Design and Operation of a Small Solar Seasoning Kiln"; and Sharma, Nath, and Bali, "A Solar Timber Seasoning Kiln."

71. M.A. Rehman and O.P. Chawla, Seasoning of Timber Using Solar Energy, Indian Forest Bulletin No. 229, (Dehra Dun, India: Forest Research Institute, 1961).

72. Ibid.

73. Sharma, Nath, and Bali, "A Solar Timber Seasoning Kiln."

74. Rehman and Chawla, "Seasoning of Timber."

75. Curtis L. Johnson, "Wind-Powered Solar-Heated Lumber Dryer," Southern Lumberman 203 (October 1961).

76. Edmund F. Rasmussen, Dry Kiln Operator's Manual, Agricultural Handbook No. 188 (Madison, Wisc.: USDA Forest Products Laboratory, March 1961), p. 178.

77. Paul J. Bois, "Constructing and Operating a Small Solar-Heated Lumber Dryer," Forest Products Utilization Technical Report No. 7, USDA Forest Service, Madison, Wisc., January 1977.

78. Edward C. Peck, "Drying 4/4 Red Oak by Solar Heat," Forest Products Journal 12 (March 1962).

79. Edwin D. Maldonado, "Solar Radiation Used to Dry Mahogany Lumber in Puerto Rico," Tropical Forest Notes No. 14, Tropical Forest Research Center, Rio Piedras, Puerto Rico, December 1962.

80. Maldonado and Peck, "Drying Lumber."

81. Chudnoff, Maldonado, and Goytia, "Solar Drying of Tropical Hardwoods."

82. Y. Tao and C. Hsiao, "Lumber Solar Drying Experiment at Taichung," National Chung Hsing University, Bulletin No. 63-N-490/C (Taichung, Taiwan, 1964).

83. H.E. Troxwell and L.A. Mueller, "Solar Lumber Drying in Central Rocky Mountain Region," Forest Products Research J. 18 (January 1968): 19-24.

84. P. Gueneau, "Une Expérience de Sechage Solaire du Bois," Bois et Forêts des Tropiques, No. 131 (1970); summarized in Brace Research Institute, "A Survey of Solar Agricultural Dryers," Technical Report T99, Quebec, December 1975.

85. Sharma, Nath, and Bali, "A Solar Timber Seasoning Kiln"; and S.N. Sharma, "Solar Timber Drying," Seminar Proceedings: Industrial Applications of Solar Energy, Madras, June 4-6, 1975, National Productivity Council, New Delhi, 1976, pp. 232-38.

86. Sharma, Nath, and Bali, "A Solar Timber Seasoning Kiln."

87. S.N. Sharma and P. Nath, "Solar Kiln Field Trials (Commercial Trials on the F.R.I. 7.1 cu. m. Solar Heating Timber Seasoning Kiln)," Final Research Report, Forestry Research Institute, Dehra Dun, India, August 1977.

88. Ibid.

89. R.F. Casin, "Solar Drying for Lumber," Technical Note No. 76, Forest Products Research Institute, Los Baños, Philippines, 1967; and R.F. Casin, E.B. Ordinario, and G.Y. Tamayo, Jr., "Solar Drying of Apitong, Narra, Red Luan and Tanguile," Philippine Lumberman 15 (1969): 23-30.

90. R.A. Plumptre, "Solar Timber Kilns: Their Suitability for Developing Countries," Technology for Solar Energy Utilization, Development and Transfer of Technology Series No. 5, UNIDO, ID/202 (New York, 1978).

91. E. Martinka, "Predrying of Some Ghanaian Timber," Technical Note No. 11, Ghanaian Forest Research Institute, Kumasi, 1969.

92. Plumptre, "The Design and Operation of a Small Solar Seasoning Kiln," and "Solar Timber Kilns."

93. Plumptre, "The Design and Operation of a Small Solar Seasoning Kiln," p. 308.

94. R.A. Plumptre, "A Simple Solar Heated Dryer for Timber," Intermediate Technology Development Group Conference on Small-Scale Energy for Developing Countries, January 1979.

95. Plumptre, "Solar Timber Kilns."

96. Plumptre, "A Simple Solar Heated Dryer for Timber."

97. Frank Hughes, "Prof. Yang's Solar-Heated Kilns," Canadian Forest Industries, (March 1978): 62-63.

98. W.R. Read, A. Choda, and P.I. Cooper, "A Solar Timber Kiln," Solar Energy 15 (April 1974): 309-16.

99. John L. Tschernitz and William T. Simpson, "Solar-Heated, Forced-Air, Lumber Dryer for Tropical Latitudes," Solar Energy 20 (1979): 563-66.

100. Tschernitz and Simpson, "Solar Kilns."

101. Peck, "Drying 4/4 Red Oak."

102. Sharma, "Solar Timber Drying."

103. Sharma, Nath, and Bali, "A Solar Timber Seasoning Kiln."

104. Sharma and Nath, "Solar Kiln Field Trials."

105. Read, Choda, and Cooper, "A Solar Timber Kiln."

106. Plumptre, "A Simple Solar Heated Dryer for Timber."

107. Plumptre, "Solar Timber Kilns."

108. Tschernitz and Simpson, "Solar Kilns," p. 26.

109. Ibid., p. 28.

110. Sharma and Nath, "Solar Kiln Field Trials."

111. Plumptre, "A Simple Solar Heated Dryer for Timber."

112. Tschernitz and Simpson, "Solar Kilns," p. 56.

113. Ibid., p. 58.

114. Plumptre, "A Simple Solar Heated Dryer for Timber."

115. Tschernitz and Simpson, "Solar-Heated, Forced-Air, Lumber Dryer."

116. Tschernitz and Simpson, "Solar Kilns."

117. Ibid., p. 2.

118. Eugene M. Wengert, "Improvements in Solar Dry Kiln Design," U.S. Forest Service Research Note FPL-0212, Forest Products Laboratory, Madison, Wisc., 1971. See also, Troxwell and Mueller, "Solar Lumber Drying."

119. Plumptre, "Solar Timber Kilns," and "A Simple Solar Heated Dryer for Timber."

120. Plumptre, "Solar Timber Kilns," p. 86.

121. See, for example, Maldonado and Peck, "Drying Lumber"; Johnson, "Wind-Powered, Solar-Heated Lumber Dryer"; Gueneau, "Une Expérience"; Plumptre, "A Simple Solar Heated Dryer"; Hughes, "Prof. Yang's Solar Heated Kilns"; and Sharma, "Solar Timber Drying."

122. See, for example, Plumptre, "The Design and Operation of a Small Solar Seasoning Kiln," and "Solar Timber Kilns"; and Sharma et al., "A Solar Timber Kiln."

123. Tschernitz and Simpson, "Solar Kilns," pp. 48-51.

124. Ibid., p. 48.

125. Wengert, "Improvements in Solar Dry Kiln Design."

126. See, for example, Casin, Ordinario, and Tamayo, "Solar Drying"; Read, Choda, and Cooper, "A Solar Timber Kiln"; Sharma, "Solar Timber Drying"; and C.H. Banks, "Solar Drying of Timber - A Development Study," CSIR Subject Survey O/HOUT 10, Pretoria, June 1969.

127. Read, Choda, and Cooper, "A Solar Timber Kiln."

128. Neil A. Duffie and D.J. Close, "The Optimisation of a Solar Timber Drier Using an Absorbent Energy Store," Solar Energy 20 (1978): 405-11.

129. Tschernitz and Simpson, "Solar Kilns," p. 26.

130. Sharma, "Solar Timber Drying," p. 234.

131. For example, Chudnoff, Maldonado and Goytia, "Solar Drying of Tropical Hardwoods"; Sharma, Nath, and Bali, "A Solar Timber Seasoning Kiln"; Sharma and Nath, "Solar Kiln Field Trials"; Casin, Ordinario, and Tamayo, "Solar Drying"; and Tschernitz and Simpson, "Solar-Heated, Forced-Air, Lumber Drying."

132. Sharma, Nath, and Bali, "A Solar Timber Seasoning Kiln"; and Tao and Hsiao, "Lumber Solar Drying."

133. Plumptre, "The Design and Operation of a Small Solar Seasoning Kiln," and "A Simple Solar Heated Dryer"; Johnson, "Wind-Powered Solar-Heated Lumber Dryer"; and Peck, "Drying 4/4 Red Oak."

134. Plumptre, "Solar Timber Kilns."

135. Plumptre, "A Simple Solar Heated Dryer."

136. Chudnoff et al., "Solar Drying of Tropical Hardwoods."

137. Troxwell and Mueller, "Solar Lumber Drying."

138. Sharma and Nath, "Solar Kiln Field Trials."

139. Tschernitz and Simpson, "Solar Kilns," p. 26.

140. Plumptre, "A Simple Solar Heated Dryer for Timber."

141. Johnson, "Wind-Powered Solar-Heated Lumber Dryer."

142. Tschernitz and Simpson, "Solar Kilns."

143. Sharma and Nath, "Solar Kiln Field Trials."

144. Plumptre, "Solar Timber Kilns," and "A Simple Solar Heated Dryer."

145. Ibid.

146. Ibid.

147. Plumptre, "A Simple Solar Heated Dryer."

148. Personal communication received from R.A. Plumptre, January 25, 1979.

149. Donald G. Cuppett, "Use of Solar Energy in Drying Lumber," and P.O. McCormick and S.J. Robertson, "Solar Industrial Process Heat for Kiln Drying Lumber," Proceedings: Practical Application of Solar Energy to Wood Processing, Virginia Tech, January 1977, Forest Products Research Society, Madison, Wisc., 1977.

150. John L. Tschernitz, VITA/CDT Solar Drying Panel.

Chapter 4

1. D. Adair, "Rice Bran Stabilization," UNIDO Ad-Hoc Expert Group Meeting, Vienna, December 1976, ID/WG.240/5.

2. E.B. Shultz, Jr., the principal author of this chapter, began work on rice bran processing concepts at Washington University in 1976. An initial report was: E.B. Shultz, Jr., "Concepts for Greater Utilization of Rice Bran in Developing Countries," Report No. CDT-77/9, Center for Development Technology, Washington University, August 1977. See also E.B. Schultz, Jr., and R.P. Morgan, "Appropriate Technology for Village-Level Rice Bran Processing and Utilization in Developing Countries," Report No. CDT 79/1, Center for Development Technology, Washington University, April 1979.

3. FAO Production Yearbook, Vol. 22 (Rome: United Nations, 1968); and USDA, Foreign Agricultural Service, Foreign Agriculture Circular FR 2-71, July 1971.

4. C.R. Adair, "Production and Utilization of Rice," in Rice: Chemistry and Technology, edited by D.F. Houston (St. Paul, Minn.: American Association of Cereal Chemists, 1972), p. 2.

5. FAO Trade Yearbook, (Rome: United Nations, 1968), p. 122.

6. D.F. Houston, "Rice Bran and Polish," in Rice: Chemistry and Technology, edited by D.F. Houston (St. Paul, Minn.: American Association of Cereal Chemists, 1972), p. 273.

7. D.H. Grist, Rice, 3rd ed. (London: Longman Green, 1959).

8. Houston, "Rice Bran and Polish," p. 276.

9. Ibid., p. 277.

10. S. Barber, "Stabilization of Rice Bran," UNIDO Ad-Hoc Expert Group Meeting, Vienna, Decmeber 1976, ID/WG.240/7.

11. R.M. Saunders, A.A. Betschart, and G.O. Kohler, "Preparation of Cereal Protein Concentrates," The Bakers Digest 64 (February 1975): 49-52.

12. Houston, "Rice Bran and Polish," p. 285.

13. Ibid.

14. J. Duckworth and J.M. Dent, "The Preparation of Refined Rice Bran," Tropical Agriculture 23 (1946): 25-28.

15. H.G.R. Reddy, "Rice Bran Stabilization," UNIDO Ad-Hoc Expert Group Meeting, Vienna, December 1976, ID/WG.240/3; International Centre for Industrial Studies, "Research and Development of a Small-Scale, Low-Cost Rice Bran Stabilizing Unit," Report of Ad-Hoc Expert Group Meeting, UNIDO, Vienna, December 1976, ID/WG.240/10; R. Hawkey, "The Case for Rural Bran Stabilization," Report to UNIDO, February 1975; and Applied Scientific Research Corporation of

Thailand, "Study on the Verification and Definition of the Most Suitable Rice Bran Stabilizing Technology and Specification of its Technical Parameters," Report to UNIDO, January 1977.

16. Hawkey, "The Case for Rural Rice Bran Stabilization."

17. Central Food Technological Research Institute, "The Food Industry in India," background paper for Working Group No. 6, International Forum on Appropriate Industrial Technology, UNIDO, New Delhi, November 1978, ID/WG.282/27.

18. Report of the ECAFE Expert Team in Rice Processing Machinery (AIDC(6)/1) January 22, 1970, discussed in Reddy, "Rice Bran Stabilization."

19. D.S.C. Spencer, S.C. Dunstan, Ibi I May-Parker, and Frank S. Rose, "Employment, Efficiency and Incomes in the Rice-Producing Industry of Sierra Leone," African Rural Economy Paper No. 15, Michigan State University, East Lansing, 1976.

20. Ibid.

21. W.L. Collier, Jusuf Colter, Sinarhadi, and Robert A. Shaw, "Choice of Technique in Rice Milling in Java," <u>Bull. of Indonesian Economic Studies</u>, (March 1974).

22. Y. Koga, "Present Status of Post-Harvest Rice Processing in Thailand and Prospects for Rural Industrialization Through Cooperative Ventures in Such Processing," ESCAP/UNIDO Division of Industry, Housing and Technology, Bangkok, 1976.

23. Spencer et al., "Employment, Efficiency and Incomes."

24. Ibid.

25. Collier et al., "Choice of Technique."

26. CFTRI, "The Food Industry in India."

27. Koga, "Present Status of Post-Harvest Rice Processing in Thailand."

28. CFTRI, "The Food Industry in India."

29. Spencer et al., "Employment, Efficiency and Incomes."

30. Koga, "Present Status of Post-Harvest Rice Processing in Thailand."

31. G.C. Witte, "Conventional Rice Milling in the United States," in <u>Rice: Chemistry and Technology</u>, edited by D.F. Houston (St. Paul, Minn.: American Association of Cereal Chemists, 1972), pp. 191 and 193.

32. Ibid.

33. Ibid.

34. T.B. Wayne, U.S. Patent 3,261,690, July 19, 1966; and U.S. Patent 3,401,731, September 17, 1968.

35. L. Lynn, R.M. Anderson, and D.L. Carpenter, "New X-M Rice Bran," <u>Cereal Science Today</u> 12 (1967): 221; L. Lynn, G. Steen, and R. Anderson, "New Rice Oil," <u>Food Technology</u> 22 (1968): 10; J.W. Hunnell and J.F. Nowlin,

"Solvent Extractive Rice Milling," in Rice: Chemistry and Technology, edited by D.F. Houston (St. Paul, Minn.: American Association of Cereal Chemists, 1972). pp. 201-14; and L. Lynn, "Edible Rice Bran Foods," in Protein Enriched Cereal Foods for World Needs, edited by M. Milner (St. Paul, Minn.: American Association of Cereal Chemists, 1969).

36. Hunnell and Nowlin, "Solvent Extractive Rice Milling."

37. Personal communication from Roger E. Sande, Industrial Sales Manager, Riviana Foods, Inc., Houston, Texas, March 31, 1977.

38. C.H. Lea, "The Oxidative Deterioration of Food Lipids," in Symposium on Foods: Lipids and Their Oxidation, edited by H.W. Schultz (Westport, Conn.: Avi Publishing Co., 1962).

39. Ibid.

40. K. Yokochi, "Rice Bran Processing for the Production of Rice-Bran Oil and Rice-Bran Protein Meal," UNIDO Expert Group - Oilseed Processing Industry, Vienna, October 1972, ID/WG.120/9.

41. C.S. Viraktamath and H.S.R. Desikachar, "Inactivation of Lipase in Rice Bran in Indian Rice Mills," Journal of Food Science and Technology (India) 8 (1971): 70.

42. K. Yokochi, "Rice Bran Processing for the Production of Rice Bran Oil and Characteristics and Uses of the Oil and Deoiled Bran," in Proceedings of the Rice By-Products Utilization International Conference, Vol. III, Valencia, Spain, 1974 (Valencia: IATA, 1977).

43. Yokochi, "Rice Bran Processing for the Production of Rice-Bran Oil and Rice-Bran Protein Meal."

44. Viraktamath and Desikachar, "Inactivation of Lipase."

45. S.H.C. Lin and C.M. Cater, "Effect of Extrusion-Cooking on the Formation of Free Fatty Acids in Rice Bran," Food Protein R&D Center, Texas A&M University, College Station, Texas, unpublished, undated.

46. ASRCT, "Study on the Verification and Definition of the Most Suitable Rice Bran Stabilizing Technology."

47. Report of Ad-Hoc Export Group Meeting, 1976, ID/WG.240/10.

48. ASRCT, "Study on the Verification and Definition of the Most Suitable Rice Bran Stabilizing Technology."

49. Yokochi, "Rice Bran Processing for the Production of Rice-Bran Oil and Rice-Bran Protein Meal."

50. H.L. Burns and M.M. Cassidy, "Method of Treating Rice Bran and Polish," U.S. Patent 2,563,798, August 7, 1951.

51. P.J. Fito, A. Flors, G. Amutio, J.M. Camacho, R. Cerni, and S. Barber, "Process for the Stabilization of Rice Bran. II. Development of Pilot Plant and Industrial Equipment," in Proceedings of the Rice By-Products Utilization

International Conference, Vol. II, Rice By-Products Preservation, Valencia, Spain, 1974 (Valencia: IATA, 1977).

52. R.E. Tribelhorn, "Colorado State University Extrusion Facilities and Systems," in *Low-Cost Extrusion Cookers*, edited by J.M. Harper and G.R. Jansen (Ft. Collins, Colo.: Colorado State University, June 1976).

53. Lin and Cater, "Effect of Extrusion-Cooking."

54. Anderson Company, Division of International Basic Economy Corp., *Rice Bran "Expandolex" Process*, Bull. 108.031 (RBP-2), undated; and M. Williams and S. Baer, "The Expansion and Extraction of Rice Bran Stabilizer," *Journal of the American Oil Chemists Society* 42 (1965): 151.

55. Hawkey, "The Case for Rural Rice Bran Stabilization."

56. Ibid.

57. Viraktamath and Desikachar, "Inactivation of Lipase."

58. V. Ratanapunvorakul and W.C.J. Hermans, "Further Studies on Rice Bran Stabilization Methods," Report No. 3 on Research Project No. 13/6, Applied Scientific Research Corporation of Thailand, Bangkok, 1968.

59. Viraktamath and Desikachar, "Inactivation of Lipase."

60. Adair, "Rice Bran Stabilization."

61. Report of the Ad-Hoc Expert Group Meeting, 1976, ID/WG.240/10.

62. P.S. Barton, "Engineering Aspects of a Simple Low-Cost Rice Bran Stabilizer," UNIDO Ad-Hoc Expert Group Meeting, Vienna, December 1976, ID/WG.240/9.

63. Hawkey, "The Case for Rice Bran Stabilization."

64. Ibid.

65. Report of the Ad-Hoc Expert Group Meeting, 1976, ID/WG.240/10.

66. Barton, "Engineering Aspects of a Simple Low-Cost Rice Bran Stabilizer."

67. ASRCT, "Study on the Verification and Definition of the Most Suitable Rice Bran Stabilizing Technology."

68. Ibid.

69. International Centre for Industrial Studies, "Research and Development of a Small-Scale, Low-Cost Rice Stabilizing Unit."

70. Houston, "Rice Bran and Polish," p. 292.

71. E.R. Cousins, S.P. Fore, H.J. Janssen, and R.O. Feuge, "Rice Bran Oil," *Journal of the American Oil Chemists' Society* 30 (1953): 9.

72. Lynn et al., "New Rice Oil."

73. Yokochi, "Rice Bran Processing for the Production of Rice Bran Oil and Characteristics and Uses of the Oil and Deoiled Bran."

74. Ibid.

75. Yokochi, "Rice Bran Processing for the Production of Rice-Bran Oil or Rice-Bran Protein Meal."

76. Ibid.

77. Ibid.

78. Ibid.

79. M.A. Connor, R.M. Saunders, and G.O. Kohler, "Rice Bran Protein Concentrates Obtained by Wet Alkaline Extraction," Cereal Chemistry 53 (1976): 488.

80. A.A. Betschart, R.Y. Fong, and R.M. Saunders, "Rice By-Products: Comparative Extraction and Precipitation of Nitrogen from U.S. and Spanish Bran and Germ," J. Food Science 42 (1977): 1088.

81. S. Barber, A. Flors, E. Tortosa, J.M. Camacho, and R. Cerni, "High Protein Flours from Rice Bran by Wet Fractionation Process," in Proceedings of the Rice By-Products Utilization International Conference, Vol. IV, Rice Bran Utilization: Food and Feed, Valencia, Spain, 1974 (Valencia: IATA, 1977).

82. Ibid.

83. S. Mihara, "Nakataki Water Process Separating Rice Bran Components," Chemical Economy and Engineering Review 42 (1970): 36-38.

84. D.F. Houston and A. Mohammad, "Air Classification and Sieving of Rice Bran and Polish," Rice Journal 69 (1966): 20.

85. D.F. Houston and A. Mohammad, "Recovery of Low-Fiber Products from Defatted Rice Bran by Remilling and Sieving," Proceedings of the 13th Rice Technical Working Group, Beaumont, Texas, February 24-26, 1970, reported in Houston, "Rice Bran and Polish," p. 291.

86. S. Barber and C.B. de Barber, "Basic and Applied Research Needs for Optimizing Utilization of Rice Bran as Food and Feed," in Proceedings of the Rice By-Products Utilization International Conference, Vol. IV, Rice Bran Utilization: Food and Feed, Valencia, Spain, 1974 (Valencia: IATA, 1977)

87. F. Gariboldi, "Parboiled Rice," in Rice: Chemistry and Technology, edited by D.F. Houston (St. Paul, Minn.: American Association of Cereal Chemists, 1972), p. 365.

88. Ibid., p. 373.

89. Ibid., p. 374.

90. Ibid., p. 375.

91. Ibid., p. 361.

92. Ibid., p. 376.

93. Ibid., p. 361.

94. Ibid., p. 369.

95. Houston, "Rice Bran and Polish," p. 289.

96. Ibid., p. 280.

97. Hunnell and Nowlin, "Solvent Extractive Rice Milling."

98. Barber and de Barber, "Basic and Applied Research Needs."

99. A.O. Cruz and A.P. West, "Utilization of Rice By-Products," Revista Filipina de Medicina y Farmacia 32 (1941): 169, reported in Houston, "Rice Bran and Polish."

100. M.N. Wood, Gourmet Food on a Wheat-Free Diet (Springfield, Ill.: C.C. Thomas, 1967).

101. Houston, "Rice Bran and Polish," pp. 293-95.

102. Ibid.

103. "Rice in India: A Techno-Economic Review," Central Food Technological Research Institute, Mysore, India, 1965, reported in Houston, "Rice Bran and Polish," p. 292.

104. J.T. Hogan, "Rice Bran, Oil and Wax," FAO Agr. Eng. Informal Working Bull. 30 (1967), p. 1.

105. T.A. Printon, U.S. Patent 2,180,256, November 14, 1939.

106. Houston, "Rice Bran and Polish," pp. 293-295.

107. Lynn, "Edible Rice Bran Foods,"

108. Ibid.

109. Houston, "Rice Bran and Polish," p. 288.

110. Ibid., p. 292.

111. Lynn et al., "New X-M Rice Bran."

112. R.T. Talwalkar, N.K. Garg, and C.R. Krishna Murti, "Rice Bran: A Source Material for Pharmaceuticals," J. Food Sci. Technol. (India) 2 (Oct.-Dec. 1965): 117.

113. Houston, "Rice Bran and Polish," p. 282.

114. Ibid., p. 286.

115. "Rice Bran: Utilization and Trade," FAO Monthly Bulletin Agr. Econ. and Stat. 13 (1964): 9.

116. Houston, "Rice Bran and Polish," p. 285.

117. "Rice Bran."

118. Yokochi, "Rice Bran Processing for the Production of Rice Bran Oil and Characteristics and Uses of the Oil and Deoiled Bran."

119. S.N. Sharma and S.V. Pingale, Paddy Milling Technology in India, Part II, Rice Bran and its Utilisation (New Delhi: Food Corporation of India, 1974).

120. Ibid.

121. Ibid.

122. Ibid.

123. Gariboldi, "Parboiled Rice," p. 376.

124. D.F. Houston, "Rice Hulls," in Rice: Chemistry and Technology, edited by D.F. Houston (St. Paul, Minn.: American Association of Cereal Chemists, 1972).

125. Ibid.

126. S.N. Sharma and S.V. Pingale, Paddy Milling Technology in India, Part I, Husk and its Utilisation (New Delhi: Food Corporation of India, 1974).

127. Houston, "Rice Hulls."

128. Ibid.

129. Sharma and Pingale, Husk and its Utilisation.

130. Houston, "Rice Hulls."

131. Ibid.

132. Italian Patent 484,118, August 1955, reported in Chem Abstracts 49 (1955): 11258, and discussed in Houston, "Rice Hulls," p. 326.

133. Tropical Products Institute, "The Use of Rice Hulls in the Manufacture of Building Blocks," 1967, reported in Houston, "Rice Hulls," pp. 326-27.

134. Sharma and Pingale, Husk and its Utilization.

135. Ibid.

136. Houston, "Rice Hulls."

137. Sharma and Pingale, Husk and its Utilisation.

138. Houston, "Rice Hulls."

139. Sharma and Pingale, Husk and its Utilisation.

140. Houston, "Rice Hulls."

141. Sharma and Pingale, Husk and its Utilisation.

142. Houston, "Rice Hulls."

143. Sharma and Pingale, Husk and its Utilisation.

144. Ibid.

145. Ibid.

146. J. Keddie, "Technological Choice and Employment in Food Processing and Storage and Related Policy Issues," background paper for Working Group No. 6, International Forum on Appropriate Industrial Technology, UNIDO, New Delhi, November 1978, ID/WG.282/97.

147. "Edible Rice Bran Oil to be Produced," Hindustan Times, November 27, 1978, p. 6.

148. C.P. Timmer, "The Choice of Technique in Indonesia," in The Choice of Technology in Developing Countries, by C.P. Timmer, J.W. Thomas, L.T. Wells, and D. Morawetz, Center for International Affairs, Harvard University, 1975.

149. Appropriate Technology Development Association, Lucknow, India, "Proposal for Development of Appropriate Technology in India, A Five-Year Programme," undated.

150. "Draft Report of Working Group No. 6, Appropriate Technology for Food Storage and Processing," J. Keddie, Rapporteur, International Forum on Appropriate Industrial Technology, UNIDO, New Delhi, November 23, 1978.

151. Ibid.

152. Barber et al., "High Protein Flours from Rice Bran."

153. Houston and Mohammad, "Air Classification and Sieving of Rice Bran and Polish.'

154. Connor et al., "Rice Bran Protein Concentrates"; and Betschart et al., "Rice By-Products."

155. Barber et al., "High Protein Flours from Rice Bran"; and Mihara, "Nakataki Water Process."

156. Duckworth and Dent, "Preparation of Refined Rice Bran."

157. Ibid.

158. F.B. Morrison, Feeds and Feeding, 22nd ed. (Ithaca, N.Y.: Morrison Publishing Co., 1956).

159. Adair, "Rice Bran Stabilization."

160. M. Van Peursem, "Manufacturers' Experiences with LEC's - The Brady Crop Cooker," in Low-Cost Extrusion Cookers, edited by J.M. Harper and G.R. Jansen (Ft. Collins, Colo.: Colorado State University, June 1976).

161. Duckworth and Dent, "Preparation of Refined Rice Bran."

162. L.J. Hanson, "Manufacturers' Experiences with LEC's - The Insta-Pro Extruder," in Low-Cost Extrusion Cookers, edited by J.M. Harper and G.R. Jansen (Ft. Collins, Colo.: Colorado State University, June 1976).

163. M. Sterner, "Exploration of Potential for Low-Cost Extrusion Cooker - Worldwide," in Low-Cost Extrusion Cookers, edited by J.M. Harper and G.R. Jansen (Ft. Collins, Colo.: Colorado State University, June 1976).

164. Hanson, "The Insta-Pro Extruder."

165. Van Peursem, "The Brady Crop Cooker."

166. Sterner, "Exploration of Potential for Low-Cost Extrusion Cooker."

167. Ibid.

168. Mihara, "Nakataki Water Process."

169. Barber et al., "High Protein Flours from Rice Bran."

170. Houston and Mohammad, "Air Classification and Sieving of Rice Bran and Polish."

171. Yokochi, "Rice Bran Processing for the Production of Rice Bran Oil and Characteristics and Uses of the Oil and Deoiled Bran."

172. Duckworth and Dent, "Preparation of Refined Rice Bran."

173. Barton, "Engineering Aspects of a Low-Cost Rice Bran Stabilizer."

Chapter 5

1. Use of Agricultural and Industrial Wastes in Low Cost Construction, UN Dept. of Economic and Social Affairs, New York, 1976, p. 1; Letter from Bernard W. Burgess, Executive Vice President of the Pulp and Paper Research Institute of Canada, to Prof. John Kardos, Washington University, November 17, 1978; and Production of Panels from Agricultural Residues, UNIDO, Vienna, 1972, p. 5.

2. UNIDO has held at least one: The Expert Working Group on the Production of Panels from Agricultural Wastes met in Vienna, December 14-18, 1978. The UN Secretariat (ESA) held at least one: The Expert Working Group on the Use of Agricultural, Industrial, and Consumer Wastes in

Low-Cost Construction, Rio Piedras, Puerto Rico, December 9-13, 1974. And the FAO/UNEP held a Seminar on Residue Utilization - Management of Agricultural and Agro-Industrial Wastes in Rome, January 1977.

3. Use of Agricultural and Industrial Wastes, p. 1.

4. Building Materials Industry, UNIDO, Vienna, 1969, p. 11.

5. A.T. DiBenedetto and R.P. Morgan, "A Proposal to Establish a Cooperative International Program on the Development and Utilization of Indigenous Materials for Low-Cost Housing and Economic Development," Washington University, St. Louis, Missouri, October 1979, p. 2.

6. Ann Kestenbaum, ed., Technology for Development (London: VCOAD, November 1975), p. 67.

7. Building Materials Industry, p. 52.

8. Ibid., p. 13.

9. Agricultural Residues: Compendium of Technologies, FAO, Rome, 1978.

10. A.E. Crittenden, Historical Outline of Past Research on the Production of Boards from Agricultural Wastes and Future Research Trends, UNIDO, Vienna, 1970, p. 3.

11. Bibliography of Agricultural Residues, FAO, Rome, 1978.

12. Production of Panels from Agricultural Residues, p. 18.

13. C.G. Kick III, "Appropriate Technology for Building Materials from Agro-Wastes and Natural Fibers," Report No. THA/CDT 80/2, Center for Development Technology, Washington University, St. Louis, Mo., February 1980.

14. "Rice Hull Ash Cement . . . High Quality, Acid Resisting," Journal of the American Concrete Institute, May 1975, pp. 235-36.

15. Use of Agricultural and Industrial Wastes, p. 6.

16. Ibid., p. 7.

17. Report of VITA Panel on Wood Waste Paper Utilization, May 7, 1976.

18. Chemical Engineering, December 10, 1973, p. 67.

19. "Rice Hull Ash Cement," pp. 235-36.

20. David Homstrom, "Professor Makes Rice-Hull Cement," The Christian Science Monitor, December 7, 1973, p. 1.

21. Ibid.

22. Use of Agricultural and Industrial Wastes, p. 7.

23. "Rice Hull Ash Cement."

24. D.J. Cook, R.P. Pama, and S.A. Damer, "Rice Husk Ash as a Pozzolanic Material," in New Horizons in Construction Materials, edited by H.Y. Fang, Vol. 1 (Lehigh Valley, Pa.: Envo, 1976), pp. 431-42.

25. Use of Agricultural and Industrial Wastes, p. 11; Agricultural Residues: Compendium of Technologies, p. 119; Agricultural Residues: World Directory of Institutions, 1st

rev., FAO, Rome, 1978, p. 121; CSIR Technologies in the Field of Agro-Based Industries, CSIR, New Delhi, August 1976, p. 4; S.S. Rehsi, "Pozzolana Based Binders," Appropriate Technology 5 (February 1979): 22-23; and Marendra Verma, "Materials and Technology for Rural Buildings," Rural Buildings Division, CBRI, undated.

26. Rehsi, "Pozzolana Based Binders."

27. Verma, "Materials and Technology for Rural Buildings."

28. CSIR Technologies, p. 4.

29. Observations by C. Kick during four years (1971-1974) in Labasa on the outer large island of Fiji. Cement could be had in the capital, Suva, where it was produced, for F$20/ton, while the price in Labasa was over F$40/ton.

30. Use of Agricultural and Industrial Wastes, p. 7.

31. Agricultural Residues: Compendium of Technologies, p. 108; Agricultural Residues: World Directory, p. 121; and CSIR Technologies, p. 4.

32. Use of Agricultural and Industrial Wastes, p. 11.

33. Ibid., p. 7.

34. Ibid., pp. 7 and 8.

35. F.M. Lauricio, J.O. Simpongco, and F.N. Tamalang, "Fabrication of Hollow Blocks from Agri-forestry Materials for Low-Cost Housing," Appropriate Technology 5 (August 1978): 18-19.

36. "Agricultural Waste as a Civil Engineering Material," Small Industry Journal 7 (March 1975): 30-33.

37. Use of Agricultural and Industrial Wastes, p. 9.

38. Ibid., p. 10.

39. "Agricultural Wastes as a Civil Engineering Material."

40. Use of Agricultural and Industrial Wastes, p. 10.

41. Ibid.

42. D.J. Cook, "Waste Products Utilization - An Investigation of Materials Applications for Rice Hulls," Studies from the School of Civil Engineering, University of New South Wales, Kensington, UNICIV Report #100, December 1972.

43. B.J. Van der Kamp and A.A. Gokhale, "Resistance of Rice Husk Board to Decay Fungi," Journal of the Indian Academy of Wood Science 5 (July/Dec. 1974): 106-07.

44. Cook, "Waste Products Utilization."

45. R.C. Vasishth, "Manufacture of Composite Boards from Rice Husks Using a Batchwise, Labour Oriented Plant," 1974, cited in Van der Kamp and Gokhale, "Resistance of Rice Husk Board to Decay Fungi." See also Vasishth, "Project Proposal: Rice Husk Board Manufacture in Malaysia," 1974.

46. Use of Agricultural and Industrial Wastes, p. 9.

47. "ITDG Makes Olympics?" Appropriate Technology 5 (November 1978): 30.

48. Richard J. Frankel, "The Design and Operation of a Water Filter Using Local Materials in Southeast Asia," in

Appropriate Technology: Problems and Promises, edited by Nicolas Jequier (Paris: OECD, 1976), pp. 238-52.

49. "Coconut - The All-around Provider," Development Forum, (January-February 1979): p. 4.

50. Uses of Agricultural and Industrial Wastes, p. 12.

51. Ibid.

52. "Coconut Husk Chipping Machine," Building Practice 4 (October 1973): 14.

53. Ibid.

54. Uses of Agricultural and Industrial Wastes, p. 12.

55. CSIR Technologies, p. 6.

56. Uses of Agricultural and Industrial Wastes, p. 13; and CBRI Technical Note #68, 2nd rev., January 1978.

57. Uses of Agricultural and Industrial Wastes, p. 13.

58. CBRI Building Materials Note #9, September 1968.

59. CSIR Technologies, p. 6.

60. Uses of Agricultural and Industrial Wastes, p. 13; Agricultural Residues: World Directory, p. 107; and CSIR Technologies, p. 6.

61. CSIR Technologies, p. 6.

62. Agricultural Residues: Compendium of Technologies, p. 271.

63. "Coconut Water as an Intravenous Fluid," South Pacific Bulletin 22 (1972): 17.

64. Lauricio et al., "Fabrication of Hollow Blocks."

65. Uses of Agricultural and Industrial Wastes, p. 14.

66. "The All-Coconut House," FIJI 1 (Sept.-Oct. 1978): 2.

67. "Coconut - The All-Around Provider."

68. See George L. Ball III, Ival O. Salyer, D.W. Werkmeister, R.A. Cass, A.M. Usmani, B.S. Bryant, and J.P.R. Falconer, Development of Low-Cost Roofing from Indigenous Materials in Developing Nations, (Dayton, Ohio: Monsanto Research Corporation, 1978).

69. "Fiji's Search for Energy," FIJI 1 (July-August 1978): 10.

70. Uses for Agricultural and Industrial Wastes, p. 16; and D.N. Daruvalla and V.G. Kamath, "Utilization of Sugar Industry Waste: Mineral Wool from Filter Press Mud," Shri Ram Institute for Industrial Research, Delhi, India, 1952.

71. J.W. Powell, Appropriate Technology in India (Kumasi, Ghana: University of Science and Technology, 1978), pp. 29-30.

72. Uses of Agricultural and Industrial Wastes, p. 16.

73. Agricultural Residues: Compendium of Technology, p. 11.

74. Ibid., p. 34.

75. Letter from Burgess, November 17, 1978.

76. Agricultural Residues: Compendium of Technology, pp. 77 and 203; Agricultural Residues: World Directory, pp. 109 and 112; and CSIR Technologies, pp. 3 and 7.

77. _Agricultural Residues: Compendium of Technology_, p. 265.

78. S.M. Singh and L.K. Aggarwal, "Binder for Many Wastes," _Research and Industry_ 22 (1977): 242-45.

79. B.A. Ntim and J.W. Powell, "Appropriate Technology Consultancy Center," in _Appropriate Technology: Problems and Promises_, edited by Nicolas Jequier (Paris: OECD, 1976), pp. 276-96.

80. Powell, _Appropriate Technology in India_.

81. Various occasional reports published by the Faculty of Architecture, University of Science and Technology, Kumasi, Ghana.

82. The Forest Products Laboratory, Madison, Wisc., is a good source of information on these methods.

83. R. Austin, D. Levy, and K. Ueda, _Bamboo_ (New York and Tokyo: Walker/Weatherhill Books, 1970).

84. _The Use of Bamboo and Reeds in Building Construction_, United Nations Publication No. ST/SAO/113; Sales No. E.72.IV.3 (1972).

85. _Production of Panels from Agricultural Residues_; and Crittenden, _Historical Outline_.

86. FAO, "World Survey of Production Capacity," FAO Committee on Wood Based Panel Products, December 1970. Cited in Crittenden, _Historical Outline_.

87. UNIDO, "Technologies from Developing Countries," No. 7 in the UNIDO Transfer of Technologies Series, UNIDO, New York, 1978.

88. Wen-yuan Huang, "A Framework for Economic Analysis of Livestock and Crop By-products Utilization," _American Journal of Agricultural Economics_ 61 (February 1979): 91-96.

89. Ibid., p. 96

90. _Production Panels from Agricultural Residues_, pp. 8, 9.

91. _Use of Agricultural and Industrial Wastes_, p. 39.

92. _Building Materials Industry_, p. 9.

93. W. Koppen and R. Geiger, _Klima der Erde_ (map), revised by R. Geiger and W. Pohl (Darmstadt: Justus Perthes, 1953).

94. DiBenedetto and Morgan, "A Proposal to Establish a Cooperative International Program," p. 9.

95. CBRI, "Appropriate Technologies and Materials for Housing and Building," a background paper presented at UNIDO's International Forum on Appropriate Industrial Technology, New Delhi/Anand, India, November 20-30, 1978.

96. "Draft Report of the Technical/Official Level Meeting to the Ministerial Level Meeting," International Forum on Appropriate Industrial Technology, UNIDO, New Delhi/Anand, India, November 20-30, 1978.

97. Ibid.

98. The building regulations in Fiji made pursuant to Section 39 of Chapter 91 provide an example. Regulation 76 of Part 10 (Materials and Workmanship) reads as follows:

> All aggregates used for concrete work shall consist of clean natural river gravel or crushed stone of approved grades and sizes, and shall otherwise conform to the requirement of Part 18 [which deals with concrete construction] of these Regulations (The Laws of Fiji, 1967 Edition, Government Printer, Suva, p. 4310).

99. Recommendation #9 of the Africa Regional Meeting of the UN Habitat and Human Settlements Foundation, November 1968; and "Draft Report of the Technical/Official Level Meeting."

100. "Coconut - All-around Provider."

101. This theme recurs in M.K. Garg's, "The Scaling-Down of Modern Technology: Crystal Sugar Manufacturing in India"; "The Upgrading of Traditional Technologies in India: Whiteware Manufacturing and the Development of Home Living Technologies," in Appropriate Technology: Problems and Promises, edited by Nicolas Jequier (Paris: OECD, 1976), pp. 156-88; and in "India - Small Scale Is Not Necessarily Inefficient," in Technology for Development, edited by Ann Kestenbaum (London: VCOAD, November 1975), p. 66.

102. In 1975, at Washington University's Center for Development Technology, a small laboratory experiment was directed at the fire protection of palm leaves (the Nypa Palm, from the Philippines). It was found that difference between wood and the thatching leaves made it difficult to transfer the wood technology directly to leaves, primarily because the waxy cuticle of the leaf made penetration difficult. Gaseous treatment appeared to be a more viable alternative. Both chlorine and bromine gases have been used as fire treatments; although, being hazardous materials, they would be difficult to adapt for use in rural situations. Ammonium sulfate, a cheap and available fertilizer, has been used to fire-treat wood and paper, and as an insecticide. The Central Building Research Institute, Roorkee, India, has reported experiments with di-ammonium phosphate and sodium fluoride for combined fire, insect, and fungal protection; flame spread was eliminated, but leaching problems required the application of additional weatherproofing.

103. Personal communication, D.L. Davies of the Koppers Company, Pittsburgh, Pa., to J.P.R. Falconer, June 8, 1979.

104. P.C. Cheo and Kenneth R. Montgomery, "The Study of Fire Retardance in Plants," Lasca Leaves, September 1970.

105. "Wheat - Puffed and Carbonized - as an Aggregate," *Concrete Construction*, November 1971.

106. In 1971, Washington University's Center for Development Technology experimented with the "carbonized wheat" approach using noncereal plant materials (bamboo and reeds). This work was directed at finding suitable, low-cost, lightweight, and environmentally inert fillers for sandwich panels having matrices of various foams, the end-use being for tropical roofing components. Increased resin costs caused the foam work to be abandoned, but the production of carbonized bamboo particles worked well at laboratory scale.

107. R.P. Morgan and A.T. DiBenedetto, "An Inter-American Program in Materials Engineering - The Utilization of Indigenous Resources for Low-Cost Materials," in *Proceedings of Second Inter-American Conference on Materials Technology, Education Volume*, Mexico City, August 1970, pp. 19-26.

108. DiBenedetto and Morgan, "A Proposal to Establish a Cooperative International Program."

109. J.P.R. Falconer, "Report on Field Trip to S.E. Asia - February 15 to March 20, 1974," for USAID/Monsanto Research Corporation/Washington University Roofing Research Project (Subcontract AID 7312).

110. J.E. Atchinson, "Agricultural Residues and Other Nonwood Plant Fibers." For wood wastes, see Edwin C. Jahn and S.B. Preston, "Timber: More Effecitve Utilization," in *Materials: Renewable and Nonrenewable Resources*, AAAS, Washington, D.C., 1976, pp. 163-67. See also, *Agricultural Residues*, pp. 142-80 for some wood wastes technologies.

111. *Encyclopedia of Chemical Technology*, 2nd ed., Vol. 10 (1966) and Vol. 12 (1967) (New York: John Wiley).

112. Patents include Nos. 2,196,277 (April 9, 1940); 2,645,587 (July 14, 1953); 3,301,800 (January 31, 1967). See also T.R. McElhinney, T.F. Clark, and D.F.J. Lynch, "Low-Cost Molding Material from Agricultural Wastes," *Modern Plastics*, March 1939.

113. *Use of Agricultural and Industrial Wastes*, p. 18.

114. R.P. Morgan and J.R.P. Falconer, "U.S. Universities and Developing Countries: Some Experiences with Appropriate Chemical and Materials Technology," paper presented at AAAS Annual Meeting, San Francisco, Calif., March 1, 1974. (Available from Center for Development Technology, Washington University, St. Louis, Mo. 63130).

115. J.P.R. Falconer, "A Building System for Ghanaian Housing," Seventh Annual AIA Research Conference, Cincinnati, Ohio, November 3, 1970.

116. Ball et al., "Development of Low-Cost Roofing." (Washington University, Center for Development Technology contributions on Subcontract to J.P.R. Falconer).

117. H.Y. Fang, ed., *New Horizons in Construction Materials*, Vol. 1 (Lehigh Valley, Pa.: Envo, 1976).

Chapter 6

1. E.B. Shultz and R.P. Morgan, "Appropriate Technology for Village Level Rice Bran Processing and Utilization in Developing Countries," Report No. CDT 79/1, Center for Development Technology, Washington University, St. Louis, Mo., April 1979.

2. M.M. Hoda, "Solar Cookers," Appropriate Technology Development Association, Lucknow, India, undated.

3. Energy for Rural Development: Renewable Resources and Alternative Technologies for Developing Countries, National Academy of Sciences, Washington, D.C., 1976, p. 138.

4. Ibid., p. 153.

5. Ibid., p. 185.

6. Ibid., p. 184.

7. A.L. Hannomd, "Alcohol: A Brazilian Answer to the Energy Crisis," Science 195 (February 11, 1977): 564.

8. R.P. Morgan with E.E. Irons, E.A. Perez, T.N. Soule, and A.S. Fried, The Role of U.S. Universities in Science and Technology for Development (New York: Pergamon Press, 1979).

9. F. Smith, ed., "The Relevance of AT Developments in the U.S. to the Third World: A Compendium of U.S. AT Organizations," TRANET, Box 567, Rangeley, Maine, Spring 1979.

10. J.H. Ashworth, "Renewable Energy Sources for the World's Poor: A Review of Current International Development Assistance Programs," Report SERI/TR-51-195, Solar Energy Research Institute, Golden, Colorado, October 1979.

11. Ibid.

12. Morgan et al., The Role of U.S. Universities. p. 201.

13. R.C. Sangster and B.K.W. Copeland, "Industrial Research and Science and Technology for Development: A Workshop in Preparation for the 1979 U.N. Conference," International Science and Technology Institute, Washington, D.C., Spring 1979.

14. Papers from UNIDO Forum on Appropriate Industrial Technology, New Delhi, November 1978; and R.P. Morgan, "UNIDO and Appropriate Industrial Technology," Science 203 (March 1978): 835.

15. The Center for Appropriate Technology, Delft Technological University, convened a conference on appropriate technology in September 1979. See J. deSchutter and G. Bemer, Fundamentals of Appropriate Technology (Delft, Netherlands: Delft University Press, 1980).

16. Ashworth, "Renewable Energy Sources for the World's Poor."

17. Morgan et al., The Role of U.S. Universities. p. 190.

18. Ibid.

19. "Analysis of AID Programs for the U.N. Conference on Science and Technology for Development," International Science and Technology Institute, Washington, D.C., January 25, 1979.

20. Ashworth, "Renewable Energy Sources for the World's Poor."

21. "Congressional Presentation," Institute for Technological Cooperation, February 23, 1979, p. 9.

22. Ibid., pp. 145-46.

23. "AT International: A Report," ATI, Washington, D.C., January 1979.

24. "Appropriate Technology: A Program Plan of the National Science Foundation," Prepared for the Committee on Science and Technology, U.S. House of Representatives, February 1979.

25. Morgan et al., The Role of U.S. Universities. p. 7.

26. Ashworth, "Renewable Energy Sources for the World's Poor."

27. Ibid.

28. John H. Ashworth and Jean W. Neuendorffer, "Matching Renewable Energy Systems to Village-Level Energy Needs," Report SERI/TR-744-514, Solar Energy Research Institute, Golden, Colorado, June 1980, p. 1.

29. "Links that Connect a Hemisphere," MOSAIC 8 (November/December 1977): 44-50.

30. Morgan et al., The Role of U.S. Universities.

31. "The United Nations University: The Third Year 1977-1978," Tokyo, Japan, 1978.

32. "Third World to Demand New Fund for Scientific Development," Nature 279 (May 3, 1979): 6; see also, Science and Government Report, September 1, 1980, pp. 4-6.

33. "Preliminary Report from a Workshop on Appropriate Technology," Winrock International Conference Center, Morrilltown, Arkansas, December 1-5, 1978.

34. "A New International Mechanism for Appropriate Technology: A Feasibility Study by a Team of Specialists," August 1978 (no other identification). See also Report of TRANET 1979 Annual Meeting, Harper's Ferry, West Virginia, December 10-15, 1978.

Author Index

Subject Index

About the Authors

ROBERT P. MORGAN is Professor and Chairman of the Department of Technology and Human Affairs, and Director of the Center for Development Technology in the School of Engineering and Applied Science at Washington University in St. Louis, Missouri. He received the bachelor's degree in chemical engineering from The Cooper Union, the master's and engineer's degrees in nuclear engineering from M.I.T., and the Ph.D. in chemical engineering from Rensselaer Polytechnic Institute. He is a corporate member of Volunteers in Technical Assistance (VITA) and has served on the Committee on Science, Engineering, and Public Policy of the American Association for the Advancement of Science. He is the author or coauthor of some eighty papers, reports, and articles spanning the fields of heat and mass transfer, nuclear reactor analysis, educational telecommunications, earth observation satellite applications, technology and international development, renewable energy resources, and the societal role and impact of technology. In 1978, Dr. Morgan received the Chester F. Carlson Award of the American Society for Engineering Education for his creative leadership and success in uniting elements of the social sciences, natural sciences and engineering to provide a new kind of education for technology and human affairs. Currently, his main professional interests are focused on science and technology for development and on renewable energy sources. He has served as principal investigator of two studies as part of U.S. preparations for the 1979 UNCSTD and is the principal author of *Science and Technology for Development: The Role of U.S. Universities*, published by Pergamon Press.

LARRY ICERMAN is Associate Professor of Technology and Human Affairs and Faculty Associate, Center for Development

Technology, Washington University, St. Louis, Missouri. Dr. Icerman is currently Director of the New Mexico Energy Institute at New Mexico State University. He holds a bachelor's degree in Aeronautics and Astronautics from the Massachusetts Institute of Technology, master's degree in Applied Mechanics from the University of California, San Diego, an M.B.A. degree from San Diego State University, and the Ph.D. degree in Engineering Sciences from the University of California, San Diego. Dr. Icerman is a member of the American Institute of Aeronautics and Astronautics, the International Solar Energy Society, the International Association for Hydrogen Energy, AAAS, and has published more than 40 scientific papers. He is the co-author of two widely read books on energy technology and policy, and serves as an Associate Editor of Energy: The International Journal, published by Pergamon Press.

ELIZABETH S. ANDREWS is Research Director with Friends of the River Foundation, a non-profit organization formed to further public education on water issues related to rivers. She received the bachelor's degree in technology and human affairs at Washington University, specializing in technology and international development. Her work experience includes a staff position with the Center for Development Technology, Washington University and an internship with Volunteers in Technical Assistance.

JAMES A. CHAPMAN is currently finishing work on the master's degree in technology and human affairs at Washington University in the area of technology and international development. He received a bachelor's degree in forest biology from the State University of New York, College of Environmental Science and Forestry at Syracuse, New York. From 1975 to 1977, he served as a U.S. Peace Corps Volunteer in Nepal, working in fisheries extension activity.

J. P. R. FALCONER is an architect-builder, living and working in Culebra, Puerto Rico and St. Louis, Missouri. He graduated with bachelor's degrees in engineering and architecture from Brown and Cornell Universities, respectively. A former faculty member of the University of Science and Technology, Kumasi, Ghana, and Washington University, St. Louis, he has had extensive experience as a practicing architect and as consultant to national and international organizations. His research interests are in materials performance and the building process, and in the transfer of architectural technologies for international development.

HORACE HERRING is a Senior Research Associate at the Lawrence Berkeley Laboratory, California, working in energy